Chromatographic Fingerprint Analysis of Herbal Medicines

Hildebert Wagner • Rudolf Bauer • Dieter Melchart
Anton Staudinger
Editors

Chromatographic Fingerprint Analysis of Herbal Medicines

Thin-Layer and High Performance Liquid Chromatography of Chinese Drugs

Vol. 4

TCM-KLINIK BAD KÖTZTING
Erste deutsche Klinik für Traditionelle Chinesische Medizin
Fachklinik für Psychosomatik und Psychotherapie

Springer

University Hospital at Beijing University of Chinese Medicine

Editors
Hildebert Wagner
Center of Pharma Research
Ludwig-Maximillians-University
Munich
Germany

Dieter Melchart
Klinikum rechts der Isar
Kompetenzzentrum Komplementärmedizin
Munich
Germany

Rudolf Bauer
Institute of Pharmaceutical Science
University of Graz
Graz
Austria

Anton Staudinger
TCM-Klinik Bad Kötzting
Bad Kötzting
Germany

ISBN 978-3-319-32326-8 ISBN 978-3-319-32328-2 (eBook)
DOI 10.1007/978-3-319-32328-2

Library of Congress Control Number: 2014945949

Printed on acid-free paper

This Springer imprint is published by Springer Nature
The registered company is Springer International Publishing AG Switzerland

Contents

Vol. I: Monographs No. 1 – 40
Vol. II: Monographs No. 41 – 80
Vol. III: Monographs No. 81 – 103

Table of Contents Vol. I

Contents alphabetically (lat. names) xi
Contents alphabetically (chin. names) xv
Acknowledgements xix
Introduction xxi
Practical work guidelines xxv

TCM-Analytical Monographs Vol. I

1. Bupleuri, Radix 1
2. Frittilariae, Bulbus 13
3. Rehmanniae, Radix 23
4. Schisandrae, Fructus 37
5. Asari, Radix et Rhizoma 45
6. Houttuyniae cordatae, Herba 59
7. Pinelliae, Rhizoma 71
8. Astragali, Radix 83
9. Angelicae pubescentis, Radix 99
10. Atractylodis macrocephalae, Rhizoma 113
11. Belamcandae sinensis, Rhizoma 127
12. Lycopi lucidi, Herba 141
13. Notopterygii, Rhizoma seu Radix 151
14. Angelicae sinensis, Radix 161
15. Angelicae dahuricae, Radix 171
16. Ligustici chuanxiong, Radix 181
17. Zanthoxyli, Pericarpium 191
18. Magnoliae officinalis, Cortex 203
19. Drynariae, Rhizoma 211
20. Puerariae, Radix 221
21. Codonopsis pilosulae, Radix 233
22. Gardeniae, Fructus 245
23. Gastrodiae, Rhizoma 255
24. Ecliptae, Herba 263
25. Andrographis, Herba 273
26. Paeoniae albae/rubrae, Radix 281
27. Sophorae, Flos 291
28. Coptidis, Rhizoma 301

29. Stephaniae tetrandrae, Radix ... 311
30. Ziziphi spinosae, Semen ... 325
31. Amomi rotundus, Fructus ... 335
32. Uncariae cum Uncis, Ramulus ... 343
33. Clematidis, Radix ... 355
34. Sinomenii, Caulis ... 369
35. Forsythiae, Fructus ... 381
36. Evodiae, Fructus ... 391
37. Anemarrhenae, Rhizoma ... 403
38. Acanthopanacis senticosi, Radix ... 415
39. Scrophulariae, Radix ... 427
40. Polygoni multiflori, Radix ... 439

Appendix:

Basic Solvent Systems, reagents and columns for the TLC-, GC- and HPLC-fingerprint
Analysis of main structure types of natural products. ... 451
Index ... 457
Drug monograph, Marker compounds, Chemical classification, Processing ... 461

Table of Contents Vol. II

Vol. I: Monographs No. 1 – 40
Vol. II: Monographs No. 41 – 80
Vol. III: Monographs No. 81 – 103

TCM-Analytical Monographs Vol. II

41. Alismatis, Rhizoma ... 467
42. Carthami, Flos ... 475
43. Epimedii, Herba ... 485
44. Cnidii, Fructus ... 499
45. Lycii radicis, Cortex ... 509
46. Lycii, Fructus ... 521
47. Mori radicis, Cortex ... 535
48. Mori, Folium ... 549
49. Cimicifugae, Rhizoma ... 559
50. Phellodendri amurensis, Cortex
Phellodendri chinensis, Cortex ... 573
51. Lonicerae, Flos
Lonicerae japonicae, Flos
Lonicerae japonicae, Caulis ... 587
52. Curcumae, Radix Curcumae longae, Rhizoma Curcumae, Rhizoma ... 601
53. Dioscoreae oppositae, Rhizoma
Dioscoreae hypoglaucae, Rhizoma
Dioscoreae nipponicae, Rhizoma
Dioscoreae septemlobae, Rhizoma ... 615
54. Ganoderma ... 633
55. Citri reticulatea, Pericarpium Citri reticulatea viride, Pericarpium ... 647
56. Corydalis, Rhizoma ... 665
57. Dipsaci, Radix ... 677
58. Atractylodis lanceae, Radix ... 691
59. Leonuri, Herba ... 707
60. Magnoliae, Flos ... 719
61. Piperis longi, Fructus ... 729
62. Sophorae flavescentis, Radix ... 743
63. Scutellariae, Radix ... 755
64. Chaenomelis, Fructus ... 767
65. Acori calami, Rhizoma Acori tatarinowii, Rhizoma. ... 777
66. Isatidis, Radix ... 791
67. Tribuli, Fructus ... 805
68. Ophiopogonis, Radix ... 819
69. Eucommiae, Cortex ... 831
70. Notoginseng, Radix et Rhizoma ... 843
71. Rhei, Radix et Rhizoma ... 857
72. Ginseng, Radix et Rhizoma
Panacis Quinquefolii, Radix ... 875

73. Siegesbeckiae, Herba ... 893
74. Salviae miltiorrhizae, Radix et Rhizoma ... 903
75. Poria ... 923
76. Cassiae, Semen ... 935
77. Camelliae, Folium ... 951
78. Artemisiae Scopariae, Herba ... 967
79. Aconiti lateralis praeparata, Radix ... 977
Aconiti kusnezoffii praeparata, Radix
80. Cinnamomi, Cortex ... 991

Appendix:

Basic Solvent Systems, reagents and columns for the TLC-, GC- and HPLC-fingerprint Analysis of main structure types of natural products ... 1009
Index ... 1015
Drug monograph, Marker compounds, Chemical classification, Processing ... 1019

Table of Contents Vol. III

Vol. I: Monographs No. 1 – 40
Vol. II: Monographs No. 41 – 80
Vol. III: Monographs No. 81 – 103

TCM-Analytical Monographs Vol. III

81. Crataegi, Folium/Fructus 1
82. Cyperi, Rhizoma 17
83. Lycopodii, Herba 27
84. Saposhnikoviae, Radix 35
85. Glycyrrhizae, Radix et Rhizoma 43
86. Gynostemmatis, Herba 55
87. Sarcandrae, Herba 69
88. Ligustri lucidi, Fructus 79
89. Moutan, Cortex 91
90. Peucedani, Radix 105
91. Achyranthis, Radix 119
92. Bambusae in Taenia, Caulis 131
93. Lysimachiae christiniae, Herba 145
94. Desmodii styracifolii, Herba 159
95. Retinervus Luffae, Fructus 171
96. Oldenlandiae, Herba 185
97. Siraitiae/Momordicae, Fructus 197
98. Morindae officinalis, Radix 205
99. Apocyni veneti, Folium 217
100. Eriocauli, Flos 229
101. Spatholobi, Caulis 235
102. Aucklandiae, Radix 243
103. Platycodonis, Radix 255

Index 267

Table of Contents Vol. IV

Vol. I: Monographs No. 1 – 40
Vol. II: Monographs No. 41 – 80
Vol. III: Monographs No. 81 – 103

TCM-Analytical Monographs Vol. IV

104. Cortex, Albiziae 1
105. Cortex, Fraxini 11
106. Fructus, Arctii 19
107. Fructus, Corni 27
108. Fructus, Kochiae 37
109. Fructus, Psoraleae 47
110. Fructus/Semen, Trichosanthis 59
111. Fructus, Viticis 71
112. Fructus, Xanthii 79
113. Herba Artemisiae annuae/Folium Artemisiae argyi 91
114. Herba, Ephedrae 107
115. Herba, Violae 115
116. Lignum, Sappan 125
117. Radix, Gentianae macrophyllae 137
118. Radix, Trichosanthis 147
119. Ramulus, Mori 157
120. Semen, Celosiae 169
121. Semen Nigrum, Sesami 181
122. Semen, Sinapis 191
123. Semen, Vaccariae 203
124. Semen/Herba, Plantaginis 213
125. Spica, Prunellae 225

Index 239

Contents Alphabetically

Lat. name	Chapter	Page
Albizia julibrissin	104	1
Arctium lappa	106	19
Artemisia annua	113	91
Artemisia argyi	113	91
Caesalpinia sappan	116	125
Celosia argentea	120	169
Cornus officinalis	107	27
Ephedra sinica	114	107
Fraxinus chinensis	105	11
Fraxinus excelsior	105	11
Fraxinus rhynchophylla	105	11
Fraxinus szaboana	105	11
Gentiana macrophylla	117	137
Kochia scoparia	108	37
Morus alba	119	157
Plantago asiatica	124	213
Plantago depressa	124	213
Prunella vulgaris	125	225
Psoralea corylifolia	109	47
Sesamum indicum	121	181
Sinapis alba	122	191
Trichosanthes kirilowii	110/118	59/147
Trichosanthes rosthornii	110/118	59/147
Vaccaria segetalis	123	203
Viola tricolor	115	115
Viola yedoensis	115	115
Vitex agnus-castus	111	71
Vitex trifolia	111	71
Xanthium sibiricum	112	79

Acknowledgements

- The editors wish to express their deep gratitude to the TCM Clinic Bad Kötzting Mr. A. Staudinger for financial support and Prof. A. Vollmar, LMU Munich, Department of Pharmacy, for the supply of laboratory space and various facilities for the chemical and technical investigations of the TCM drugs.
- We are deeply indebted to our technical assistants Mrs. Stefanie Püls and Mrs. Talee Barghouti.

Authors' Contributions

- Supervisor and responsible for the correct performance of the Analytical Monographs of Chinese Herbal Drugs Prof. Dr. Dr. h.c. Hildebert Wagner, Special Analytical Laboratory of the Department of Pharmacy of the University of Munich, Butenandtstr. 5, 81377 Munich; e-mail. h.wagner@cup.uni-muenchen.de

Introduction

- **Legislation**
  Among the various prerequisites for a perfect quality proof of these herbal drugs, authentication and safety proof take first precedence. Identification was in former times primarily synonymous with the macroscopic and microscopic botanical authenticity. Since that time, however, chemical composition and particularly the complex entities of the low molecular constituents have become of greater interest for oral medicinal application and thus in evaluating the pharmacological effects and therapeutic efficacy of the plant drug extracts obtained by decoction or other extraction processes.

- Independent of the specific national drug regulations for countries around the world, there is also an international consensus that all TCM drugs must meet certain, stipulated high-quality standards. Additionally, it must be guaranteed that all TCM drugs prescribed by physicians are safe for patients. The safety proof aims mainly to exclude any kind of possible falsifications of the herbal drugs and the limitation of concentrations of heavy metals, aflatoxins and defined microbial adulterations.

- **Applied Methods of the Qualified Proof**
  The main method used is TLC (thin layer chromatography), which allows us to present the visualized main characteristic constituents in the form of coloured TLC photographs. The second method, used globally, is HPLC (high pressure liquid chromatography) in the form of a so-called fingerprint analysis. This technique allows us to detect the complex entities of all low molecular constituents of a plant drug extract, with the advantage that the single constituents can be made visible in the form of peak profiles. Additionally, the single constituents can be quantified by using online recordable UV spectra with the diode array technique. It is also possible to gain preliminary information as to which chemical structure type the single compounds may belong. From this year forward, LC-MS (liquid chromatography-mass spectroscopy) is also available for the analysis of plant extracts whose chemical compositions have previously been only minimally investigated.

- **Publication of the Analytical Monographs of Investigated Herbal Chinese Drugs**
  The following volumes were published by Springer, Vienna and New York, with financial support from the TCM Clinic Bad Kötzting; Wagner, H., Bauer, R., Melchart, D., Xiao, P.-G., Staudinger, A. (Eds.)

  - Vols. I and II (2011) containing 80 analytical monographs
  - Vol. III (2015) containing 23 analytical monographs
  - Vol. IV (2016) containing 22 analytical monographs
  - Vol. V (in preparation; publication scheduled for February/March 2017)

- Note: All single analytical monographs that are already edited can be downloaded at http://www.springer.com/de/book/9783709107621

Prospects for the Improvement of the Quality Proof of Chinese Herbal Drugs

1. **Authenticity of TCM drugs not definitely assessable**
   Some herbal drugs are not yet produced under controlled cultivation but originate from wild collections. Even if they are derived from cultivations, it must be taken into account that they can originate from quite varied climate zones and that they may be harvested under a variety of conditions. Therefore, their chemical authenticity and homogeneity within a defined plant species often cannot be guaranteed. We have thus investigated as many herbal drug samples as we were able to acquire from different districts, climate zones and markets in China, as well as reference drugs from some German herbal drug firms that also import herbal drugs from China.

2. For 5–10 % of imported plant drugs from China, we do not receive specific information about the plant part (Flos, Fructus, Semen, Folium, Cortex or Radix and Rhizoma) from which they were collected. Such drugs are specified as "herba" analogues. For these drug samples, it cannot be expected that the TLC and HPLC chemical fingerprints are very homogenous. Not all parts of a herbal drug contain the same chemical constituents. The documentation in the corresponding herbal analytical monographs confirms this judgement (see e.g. Herba Leonuri, Vol. II; Herba Lysimachiae, Vol. III; or Herba Violae, Vol. IV). Therefore, it will be necessary that this discrepancy has to be corrected in the future. Otherwise, it cannot be expected that the results of clinical application can be reproduced.

3. **Uncertain botanical nomenclature**
   The non-uniform nomenclature for the same plant in various regions of China can cause impermissible substitutions or falsifications. This occurred some years ago when the root of *Stephania tetrandra* (Hanfangji) was mistaken for the root of *Aristolochia fangji* (Guanfangji). The latter of both contains the carcinogenic aristolochic acid which can produce severe nephrotoxic side effects. A similar Chinese drug is the tetraploid *Acorus tatarinowii* which differs in a very high content of carcinogenic β-asarone from that of the diploid *Acorus calamus*, known officially in most western countries. Meanwhile special chromatographic methods were developed and described in the analytical monographs to avoid such falsifications.

4. **Great variability of plant species**
   Several herbal drug monographs of the Chinese Pharmacopoeia list more than two species or subspecies and sometimes up to eight species labelled as synonyms, subspecies or subvarieties. It is assumed that all species contain the same constituents in the same amount. In our 20 years running TLC- and HPLC-fingerprint investigations, we have shown that in many cases considerable differences were detectable between the single species and the main official drugs. Correspondingly, it may be suggested that a great number of the "subspecies" do not possess the same pharmacological and therapeutic efficacy. This fact must be recognized and taken into consideration!

Guidelines for the Experimental Work

Source of the Herbal Drugs

As discussed in the preceding paragraph, the herbal drugs must originate from clearly identified botanical species. Additionally, it must be taken into consideration that differences in cultivations, climatic conditions, time of harvest, drying and storing conditions can cause slight chromatographic deviations which cannot be avoided and are normal. Therefore, it is worthwhile to investigate as many herbal drug samples of one species as can be obtained from different geographic and ecological areas.

Extraction Conditions

The chosen extraction procedures should be fast but efficient according to present scientific knowledge and inclusive of the total entity of the low molecular constituents of a herbal drug. This can be achieved in most cases using alcohol (MeOH or EtOH). Additional fingerprints can be obtained by extraction using petroleum ether/hexane or chloroform (for lipophilic compounds) or water/water-acetone mixtures (for tannins, high polymeric procyanidins and amino acids) as solvents. Polysaccharides and proteins can be characterized via their sugar or amino acid fingerprints after enrichment and acidic or enzymatic hydrolysis.

Chromatographic Conditions

Plates/Columns

- For the chromatography TLC- or HPTLC-standardized Silica Gel F254 (Merck) plates, in some specific cases also aluminium oxide- or cellulose-coated plates (Merck) are used. HPTLC plates are precoated with Silica Gel of an average particle size and a narrow size distribution of 5 µm (as opposed to TLC material of 15 µm average particle size and a broader size distribution).

- For all HPLC-analysis reversed phase C-18 or C-8 columns (LiChroCART® 125-4/250-4 LiChrospher® 100 RP-18 (5 µm), Merck or LiChroCART® 125-4/250-4 LiChrospher® 60 RP select B (5 µm), Merck) can be used with a Merck HITACHI L-4500 A Diode Array Detector.

Detection/Solvent System

In the appendix of Volumes 1 and 2 (p. 451/1009), the most used reagents and basic solvent systems in TLC and HPLC are listed for the detection of main structure types of drug constituents in herbal drugs.

Reference Compounds

The availability of reference compounds which are characteristic of any herbal drug and at the same time represent the main pharmacologically active constituents of any plant facilitates the identity (quality) proof of a herbal drug and is a requirement for quantitation determination. If they cannot be isolated in the researcher's own laboratory, some of them can also be purchased from special firms. In Germany the firm PhytoLab in Vestenbergsgreuth (www.phytolab.com) offers many reference compounds which are listed as "marker compounds" in the Chinese Pharmacopoeia.

Reproducibility of the Fingerprint Analysis

If the same technical conditions described are used, it can be expected that even with the use of instruments from other firms, nearly identical TLC and HPLC fingerprints must be obtained. If, however, for any reason, the grade of separation and/or the R*f* and Rt values deviate from those stipulated in the monographs, the sequence and the overall TLC zone and HPLC peak profiles must in any case be identical.

Photography

The TLC chromatograms were developed by a Canon PowerShot G2 digital camera in a CAMAG Reprostar 3 cabinet using winCATS software (www.camag.com).

Cortex Albiziae – *Hehuanpi*

Pharmacopoeia: [1] Pharmacopoeia of the People's Republic of China, English Edition Vol. I, 2010

Official drug: [1] Silktree Albizia Bark is the dried stem of *Albizia julibrissin* Durazz. (Fam. Fabaceae). The drug is collected in summer and autumn, and dried in the sun.

Origin: [2] China, Japan, India, Iran and Ethiopia

Description of the drug: [1] Quilled and semi-quilled, 40–80 cm long, 0.1–0.3 cm thick. Outer surface greyish-brown, somewhat longitudinally wrinkled, some shallowly fissured, with dense and distinct transversal elliptical lenticels, brown or brownish-red, occasionally showing prominent transversal ribs or some large rounded branch scars, and often covered with patches of lichens; inner surface pale yellowish-white, smooth, with fine and dense longitudinal striation. Texture hard und fragile, easily broken, fracture fibrously laminated, pale yellowish-white.

Pretreatment of the raw drug: [1] Foreign matters are eliminated, washed clean, soaked well, cut into slivers or pieces, and dried in the sun.

Medicinal use: [3] It is used as tonic, stimulant and sedative for treatment of melancholy and insomnia, also for treatment of traumatic diseases.

Effects and indications of Cortex Albiziae according to Traditional Chinese Medicine [1–6]

Taste:	Sweet
Temperature:	Neutral
Channels entered:	*Orbis hepaticus, o. cardialis, o. pulmonalis, o. lienalis*
Effects (functions):	To remove depression and tranquilize the mind. Activate blood and disperse swelling, anti-inflammatory, treating of swelling and pain of the lungs, skin ulcers, and wounds
Symptoms and indications:	Disquietude of heart spirit, depression and insomnia, lung abscess, sore and swelling, pain caused by injuries from falls

H. Wagner et al. (eds.), *Chromatographic Fingerprint Analysis of Herbal Medicines*, Vol. 4,
DOI 10.1007/978-3-319-32328-2_1, © Springer International Publishing Switzerland 2016

Main constituents: [3, 6–14]

- **Lignan glucosides**

(−)-Syringaresinol-4-O-β-D-apiofuranosyl-(1 → 2)-β-D-glucopyranoside, syringaresinol tri- and tetraglycoside

- **Lignans and derivatives**

derivatives of 3,4,5- methoxyphenol (lyoniresinol, vomifoliol and icariside E5)

- **Saponins**

Julibroside I-III, A_1–A_4, B_1, C_1, J_1–J_3, J_7 and J_9, J_{16}, J_{17}, J_{21} J_{29}–J_{31}, J_{32}, J_{34}, J_{35}, saponins with prosapogenins machaerinic acid methyl ester, acacic acid lactone, acacigenin B, machaerinic acid lactone, 16-deoxyacacigenin B, julibro triterpenoidal lactone A (16β-isomer of acacic acid lactone) julibrogenin A

- **Flavone glucosides**

7, 3′,4′-trihydroxyflavone, quercetin-3-O-β-D-galactopyranoside, quercetin-3-O-α-L- rhamnopyranoside

- **Other compounds**: syringic acid, β-sitosterol, α-spinasterol-3-*O*-β -D-glucopyranoside

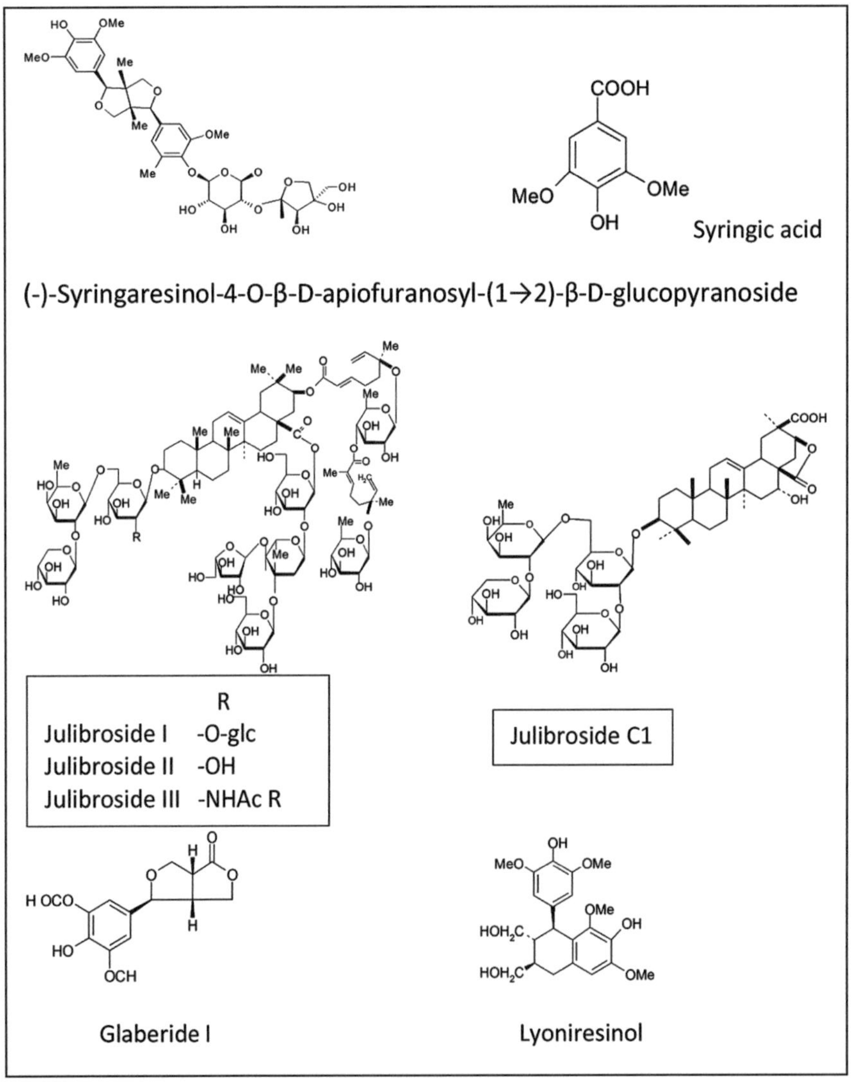

Fig. 1 Formulae of the main constituents of Cortex Albiziae [1, 3, 7, 9, 10]

Reported pharmacological effects: [3–5, 7, 10, 13, 15]

- antioxidant effect
- anti- inflammatory
- sedative effect
- anxiolytic effect
- anti-tumoral effect by induction of apoptosis type (human acute leukemia junket T-cells)

TLC Fingerprint Analysis:

	Drug samples	Origin
1	Cortex Albiziae / *Albizia julibrissin*	Province Henan (China)
2	Cortex Albiziae / *Albizia julibrissin*	Province Shanxi (China)
3	Cortex Albiziae / *Albizia julibrissin*	Province Sichuan (China)
4	Cortex Albiziae / *Albizia julibrissin*	Sample of commercial drug, obtained from HerbaSinica (origin: Hunan China)
5	Cortex Albiziae / *Albizia julibrissin*	Sample of commercial drug, obtained from China Medica (origin: Beichuan, Sichuan)
6	Cortex Albiziae / *Albizia julibrissin*	Sample of commercial drug, obtained from TCM-Clinic Bad Kötzting (Charge: K07.01.2003)

1. **TLC-fingerprint analysis of Saponins and Lignans:**

Reference compounds		R*f* of Fig. 2a	R*f* of Fig. 2b
T1	Syringaresinol diglucoside	0.61	0.71
T2	Oleanolic acid	0.98	–
T3	Saccharose	0.16	–
T4	Syringic acid	–	0.97

1. Extraction: 2 g powdered drug is extracted with 20 ml of 80 % methanol under reflux for 1 h. The extract is cooled, filtrated and evaporated to dryness. The residue is dissolved in 2 ml methanol and filtered over Chromafil® filtration unit, type 0–20 µm/25 mm.
2. Reference compounds: Each 0.5 mg is dissolved in 0.5 ml ethanol
3. Separation parameters:

Plate: HPTLC Silica gel 60 F_{254}, Merck

Applied amounts: Cortex Albiziae extracts: 10 µl each

Reference compounds: 10 µl each

Solvent system: Chloroform + methanol + glacial acetic acid + water

(7.5 + 4 + 1.5 + 1.25)

Detection: a) Anisaldehyde – sulphuric acid reagent

0.5 ml anisaldehyde is mixed with 10 ml glacial acetic acid, followed by 85 ml methanol and 5 ml sulphuric acid in that order.

The plate is sprayed with 10 ml and then heated at 110 °C for 7–10 min. Evaluation is carried out under visible light.

b) Trichloroacetic acid- potassium hexacyanoferrate-iron-III-chloride reagent (TPF)

1) 25 % trichloroacetic acid in chloroform.

2) 1 % aqueous potassium hexacyanoferrate mixed with an equal volume of 5 % aqueous iron-III- chloride.

The plate is sprayed with solution (1) and heated at 110 ° C for 10 min. It is then sprayed with solution (2) and evaluated in VIS.

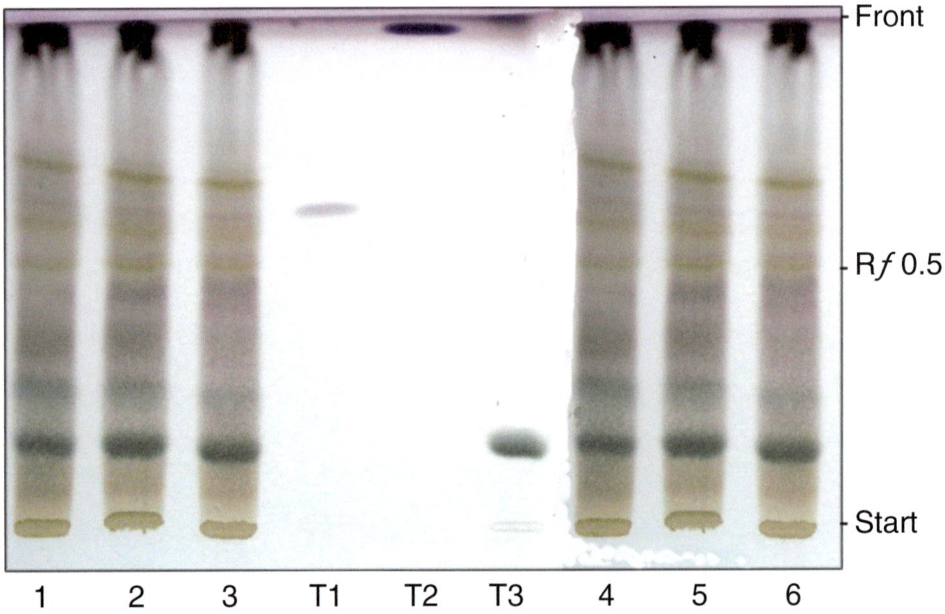

Fig. 2a Thin layer chromatogram of the 80 % methanol extracts of Cortex Albiziae, sprayed with anisaldehyde-sulphuric acid reagent (VIS)

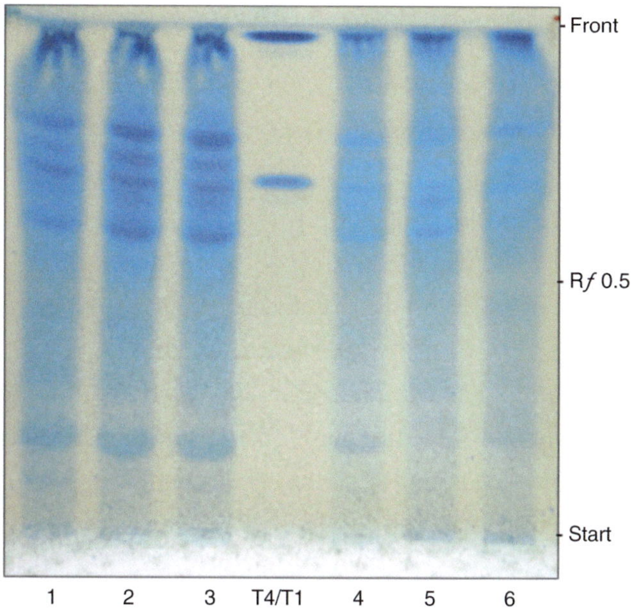

Fig. 2b Thin layer chromatogram of the 80 % methanol extracts of Cortex Albiziae, sprayed with Trichloroacetic acid- potassium hexacyanoferrate-iron-III-chloride reagent (VIS)

4) Description of TLC-fingerprint

Fig. 2a: The TLC is characterised by one dark brown zone at R*f*=0.98 (Oleanolic acid) and 7–8 weak brown zones from ~R*f* =0.70 down to R*f*=0.25. In the first R*f*-range from 0.70 till ~R*f*=0.55 violet brown coloured lignan (glycosides) appear with the reference compound syringaresinol diglucoside at R*f*=0.61. In the second R*f*-range from R*f*=0.55 till ~R*f*=0.15 some of the yellow- brown zones above saccharose could be not assigned to the julibrosides with 3-4 sugar moites. The genuine julibrosides I, II or III (see formula Fig. 1) could not be identified. According to the literature they were never isolated in genuine form and only registered as "prosapogenins" after partial hydrolysis by NMR and MS- spectroscopy!

Fig. 2b: The Cortex Albiziae extracts were developed in the same solvent system but sprayed with the TPF-reagent. The lignan (glucosides) appear in the upper R*f*- range in pigeon blue coloured zone. The saponins cannot be detected with the TPF- reagent.

HPLC-Fingerprint Analysis:

1. Sample preparation: The same extracts used for TLC-fingerprint analysis
2. Injection volume: Cortex Albiziae extracts: 7 µl each
3. HPLC parameter:

Apparatus: MERCK HITACHI D-6000 A Interface
MERCK HITACHI L-4500 A Diode Array Detector
MERCK HITACHI AS-2000 Auto sampler
MERCK HITACHI L-6200 A Intelligent Pump

Separation column: LiChroCART® 250-4 LiChrospher® 100 RP-18 (5 µm), Merck

Precolumn: LiChroCART® 4-4 LiChrospher® 100 RP-18 (5 µm), Merck

Solvent System: A: 0.1 % H_3PO_4 (Millipore Ultra Clear UV plus® filterd)
B: Acetonitrile (VWR)

Gradient: 5–18 % B in 15 min., (flow: 0.8 ml/min)
18–40 % B in 35 min., (flow: 0.8 ml/ min)
40–95 % B in 10 min., (flow: 1 ml/min)
95–100 % B in 5 min., (flow: 1 ml/min)
Total run time: 65 min

Detection: 215 nm

Retention times of the main peaks

Peak	Rt (min)	Compound
1	6.0	Syringic acid
2	11.0–18.0	Triterpene glucosides
3	21.2	(−)-Syringaresinol-4-O-β-apiofuranosyl- (1 → 2)-β-D-glucopyranosid
4	26.8	Lignan diglucoside
5,6	37.4, 47.21	Triterpene oligoside
7	62.6	α-Spinasterol
8	67.5	Oleanolic acid

4) Description of the HPLC-Figures 3a, 3b and 3c

The extract samples 2,3 and 6 show a very similar peak profile with main peaks in the Rt -range **A** between Rt = 20 and 28 with Peak No. **3** (Rt = 21.2) which could be assigned to (−)-Syringaresinol-4- O-β-apiofuranosyl-(1 → 2)-β-D- glucopyranoside. The prominent peak **4** in the same range (Rt =26.8) might be related to a second lignan glycoside. The peaks in the Rt- range of 11.0–20.0 can be assigned to triterpene oligo-glycosides and the peaks between Rt = 35.0 and 70.0 to sterols inclusive oleanolic acid

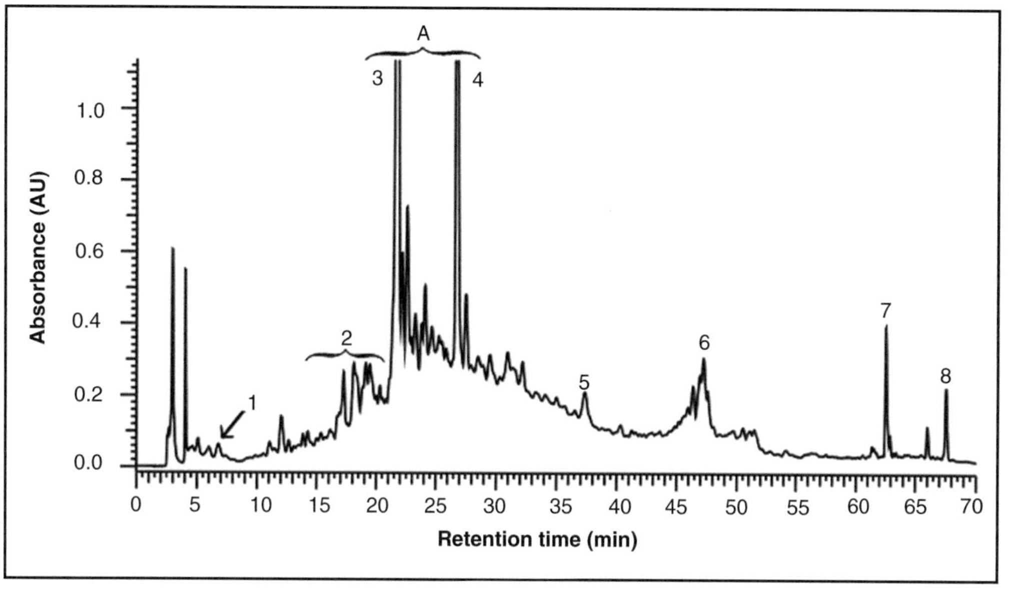

Fig. 3a HPLC-fingerprint analysis of the 80 % methanol extract of Cortex Albiziae, sample 2

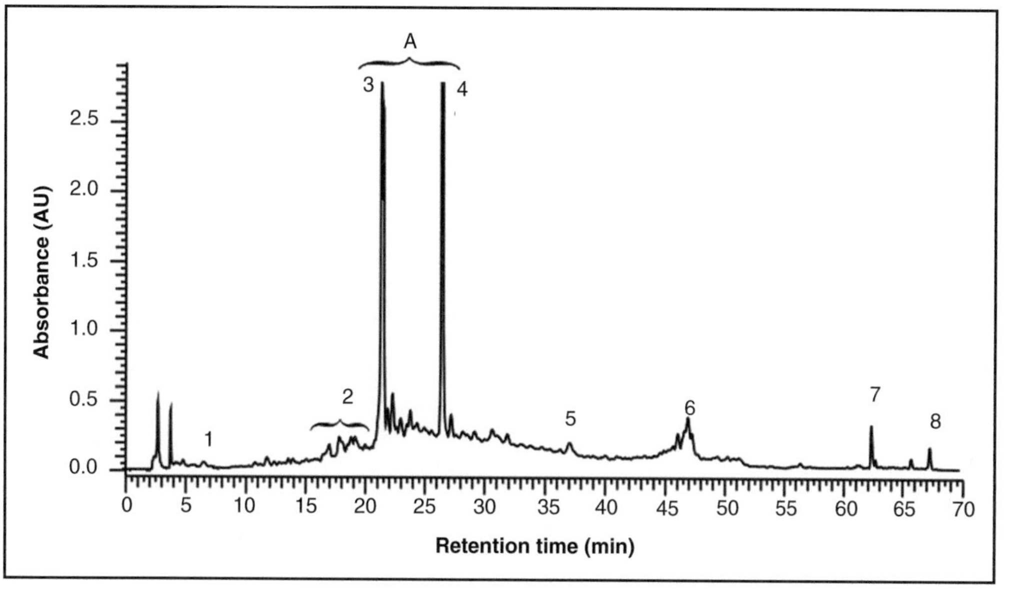

Fig. 3b HPLC-fingerprint analysis of the 80 % methanol extract of Cortex Albiziae, sample 3

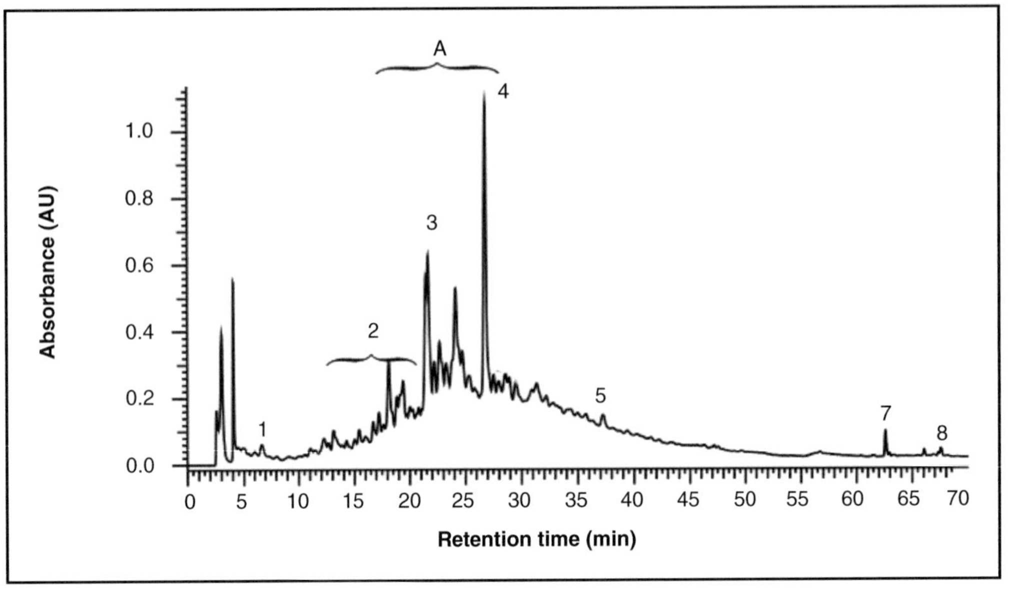

Fig. 3c HPLC-fingerprint analysis of the 80 % methanol extract of Cortex Albiziae, sample 6

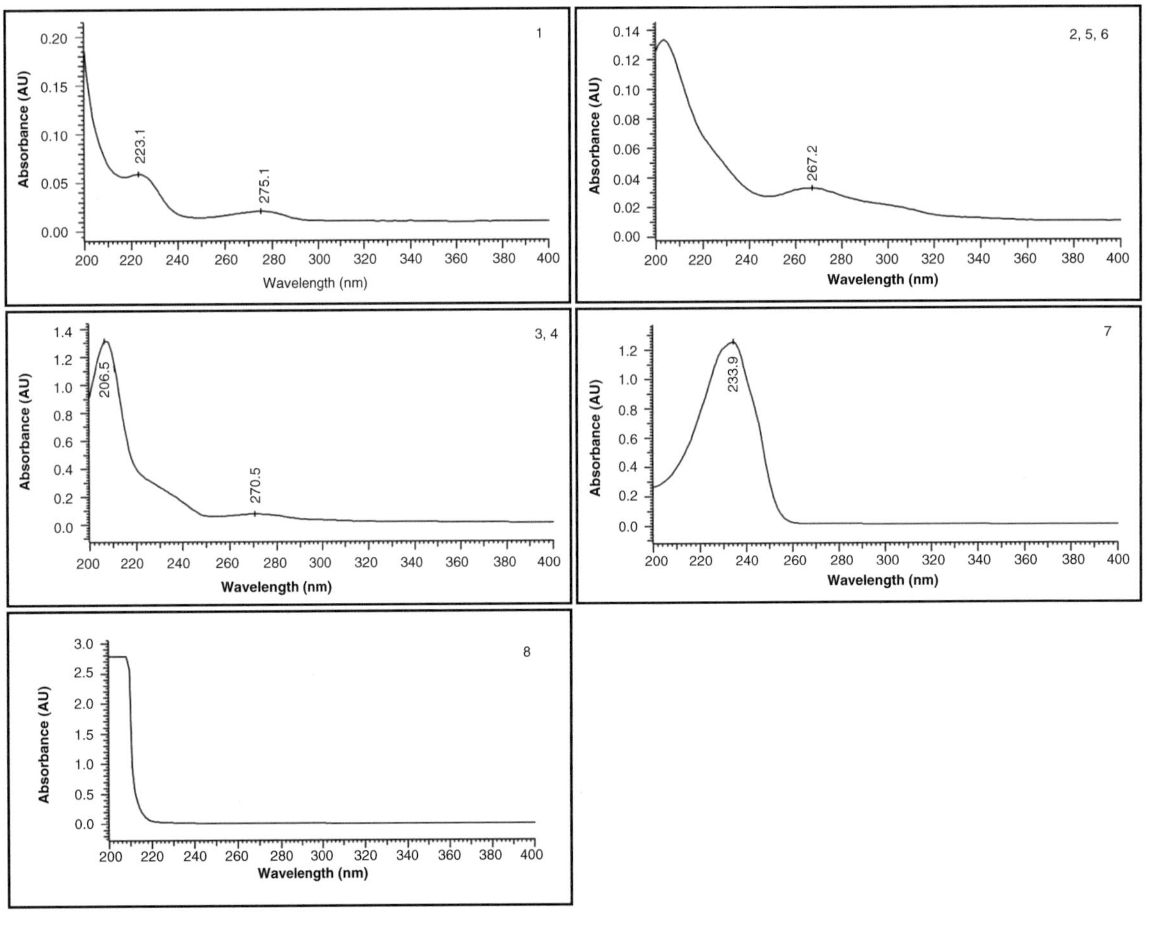

Fig. 4 On line UV-spectra of the main characteristic peaks of Cortex Albiziae

Conclusion

With TLC and HPLC the authentication of Cortex Albiziae can be easily achieved based on the lignan glycosides. The triterpene oligosids are only present in very low concentration, therefore not suitable as marker compounds.

References

1. Pharmacopoeia of the People's Republic of China. English Edition vol. 1. People's Medical Publishing House, Beijing (2010)
2. Keys, J.D.: Chinese herbs, their botany, chemistry, and pharmacodynamics. Charles E. Tuttle Company, Tokyo (1976)
3. Tang, W., Eisenbrand, G.: Handbuch of Chinese medicinal plants. Wiley-VCH Verlag, Wheinheim (2011)
4. Hempen, C.-H., Fischer, T.: Leitfaden chinesische phytotherapie 2. Auflage/Urban & Fischer, Munich (2007)
5. Liang, H., Tong, W.Y., Zhao, Y., Cue, J., Tu, G.: An antitumor compound julibroside J_{28} from *Albizia julibrussin*. Bio. Med. Chem. Letters **15**(20), 4493–4495 (2005)
6. Ikeda, T., Fujiwara, S., Araki, K., Kinjo, J., Nohara, T., Miyosh, T.: Cytotoxic glycosides from *Albizia julibrissin*. J. Nat. Prod. **60**, 102–107 (1997)
7. Hong Kong Chinese materia medica standards. vol. 6. Chinese Medicine Division – Department of Health – Government of the Hong Kong Special Administrative Region – the People's Republic of China, Hong Kong (2008)
8. Jung, Y.H., Ha, R.R., Kwon, S.H., Hong, S.I., Lee, K.H., Kim, S.Y., Lee, S.Y., Jang, C.G.: Anxiolytic effects of Julibroside C1 isolated from *Albizia julibrissin* in mice. Prog. Neuropsychopharmacol Biol. Psychiatry **44**, 184–192 (2013)
9. Kokila, K., Priyadharshini, S.D., Sujatha, V.: Phytopharmacologial properties of Albizia Species. Int. J. Pharm. Sci. **3**(5), 70–73 (2013)
10. Zou, K., Zhao, Y., Tu, G., Cui, J., Jia, Z., Zhang, R.: Two diastereometric saponins with cytotoxic from *Albizia julibrissin*. Carbohydr. Res. **324**, 182–188 (2000)
11. Ma, L., Tu, G., Chen, S., Zhang, R., Lai, L., Xu, K., Tang, Y.: NMR determination of the structure of Julibroside J_1, Carbohydr. Research **281**(1), 35–46 (1996)
12. Sun, H., He, S., Shi, M.: Adjuvant- active fraction from *Albizia julibrissin* saponins improves immune reponses by inducing cytokine and chemokine at the site of injection. Int. Immunopharmacol. **22**(2), 346–355 (2014)
13. Wagner, H., Ulrich- Merzenich, G.: Evidence and rational based research on Chinese drugs, 1st edn. Springer – Verlag, Wien (2013)
14. Haddad, M., Myamoto, T., Laurens, V., Lacaille- Dubois, M.: Two new biologically active triterpenoidal saponins acytalted with salicylic acid from *Albizia adianthifolia*. J. Nat. Prod. **66**(3), 372–377 (2003)
15. Zheng, L., Zheng, J., Zhao, Y., Wang, B., Wu, L., Liang, H.: Three anti- tumor saponins from *Albizia julibrissin*. Biol. Med. Chem. Letters **10**(16), 2765–2768 (2006)

Cortex Fraxini – *Qinpi*

Pharmacopoeia: [1]
Pharmacopoeia of the People's Republic of China, English Edition Vol. I, 2010

Official drug: [1, 2]
Ash bark is the dried branch bark or stem bark of *Fraxinus rhynchophylla* Hance, *Fraxinus chinensis* Roxb., *Fraxinus szaboana* Lingelsh. or *Fraxinus stylosa* Lingelsh. (Fam. Oleaceae)

The drug is collected in spring and autumn and dried in the sun.

Origin: [3–6]
Mainly in Chinese provinces such as Gansu, Hebei, Heilongjiang, Henan, Jilin, Liaoning, Shaanxi, Shandong, Shanxi and Sichuan. Available or cultivated also in Japan, Korea, Russia and Vietnam

Description of the drug: [1]
Branch bark:

Quilled or channelled, 10–60 cm long, 1.5–3 mm thick. Outer surface greyish-white, greyish-brown to blackish-brown, or alternated in patches, even or slightly rough, with greyish-white and rounded dotted lenticels, and fine oblique wrinkles, some with branch scars; inner surface yellowish-white or brown, smooth. Texture hard and fragile, fracture fibrous, yellowish-white. Odor, slight; taste, bitter

Stem bark:

Slat pieces, 3–6 mm thick. Outer surface greyish-brown, with rimose furrows and reddish-brown rounded or transversal lenticels. Texture hard, fracture relatively fibrous.

Medicinal use: [7]
Mainly used to treat chronic bronchitis, further inflammatory diseases and bacterial dysentery

Effects and indications of Cortex Fraxini according to Traditional Chinese Medicine [1–3, 7–11]

Taste:	Bitter, astringent
Temperature:	Cold
Channels entered:	*Orbis hepaticus et felleus, Orbis intestini crassi, Orbis stomachi*
Effects (functions):	To clear heat and dry dampness, astringe to check dysentery, check vaginal discharge, and improve vision
Symptoms and indications:	Dampness-heat diarrhea and dysentery, red or white vaginal discharge, red painful swelling eyes, nebula

H. Wagner et al. (eds.), *Chromatographic Fingerprint Analysis of Herbal Medicines,* Vol. 4,
DOI 10.1007/978-3-319-32328-2_2, © Springer International Publishing Switzerland 2016

Main and minor constituents: [2, 3, 7, 9–16]

Coumarins:	Esculin, esculetin, escuside, fraxin, fraxetin, fraxidin, fraxinol, scopoletin, isoscopoletin, scopolin, cichoriin
Phenylpropane derivatives:	Syringin (= Eleutherosid B), lariciresinol, cycloolivil, coniferyl alcohol, dehydroconiferyl alcohol, coniferyl aldehyde
Terpene:	Stylosin
Secoiridoid glycosides:	Chinensisol, oleuropein, neooleuropein, ligustroside, salidroside, frachinoside
Lignans:	Pinoresinol, acetoxypinoresinol, pinoresinol-4-*O*-β-D-glucopyranoside, 8-hydroxypinoresinol, medioresinol, syringaresinol, balanophonin, ficusal
Other compounds:	Flavonoids (rhoiflorin, cosmosiin), saponins, sterols, phenolic acids

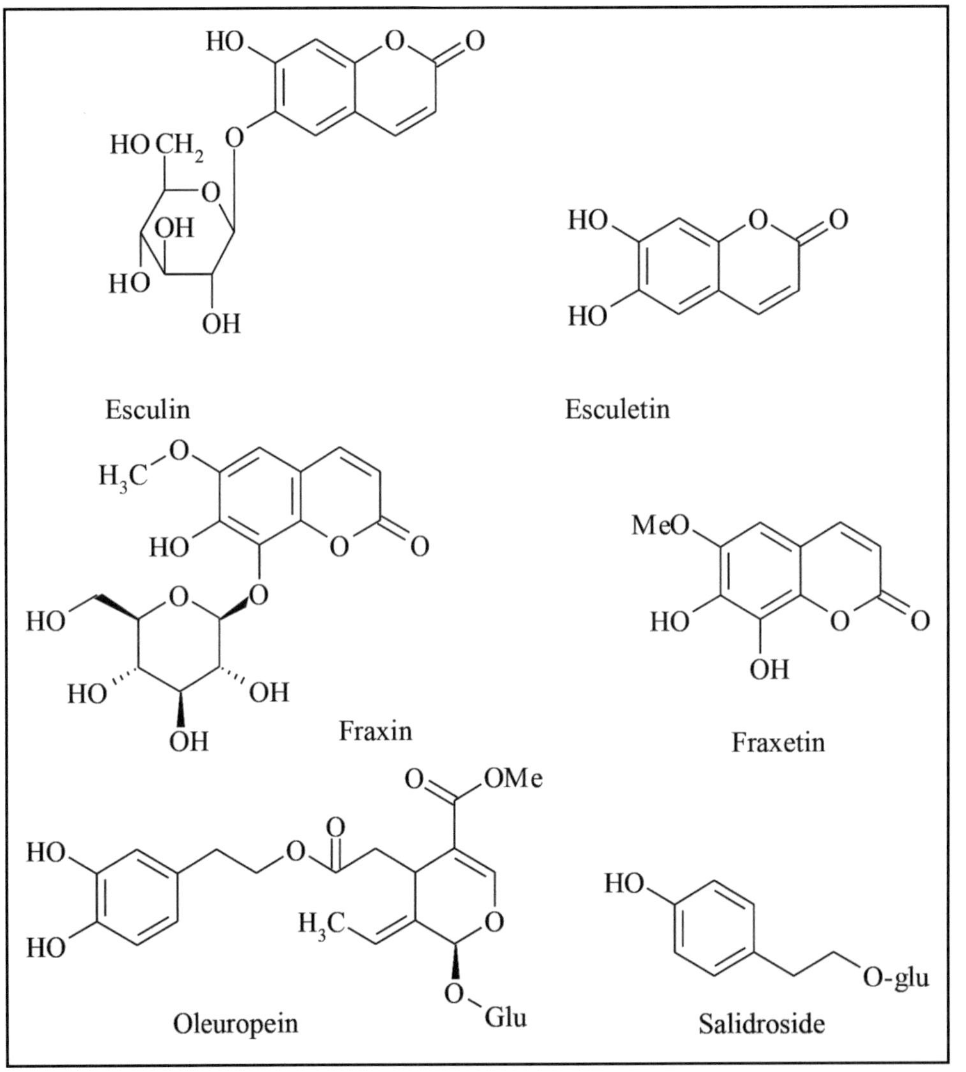

Fig. 1 Formulae of the main compounds of Cortex Fraxini [2, 10, 13]

Reported pharmacology:

- antiallergic [10, 11, 15]
- anti-asthmatic [7]
- expectorant [7, 9]
- antioxidant [10, 11]
- anti-arthritis [11, 13]
- anti-dysenteric [2]
- antitussive [7, 9]
- sedative [7]
- analgesic [7, 11, 15]
- diuretic [7, 10, 11, 15]
- anticonvulsive [7]
- immunomodulatory effects [2, 11, 13, 15]
- anticancer [10, 11, 13]
- inhibition of low-density lipoprotein oxidation (LDL) [2]
- anticoagulant [10, 11, 15]
- antibacterial [10, 11, 15]
- antibiotic [7]

<u>**TLC-Fingerprint Analysis**</u> [17]

	Drug samples	**Origin**
1	Cortex Fraxini / *Fraxinus chinensis*	Province Hunan, China
2	Cortex Fraxini / *Fraxinus rhynchophylla*	Province Liaoning
3	Cortex Fraxini / *Fraxinus szaboana*	Beijing, China
4	Cortex Fraxini / *Fraxinus excelsior*	Sample of commercial drug (firm Finzelberg)
5	Cortex Fraxini / unknown species	Sample of commercial drug (firm Galke)
6	Cortex Fraxini / unknown species	Sample of commercial drug (Munich pharmacy, origin: Hanzhong, Shaanxi)

Reference compounds of Fig. 2		**R*f***
T1	Esculetin	0.89
T2	Esculin	0.60
T3	Fraxetin	0.96
T4	Fraxin	0.44
T5	Oleuropein	0.70
T6	Syringin	0.42
T7	Lariciresinol	0.90
T8	Salidroside	0.61

1. Extraction: 1.0 g powdered drug is extracted with 10 ml methanol under reflux for 10 min. The extract is cooled, filtered and evaporated to dryness. The residue is dissolved in 1 ml methanol
2. Reference compounds: each 0.5 mg is dissolved in 0.5 ml methanol
3. Separation parameters:

   Plate: HPTLC Silica gel 60 F_{254}, Merck

   Applied amounts: Cortex Fraxini extracts: each 5 µl, Reference compounds: each 10 µl

   Solvent system: Ethyl acetate + methanol + water (15.4 + 3 + 1.6)

   Detection: Iron(III)-chloride / Potassium ferricyanide.

   a) 4.5 g ferric chloride are dissolved in 10 ml water.

   b) 1 g potassium ferricyanide is dissolved in 10 ml water.

   Solution **a** and solution **b** are mixed (1:1), the plate is sprayed with the reagent and evaluated in VIS.
4. Description:

   Apart from sample 1 and 2 all extract samples show a rather homogeneous zone profile of 6 dark blue zones which could be very well assigned to the eight references **T1** – **T8.**

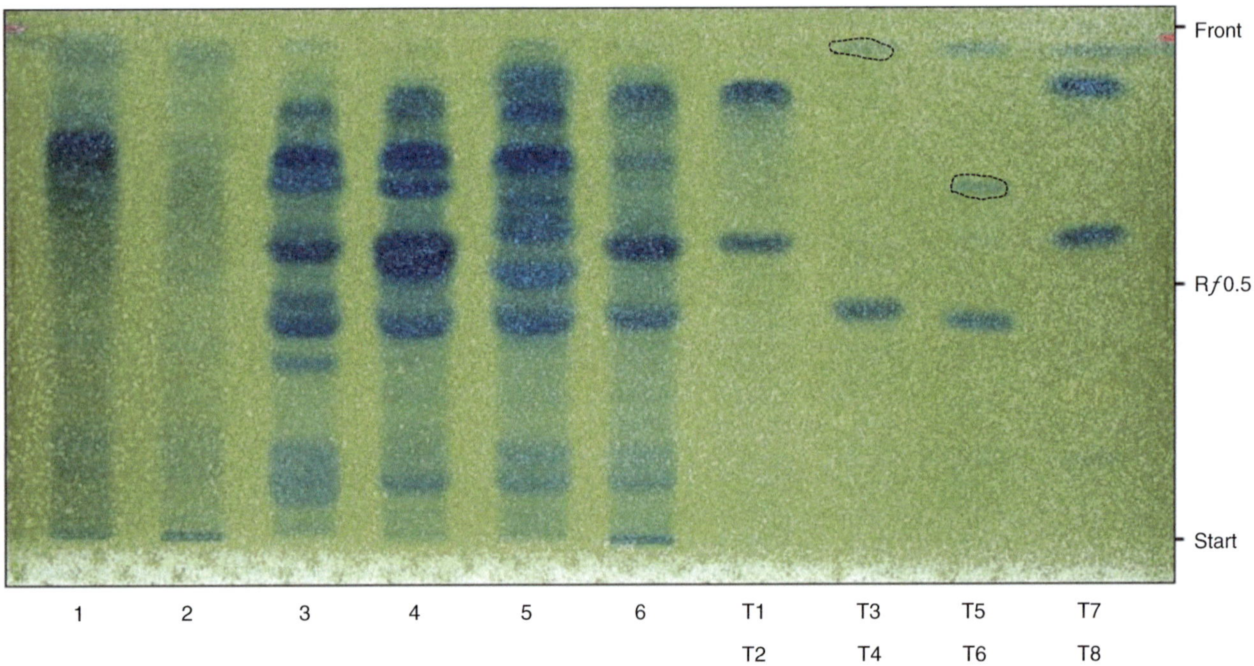

Fig. 2 Thin layer chromatogram of the methanol extracts of Cortex Fraxini, sprayed with Iron(III)-chloride / Potassium ferricyanide (VIS)

HPLC Fingerprint Analysis: [18]

1. Extraction: 1.0 powdered drug is extracted with 10 ml methanol under reflux for 10 min. The extract is cooled, filtered and evaporated to dryness. The residue is dissolved in 1 ml methanol and filtered over Chromafil®, type 0.20 µm
2. Injection volume: Cortex Fraxini extracts: each 5 µl
3. HPLC parameter:

Apparatus: MERCK HITACHI D-6000 A Interface
MERCK HITACHI L-4500 A Diode Array Detector
MERCK HITACHI AS-2000 Autosampler
MERCK HITACHI L-6200 A Intelligent Pump

Separation column: LiChroCART® 250-4 LiChrospher® 100 RP-18 (5 µm), Merck

Precolumn: LiChroCART® 4-4 LiChrospher® 100 RP-18 (5 µm), Merck

Solvent: A: 0. 1 % Phosphoric acid / Water (Millipore Ultra Clear UV plus® filtered)
B: Acetonitrile (VWR)

Gradient: 5–25 % B in 55 min, 25–55 % B in 5 min, Total runtime: 60 min

Flow: 1.0 ml/min

Detection: 210 nm

Retention times of the main peaks

peak	Rt (min)	compound
1	15.3	Esculin
2	21.8	not assignable
3	22.8	Esculetin
4	23.8	Fraxin
5	25.2	not assignable
6	27.6	Fraxetin
7	29.8	not assignable
8	32.8	
9	38.6	
10	41.7	
11	42.9	
12	50.0	Oleuropein / Lariciresinol
13	57.5	not assignable

4. Description of the HPLC-Figures

   The HPLC-fingerprint analysis shows analogue to TLC a fairly well peak profile.

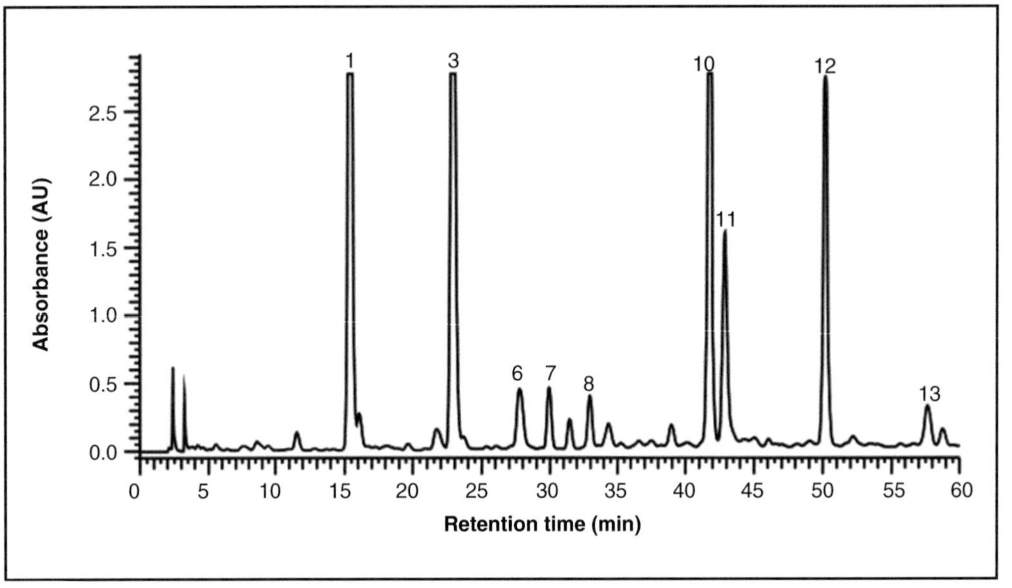

Fig. 3a HPLC fingerprint analysis of the methanol extract of Cortex Fraxini, sample 3

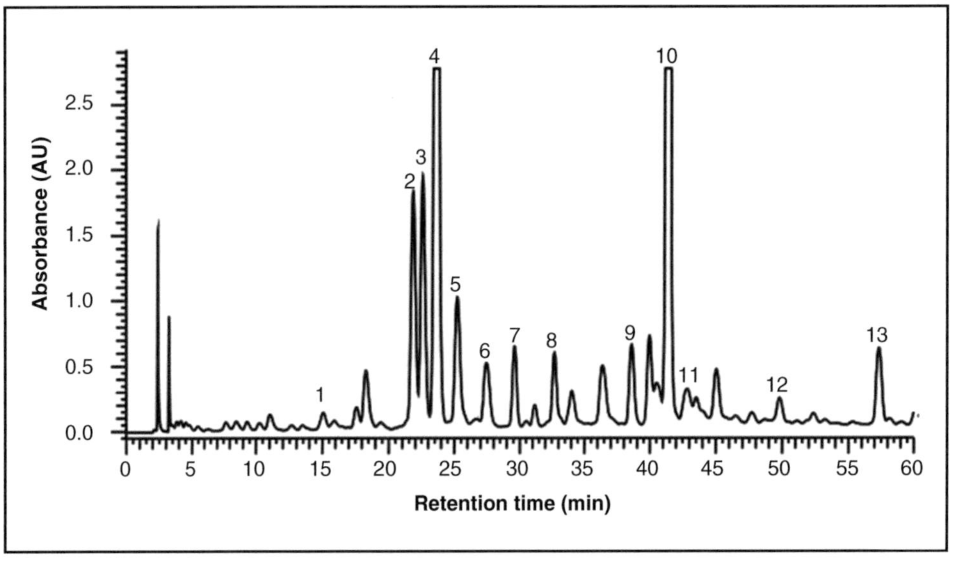

Fig. 3b HPLC fingerprint analysis of the methanol extract of Cortex Fraxini, sample 5

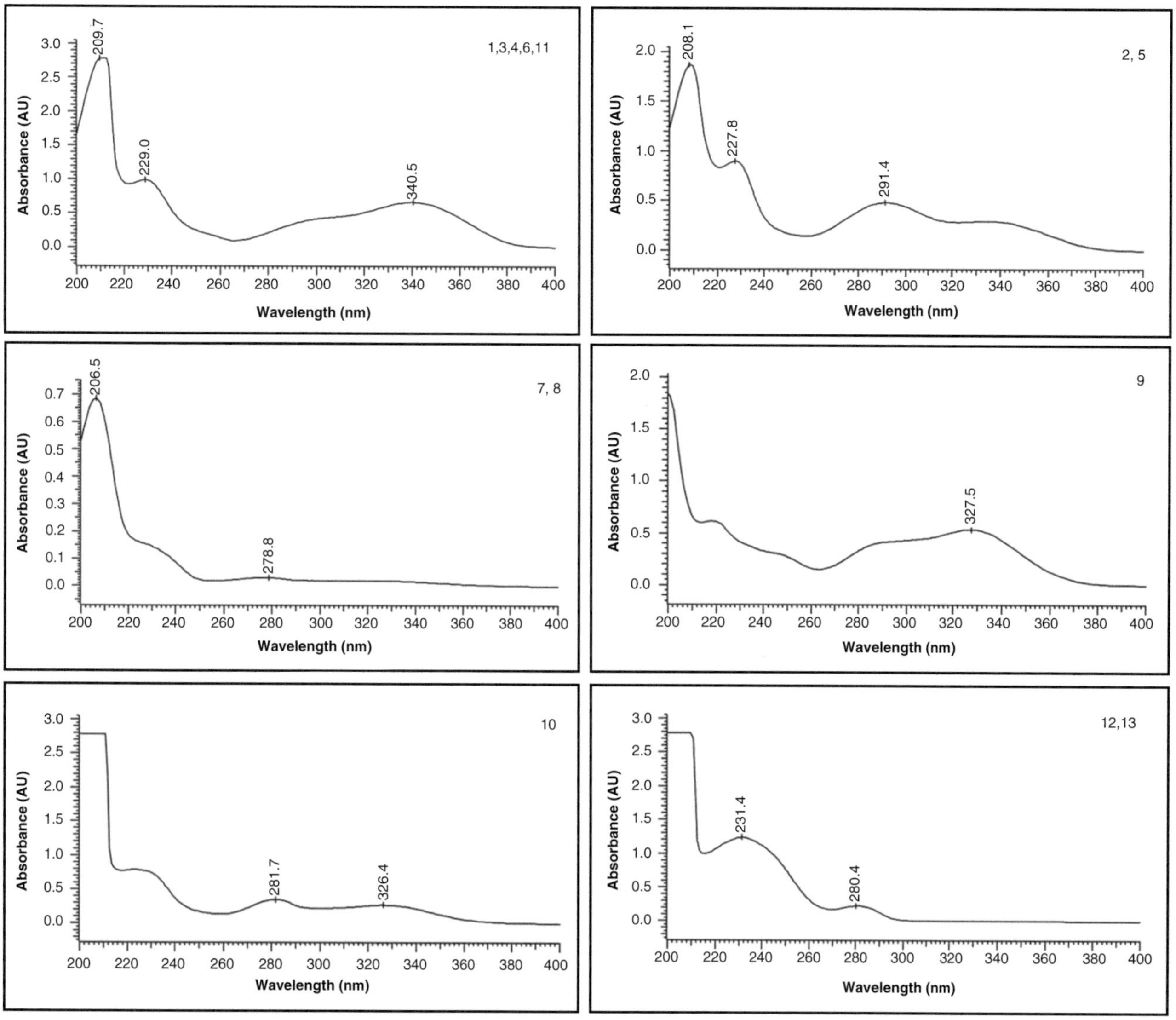

Fig. 4 On line UV-spectra of the main peaks of Cortex Fraxini

Note: According to the Chinese Pharmacopeia 2010 Cortex Fraxini contains not less than 1.0% of the total amount of esculin and esculetin, calculated with reference to the dried drug. [1]

Conclusion

The authentication of Cortex Fraxini is very well stipulated with the TLC and HPLC chromatographic fingerprints.

References

1. Pharmacopoeia of the People's Republic of China. English Edition, vol. 1. People's Medical Publishing House, Beijing (2010)
2. Tang, W., Eisenbrand, G.: Handbook of Chinese medicinal plants – chemistry, pharmacology, toxicology, vol. 1. WILEY-VCH Verlag GmbH & Co. KgaA, Weinheim (2011)
3. Zhang, Z.Z.: An illustrated Chinese materia medica in Hong Kong. School of Chinese Medicines/Hong Kong Baptist University, Hong Kong (2004)
4. Flora of China, *Fraxinus chinensis*, FOC 15, 277 (www.eFloras.org)
5. Flora of China, *Fraxinus chinensis* subsp. *rhynchophylla*, FOC 15, 278 (www.eFloras.org)
6. Flora of China, *Fraxinus stylosa*, FOC 15, 276 (www.eFloras.org)
7. Hempen, C.-H., Fischer, T.: A materia medica for Chinese medicine, English Edition. Elsevier GmbH/Urban & Fischer, Munich (2009)
8. Porkert, M.: Klinische Chinesische Pharmakologie. Verlag für Medizin Dr. Ewald Fischer, Heidelberg (1978)
9. Zhang, H., Li, Q., Shi, Z., Hu, Z., Wang, R.: Analysis of aesculin and aesculetin in Cortex Fraxini by capillary zone electrophoresis. Talanta **52**(4), 607–621 (2000)
10. Wang, L., Sun, F., Zhang, X., Ma, Z., Cheng, L.: A secoiridoid with quinone reductase inducing activity from Cortex Fraxini. Fitoterapia **81**(7), 834–837 (2010)
11. Zhou, L., Kang, J., Fan, L., Ma, X.C., Zhao, H.Y., Han, J., Wang, B.R., Guo, D.A.: Simultaneous analysis of coumarins and secoiridoids in Cortex Fraxini by high-performance liquid chromatography-diode array detection-electrospray ionization tandem mass spectrometry. J. Pharm. Biomed. Anal. **47**(1), 39–46 (2008)
12. Kim, N.Y., Pae, H.O., Ko, Y.S., Yoo, J.C., Choi, B.M., Jun, C.D., Chung, H.T., Inagaki, M., Higuchi, R., Kim, Y.C.: In vitro inducible nitric oxide synthesis inhibitory active constituents from *Fraxinus rhynchophylla*. Plante. Med. **56**(7), 656–658 (1999)
13. Fu, S., Zhang, J., Li, T., Wang, S., Ding, W., Zhao, M., Du, Y., Wang, Q., Jia, J.: Multi-responses extraction optimization based on response surface methodology combined with polarity switching HPLC-MS/MS for the simultaneous quantitation of 11 compounds in Cortex Fraxini: application to four species of Cortex Fraxini and its 3 confusable species. J. Pharma. Biomed. Anal. **91**, 210–221 (2014)
14. Kostova, I., Iossifova, T.: Chemical components of Fraxinus species. Fitoterapia **78**(2), 85–106 (2007)
15. Shin, E., Choi, K.M., Yoo, H.S., Lee, C.K., Hwang, B.Y., Lee, M.K.: Inhibitory effects of Coumarins from the Stem Barks of *Fraxinus rhynchphylla* on adipocyte differentiation in 3 T3-L1 cells. Biol. Pharm. Bull. **33**(9), 1610–1614 (2010)
16. Kuwajima, H., Morita, M., Takaishi, K., Inoue, K., Fujita, T., He, Z.D., Yang, C.R.: Secoiridoid, coumarin and secoiridoid-coumarin glucosides from *Fraxinus chinensis*. Phytochemistry **31**(4), 1277–1280 (1992)
17. Wagner, H., Bauer, R., Melchart, D., Xiao, P.G., Staudinger, A.: Chromatographic fingerprint analysis of herbal medicines, vol. 1–3. Springer, Wien/New York (2011/2015)
18. Hong Kong Chinese materia medica standards, vol. 6. Chinese Medicine Division – Department of Health – Government of the Hong Kong Special Administrative Region – the People's Republic of China, Hong Kong (2013)

Fructus Arctii – *Niubangzi*

Pharmacopoeia: [1]	Pharmacopoeia of the People's Republic of China, English Edition Vol. I, 2010
Official drug: [1]	Great Burdock Achene is the dried ripe fruit of *Arctium lappa* L. (Fam. Asteraceae). The infructescence is collected in autumn when ripe, and dried in the sun, the fruit is tapped out, removed from foreign matter, and dried again in the sun.
Synonyms: [2–5]	Fructus Bardanae, Semen Bardanae, Semen Lappae majoris, *Lappa communis* Coss et Germ., *L. edulis* Sieb., *L. major* Gaerth., *L. minor* DC.
Origin: [3, 6]	Mainly in Chinese provinces such as Zhejiang, Sichuan, Hubei, Hebei and Henan, and in north-eastern China, Asia and Europe.
Description of the drug: [1]	Long-obovate, slightly flattened, somewhat curved, 5–7 mm long, 2–3 mm wide. Externally greyish-brown, purplish-black mottled, with several longitudinal ribs, usually one to two middle ribs relatively distinct. Summit obtuse-rounded, slightly broad, with a circular ring at the top, and a pointed remain of style in the center; base slightly narrowed, bearing surface pale in colour. Pericarp relatively hard, cotyledons 2, yellowish-white, oily. Odour, slight; taste, bitter, slightly pungent and numb.
Processing: [1]	*Fructus Arctii* (*stir-baked*) The clean Fructus Arctii is stir-baked as described under the method for simple stir-baking (Appendix II D) until it becomes inflated and slightly scented. The drug is broken to pieces before use.
Medicinal use: [6, 7]	Treatment of common cold, cough, diabetes and headache.

Effects and indications of Fructus Arctii according to Traditional Chinese Medicine [1, 3–5, 8–14]

Taste:	Pungent and bitter
Temperature:	Cold
Channels entered:	*Orbis pulmonalis, o. stomachi, o. intestini crassi*
Effects (functions):	To disperse wind-heat, diffuse the lung to promote eruption, remove toxin and soothe the throat
Symptoms and indications:	Common cold caused by wind-heat, cough, profuse sputum, measles, rubella, swelling and sore of throat, mumps, erysipelas, swelling abscess, sore and toxin

H. Wagner et al. (eds.), *Chromatographic Fingerprint Analysis of Herbal Medicines,* Vol. 4,
DOI 10.1007/978-3-319-32328-2_3,

Main constituents:

- **Butyrolactone lignans and lignan glycosides** [2, 3, 5–22]

Arctiin, arctigenin, isoarctigenin, arctigenic acid, arctignan E, neoarctin A+B, diarctigenin, 7,8-didehydroarctigenin, matairesinol, matairesinoside, lappaol A-F+H, isolappaol C

- **Caffeoylquinic acids** [10, 17, 20, 23]

Chlorogenic acid, 1,5-dicaffeoylquinic acid

Minor constituents:

- Fatty acids (e.g. arachidic acid, stearic acid, palmitic acid, linoleic acid), sterols (β-sitosterol, daucosterol), gobosterin [2, 3, 8, 16–18]

Reported pharmacology:

- anti-inflammatory [7, 9, 11, 13, 14, 16–19, 23]
- hepatoprotective [5, 7, 23]
- anti-mutagenic [7, 9, 13]
- chemopreventive [7]
- cytotoxic [7, 9, 13, 22]
- inhibits platelet aggregation [5, 9, 10, 13, 14]
- antibiotic [5]
- diuretic [5, 11, 19, 23]
- antidiabetic / hypoglycemic [5, 12, 16–18, 20, 21]
- anti-proliferative [9, 10, 13, 15, 22, 23]
- anti-carcinogenesis [7, 9, 10, 14, 22]
- calcium antagonist [9, 13, 14, 22]
- anti-tumorpromoting effects [9, 17, 20]
- enhancement of anti-coagulation [9]
- anticancer [11, 13, 18]
- antioxidant [11, 14, 23]
- antibacterial [11]
- anti-viral [11, 23]
- immunosuppressive activities [11]
- antimicrobial [14, 16, 23]
- inhibitory activity of diabetic retinopathy / nephropathy [14, 21]

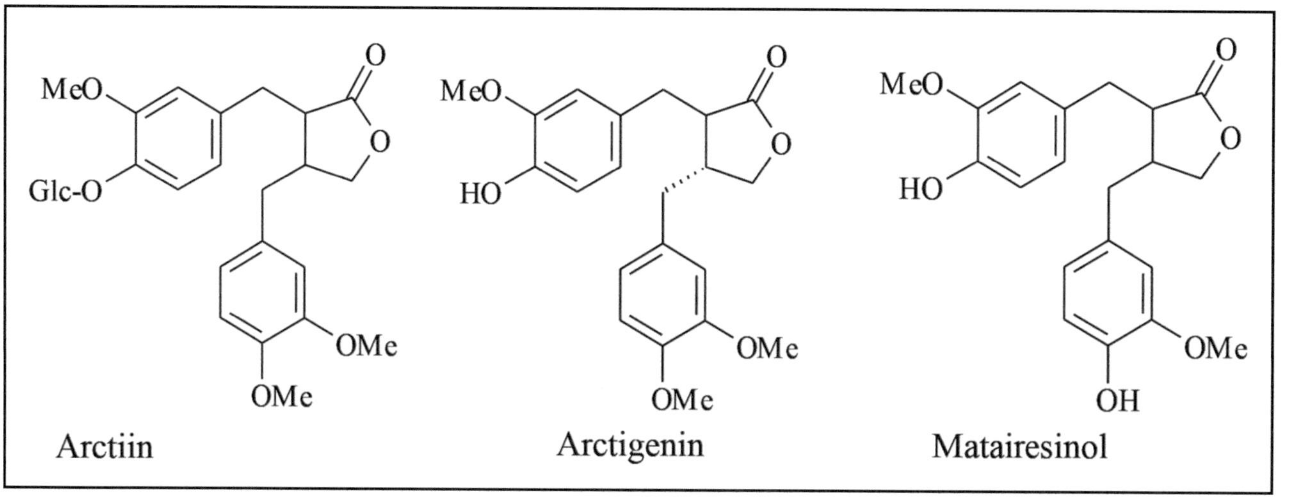

Fig. 1 Formulae of the main compounds of Fructus Arctii [2, 7]

TLC-Fingerprint Analysis

Drug samples		Origin
1	Fructus Arctii / *Arctium lappa*	Sample of commercial drug obtained from firm HerbaSinica (origin: Anhui)
2	Fructus Arctii / *Arctium lappa*	Sample of commercial drug obtained from TCM-Clinic Bad Kötzting (Charge: K14.03.2001)
3	Fructus Arctii / *Arctium lappa*	Sample of commercial drug obtained from TCM- Clinic Bad Kötzting (Charge: K01.09.2005)
4	Fructus Arctii / *Arctium lappa*	Sample of commercial drug obtained from firm China Medica (origin: Neiqiu, Hebei)

Reference compounds		R*f*
T1	Arctigenin	0.89
T2	Arctiin	0.46

1. Extraction: 1.0 g powdered drug is extracted under reflux with 10 ml methanol for 30 min. The extract is filtered, evaporated to dryness and the residue is dissolved in 1 ml methanol
2. Reference compounds: 1.0 mg is dissolved in 1.0 ml methanol

3. Separation parameters:

Plate:	HPTLC Silica gel 60 F_{254}, Merck
Applied amounts:	Fructus Arctii extracts: each 10 µl, Reference compounds: 10 µl
Solvent system:	Dichloromethane + methanol + water (10 + 2 + 0.25)
Detection:	**1. 10 % ethanolic sulphuric acid**
	The plate is sprayed with 8 ml reagent and heated at 110 °C for 10 min. The plate is evaluated in VIS (Fig. 2a)
	2. For visualization of Arctigenin
	Solution **I**: 1 % aqueous potassium hexacyanoferrate (III)
	Solution **II**: 5 % aqueous iron(III) chloride
	Solution I and II are mixed (1:1); the plate is sprayed with 8 ml reagent and evaluated in VIS (Fig. 2b)

4. Description of Fig. 2a:

Arctiin (**T2**) as dominant constituent of Frcutus Arctii appears in VIS as strong brown-purple zone at R*f*=0.46. Arctigenin (**T1**) is not clearly detectable with this spray reagent. The other brown zones in the upper and deep R*f*-range could be not identified but can be assigned to other butyro-lignans and resp. lignan glycosides.

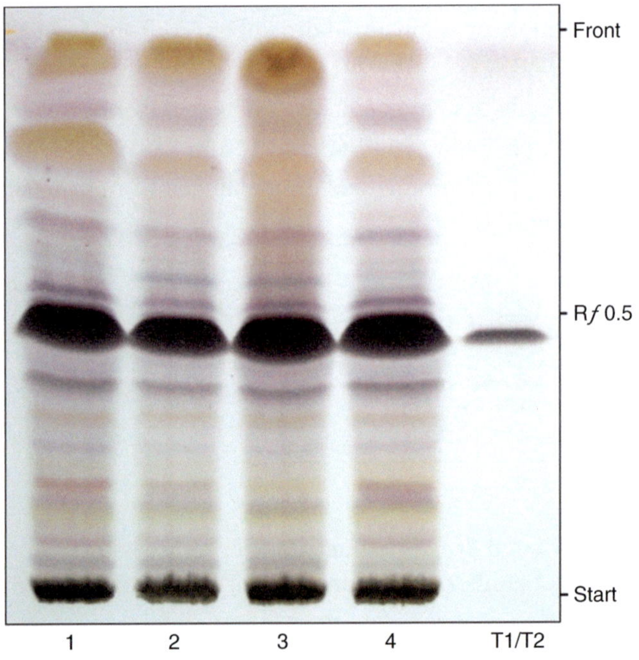

Fig. 2a Thin layer chromatogram of the methanol extracts of Fructus Arctii, sprayed with 10 % ethanolic sulphuric acid (VIS)

Fig. 2b Thin layer chromatogram of the methanol extracts of Fructus Arctii, sprayed with $C_6FeK_3N_6/FeCl_3$ solution (VIS)

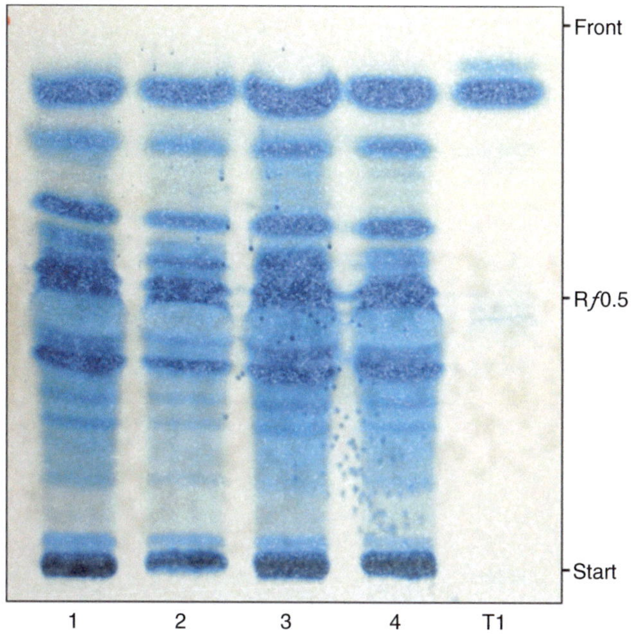

Description of Fig. 2b:

With the $C_6FeK_3N_6/FeCl_3$ reagent all characteristic constituents of Fructus Arctii can be detected as blue zones divided over the whole R*f*-range. Arctigenin (**T1**) is also detectable as distinct blue zone.

HPLC-Fingerprint Analysis: [24]

1. Extraction: 1.0 g powdered drug is extracted under reflux with 10 ml methanol for 30 min. The extract is filtered, evaporated to dryness and the residue dissolved in 1 ml methanol. The extract is filtered over Chromafil®, Type 0.20 µm
2. Injection volume: Fructus Arctii extracts: each 2.5 µl
3. HPLC parameter:

Apparatus:	MERCK HITACHI D-6000 A Interface
	MERCK HITACHI L-4500 A Diode Array Detector
	MERCK HITACHI AS-2000 Autosampler
	MERCK HITACHI L-6200 A Intelligent Pump
Separation column:	LiChroCART® 250–4 LiChrospher® 60 RP select B (5 µm), Merck
Precolumn:	LiChroCART® 4–4 LiChrospher® 60 RP select B (5 µm), Merck
Solvent system:	A: 0.1 % phosphoric acid/water (Millipore Ultra Clear UV plus® filtered)
	B: acetonitrile (VWR)
Gradient:	5–30 % B in 5 min, 30 % B for 20 min, 30–95 % B in 10 min, 95 % B for 20 min
	total runtime: 55 min
Flow:	0.6 ml/min
Detection:	280 nm

Retention times of the main peaks

Peak	Rt (min)	Compound
1	11.6	Chlorogenic acid
2	14.2	Caffeoylquinic acid isomer
3	18.3	Arctiin
4	37.1	Arctigenin

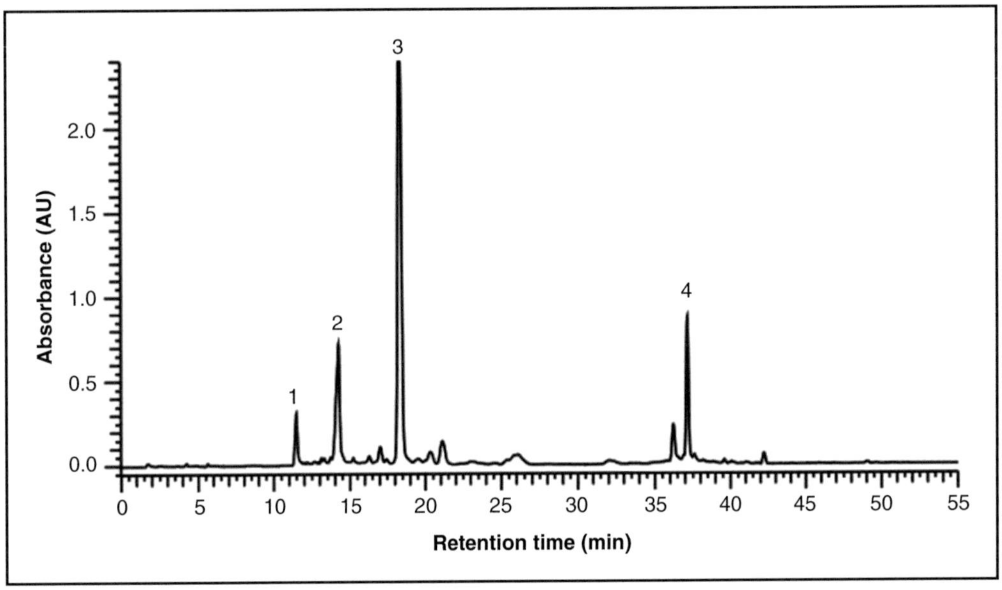

Fig. 3a HPLC-fingerprint analysis of the methanol extract of Fructus Arctii, sample 3

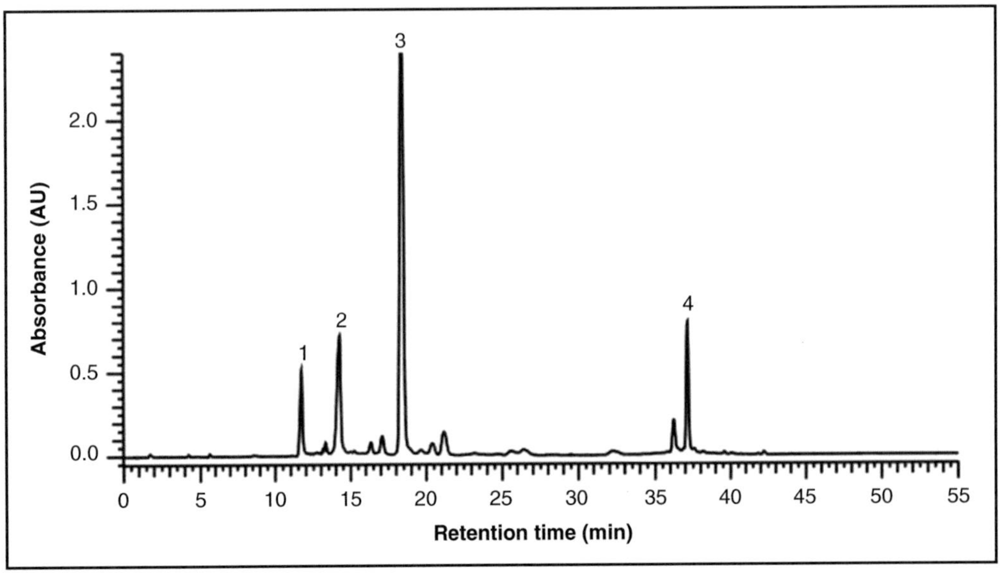

Fig. 3b HPLC-fingerprint analysis of the methanol extract of Fructus Arctii, sample 4

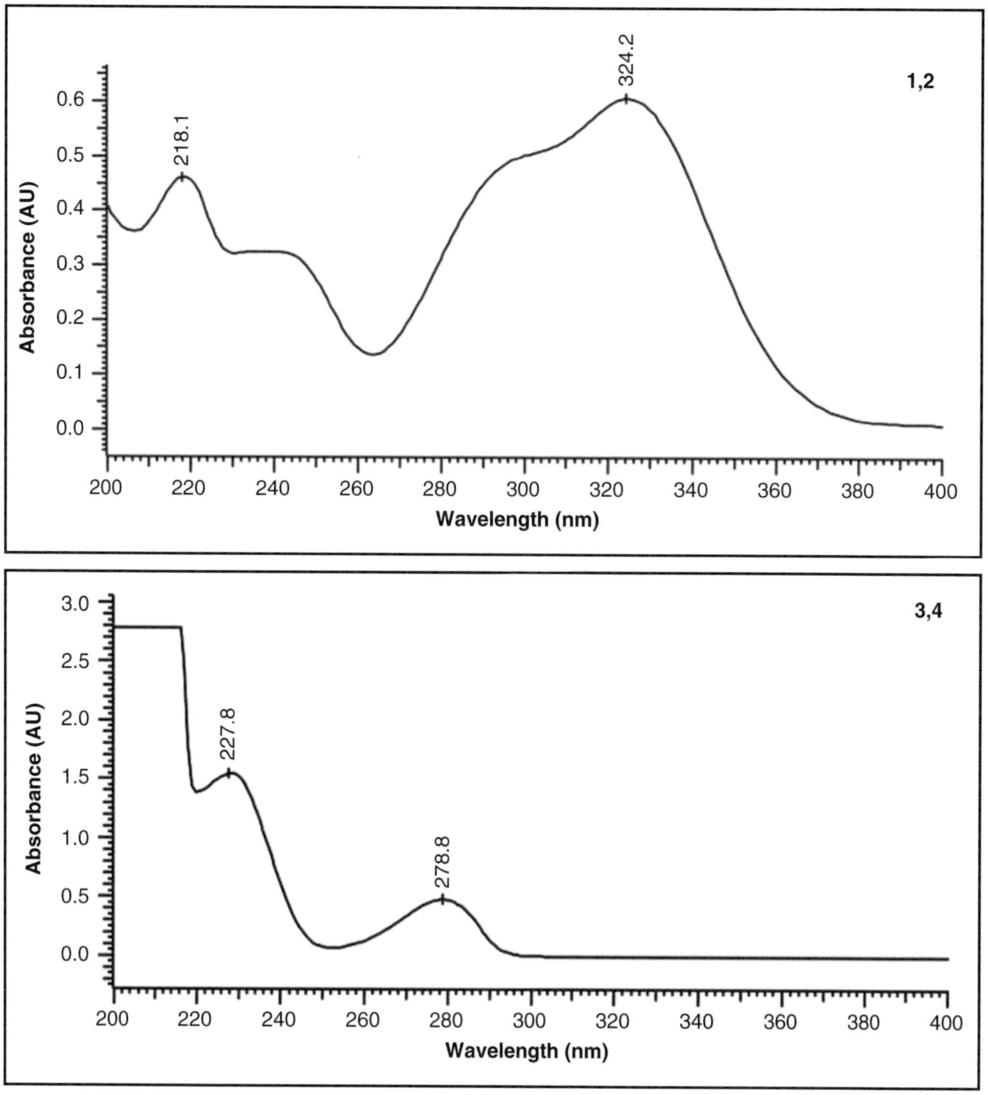

Fig. 4 On line UV-spectra of the main peaks of Fructus Arctii extracts

4. Description of the HPLC-Figures

   The Fructus Arctii extract samples 3 and 4 provide a characteristic peak profile which shows besides the caffeoylquinic acids (Peak **1**=chlorogenic acid, **2**), arctiin (**3**) at Rt=18.3 and arctigenin (**4**) at Rt=37.1.

Note: According to the Chinese Pharmacopeia 2010 Fructus Arctii contains not less than 5.0% of arctiin, calculated with reference to the dried drug [1].

Conclusion

The methanol extracts of Fructus Arctii show in the chosen TLC- and HPLC-analysis the same distinct authentic chromatographic profiles.

References

1. Pharmacopoeia of the People's Republic of China. English Edition, vol. 1. People's Medical Publishing House, Beijing (2010)
2. Blaschek, W., Hänsel, R., Keller, K., Reichling, J., Rimpler, H., Schneider, G.: Hagers Handbuch der Pharmazeutischen Praxis, Folgeband 2, Drogen A – K. Springer, Berlin/Heidelberg (1998)
3. Keys, J.D.: Chinese herbs: their botany, chemistry, and pharmacodynamics. Charles E. Tuttle Company, Rutland/Tokyo (1987)
4. Porkert, M.: Klinische Chinesische Pharmakologie. Verlag für Medizin Dr Ewald Fischer, Heidelberg (1978)
5. Hempen, C.-H., Fischer, T.: A materia medica for Chinese medicine, english edition. Elsevier GmbH, Urban & Fischer, Munich (2009)
6. Zhang, Z.Z.: An illustrated Chinese materia medica in Hong Kong. School of Chinese Medicine/Hong Kong Baptist University, Hong Kong (2004)
7. Tang, W., Eisenbrand, G.: Handbook of Chinese medicinal plants: chemistry, pharmacology, toxicology, vol. 1. WILEY-VCH Verlag GmbH & Co. KgaA, , Weinheim (2011)
8. Paulus, E., Ding, Y.-H.: Handbuch der traditionellen chinesischen Heilpflanzen. Karl F. Haug Verlag, Heidelberg (1987)
9. Wu, J.G., Wu, J.Z., Sun, L.N., Han, T., Du, J., Ye, Q., Zhang, H., Zhang, Y.G.: Ameliorative effects of arctiin from *Arctium lappa* on experimental glomerulonephritis in rats. Phytomedicine **16**(11), 1033–1041 (2009)
10. Qin, K., Liu, Q., Cai, H., Cao, G., Lu, T., Shen, B., Shu, Y., Cai, B.: Chemical analysis of raw and processed Fructus Arctii by high-performance liquid chromatography/diode array detection-electrospray ionization-mass spectrometry. Pharmacogn. Mag. **10**(40), 541–546 (2014)
11. Liu, J., Cai, Y.Z., Wong, R.N.S., Lee, C.K.F., Tang, S.C.W., Sze, S.C.W., Tong, Y., Zhang, Y.: Comparative analysis of caffeoylquinic acids and lignans in roots and seeds among various burdock (arctium lappa) genotypes with high antioxidant activity. J. Agric. Food Chem. **60**(16), 4067–4075 (2012)
12. Xu, Z., Ju, J., Wang, K., Gu, C., Feng, Y.: Evaluation of hypoglycemic activity of total lignans from fructus arctii in the spontaneously diabetic goto-kakizaki rats. J. Ethnopharmacol. **151**(1), 548–555 (2014)
13. Guo, M., Liang, J., Wu, S.: On-line coupling of counter-current chromatography and macroporous resin chromatography for continuous isolation of arctiin from the fruit of *Arctium lappa* L. J. Chromatogr. A **1217**(33), 5398–5406 (2010)
14. Wang, W., Pan, Q., Han, X.Y., Wang, J., Tan, R.Q., He, F., Dou, D.Q., Kang, T.G.: Simultaneous determination of arctiin and its metabolites in rat urine and feces by HPLC. Fitoterapia. **86**, 6–12 (2013)
15. Matsumoto, T., Hosono-Nishiyama, K., Yamada, H.: Antiproliferative and apoptotic effects of butyrolactone lignans from *Arctium lappa* on leukemic cells. Planta Med. **72**(3), 276–278 (2006)
16. Zhao, J., Evangelopoulos, D., Bhakta, S., Gray, A.I., Seidel, V.: Antitubercular activity of arctium lappa and tussilago farfara extracts and constituents. J. Ethnopharmacol. **155**(1), 796–800 (2014)
17. Yang, Y.N., Huang, X.Y., Feng, Z.M., Jiang, J.S., Zhang, P.C.: hepatoprotective activity of twelve novel 7′-hydroxy lignan glucosides from arctii fructus. J. Agric. Food Chem. **62**(37), 9095–9102 (2014)
18. Ming, D.S., Guns, E.S., Eberding, A., Towers, G.H.N.: Isolation and characterization of compounds with anti-prostate cancer activity from arctium lappa L. Using bioactivity-guided fractionation. Pharm. Biol. **42**(1), 44–48 (2004)
19. Park, S.Y., Hong, S.S., Han, X.H., Hwang, J.S., Lee, D., Ro, J.S., Hwang, B.Y.: Lignans from arctium lappa and their inhibition of LPS-induced nitric oxide production. Chem. Pharm. Bull. **55**(1), 150–152 (2007)
20. Qin, K., Wang, B., Li, W., Cai, H., Chen, D., Liu, X., Yin, F., Cai, B.: Quality assessment of raw and processed Arctium lappa L. through multicomponent quantification, chromatographic fingerprint, and related chemometric analysis. J. Sep. Sci. **38**(9), 1491–1498 (2015)
21. Xu, Z., Gu, C., Wang, K., Ju, J., Wang, H., Ruan, K., Feng, Y.: Arctigenic acid, the key substance responsible for the hypoglycemic activity of Fructus Arctii. Phytomedicine **22**(1), 128–137 (2015)
22. Liu, H., Zhang, Y., Sun, Y., Wang, X., Zhai, Y., Sun, Y., Sun, S., Yu, A., Zhang, H.: Determination of the major constituents in fruit of Arctium lappa L. by matrix solid-phase dispersion extraction coupled with HPLC separation and fluorescence detection. J. Chromatogr. B **878**(28), 2707–2711 (2010)
23. Haghi, G., Hatami, A., Mehran, M.: UPLC and HPLC of caffeoyl esters in wild and cultivated *Arctium lappa* L. Food Chem. **138**(1), 321–326 (2013)
24. Wagner, H., Bauer, R., Melchart, D., Xiao, P.G., Staudinger, A.: Chromatographic fingerprint analysis of herbal medicines: thin layer and high performance liquid chromatography of Chinese drugs, volume 1: fructus forsythiae. Springer, Wien (2011)

Fructus Corni – *Shanzhuyu*

Pharmacopoeia: [1]	Pharmacopoeia of the People's Republic of China, English Edition Vol. I, 2010
Official drug: [1]	Asiatic Cornelian Cherry Fruit is the dried ripe sarcocarp of *Cornus officinalis* Sieb. et Zucc. (Fam. Cornaceae). The fruit is collected in later autumn and early winter when the pericarp turns red, baked over a soft fire or treated with boiling water for a moment, removed from kern in time and dried
Origin: [2–7]	Eastern China (especially in the area between Huangho and Yangtse, Anhui, Gansu, Henan, Hubei, Hunan, Jiangsu, Jinagxi, Shaanxi, Shandong, Shanxi, Sichuan, Zheijiang), Korea and Japan.
Description of the drug: [1]	Irregular flaky or bladdery, 1–1.5 cm long and 0.5–1 cm wide. Externally purplish-red to purplish-black, shrunken, lustrous. Sometimes with a rounded scar of persistent calyx at the apex and a scar of fruit stalk at the base. Texture soft. Odour, slight; taste, sour, astringent and slightly bitter.
Pretreatment of the raw drug: [1]	*Pulp of Fructus Corni:* Foreign matters are eliminated and kerns are remained.
Processing: [1]	*Pulp of Fructus Corni (processed with wine)* The pulp of Fructus Corni is stewed or steamed as described under the method for stewing or steaming with wine (Appendix II D) until the wine is absorbed entirely.
Medicinal use:	For treating allergy, hepatitis, rheumatoid arthritis, nephritis and for external treatment of itchy skin conditions.

H. Wagner et al. (eds.), *Chromatographic Fingerprint Analysis of Herbal Medicines*, Vol. 4,
DOI 10.1007/978-3-319-32328-2_4, © Springer International Publishing Switzerland 2016

Effects and indications of Fructus Corni according to Traditional Chinese Medicine [1–5, 7–16]

Taste:	Sour and astringent
Temperature:	Warm tendency
Channels entered:	*Orbis hepaticus, orbis renalis*
Effects (functions):	To tonify and nourish kidney and liver, astringe to secure collapse
Symptoms and indications:	Dizziness, tinnitus, sore pain in the low back and knees, impotence, seminal emission, uroclepsia, enuresis, frequent urination, flooding and spotting, vaginal discharge, profuse sweating, collapse, interior heat and wasting-thirst. Spermatorrhea, hemorrhage, thamurgia, lumbago, vertigo, and night sweats. Reducing blood glucose, treating osteoporosis. Metrorrhagia and leukorrhea

Main constituents:

Iridoid glycoside / aglycones [4, 6, 7, 9, 11–30]	Loganin, secoxyloganin, loganic acid, logmalicid A+B, cornuside, morronoside, 7-α-morroniside, 7-β-morroniside, 7-α-O-methylmorroniside, 7-β-O-methylmorroniside, 7-α-O-ethylmorroniside, 7-β-O-ethylmorroniside, morroniside acetate, sweroside cornel iridoid glycoside, cornin (verbenalin)
Tannins, tannic acid and Gallotannins [4, 6, 7, 9, 11, 12, 15, 16, 18, 19, 21, 24–26, 29]	2,3-di-O-galloyl-D-glucose; 1,2,3-tri-O-galloyl-β-D-glucose; 1,2,6-tri-O-gallyol-β-D-glucose; 1,2,3,6-tetra-O-gallyol-β-D-glucose; cornusiin A-C,G; 7-O-galloyl-D-sedoheptulose; gemin D; isoterchebin; tellimagrandins I+II; oenothein C; isorugosin B; gallic acid; tannic acid; ellagic acid
Organic / phenolic acids [4, 6, 11, 12, 17, 30]	Caffeic acid, tartaric acid, malic acid, succinic acid, citric acid, p-coumaric acid, protocatechuic acid
Triterpenoids [6, 9, 15, 21, 30]	Oleanolic acid, ursolic acid
Flavonoids [6, 15]	Kaempferol, kaempferol-3-O-β-D-galactopyranoside; kaempferol-3-O-β-D-rutinoside (**nicotiflorin**), quercetin, quercetin-3-O-β-D-glucuronide methyl ester, quercetin-3-O-β-D-glucopyranoside (**isoquercitrin**), quercetin-3-O-β-D-galactopyranoside (**hyperoside**)
Volatile flavor components [15, 24]	Benzyl cinnamate, iso-butyl alcohol, iso-amyl alcohol, furfural, methyl eugenol, iso-asarone, β-phenyl ethyl alcohol, trans-linalool oxide, elemicine

Minor constituents:

Furan derivatives (e.g., 5-hydroxymethylfurfural) [9, 15, 16, 26], polysaccharides [6, 7, 11, 12] and triterpen-saponins [7, 9, 25, 30]

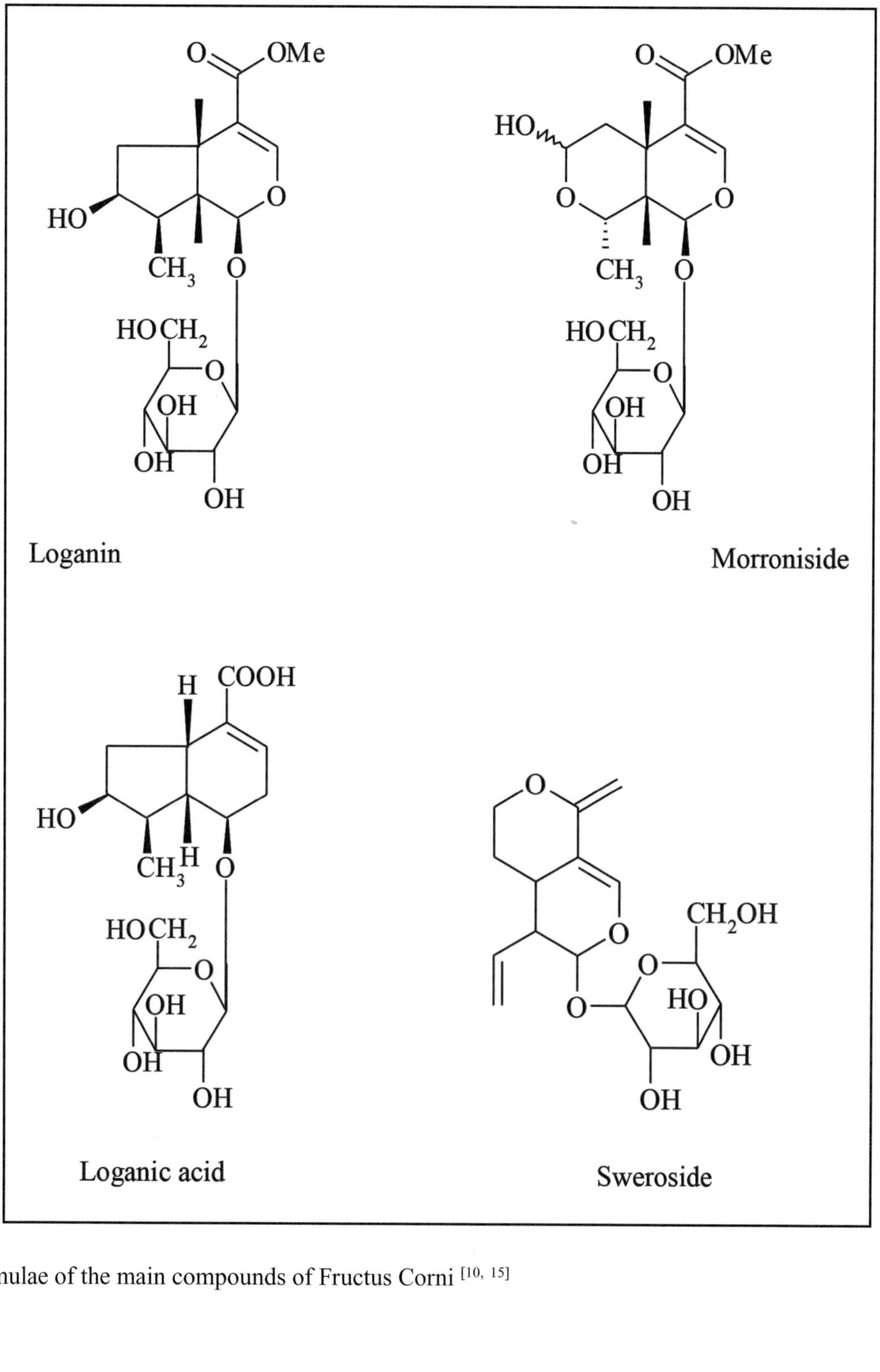

Fig. 1 Formulae of the main compounds of Fructus Corni [10, 15]

Reported main pharmacological activities:

- anti-inflammatory [7, 9, 12, 16, 17, 23, 27, 28, 30]
- anti-diabetic [13, 20, 23, 27, 28, 30]
- anti-hyperlipidemic / anti-hypertriglyceridemic activity [9, 19, 20]
- immunomodulatory [11–13, 16]
- neuroprotective [6, 14, 22]
- anti-cancer [11, 16]

TLC-Fingerprint Analysis

Drug samples		**Origin**
1	Fructus Corni / *Cornus officinalis*	Sample of commercial drug (Caelo)
2	Fructus Corni / *Cornus officinalis*	Sample of commercial drug (TCM-Clinic, Bad Kötzting)
3	Fructus Corni / *Cornus officinalis*	Sample of commercial drug (TCM-Clinic, Bad Kötzting)
4	Fructus Corni / *Cornus officinalis*	Sample of commercial drug (TCM-Clinic, Bad Kötzting)
5	Fructus Corni / *Cornus officinalis*	Sample of commercial drug (HerbaSinica, origin: Shanxi)
6	Fructus Corni / *Cornus officinalis*	Sample of commercial drug (China Medica, origin: Zheijiang)

Reference compounds of Fig. 2a/b		**R*f***
T1	Loganin	0.50
T2	Loganic acid	0.24
T3	Oleanolic acid	0.95
T4	Sweroside	0.47

1. Extraction: 1.0 g powdered drug is ultrasonicated with 10 ml ethanol for 30 min. The extract is filtered, evaporated to dryness and the residue dissolved in 1 ml ethanol.

2. Reference compounds: Each 1.0 mg is dissolved in 1.0 ml methanol

3. Separation parameters:

   Plate: HPTLC Silica gel 60 F254, Merck

   Applied amounts: Fructus Corni extracts: each 5 µl, Reference compounds: each 10 µl (T1–T3), 20 µl (T4)

   Solvent system: Ethyl acetate water + methanol + water (15 + 4 + 3.5)

   Detection: Anisaldehyde – Sulphuric acid reagent

   0.5 ml anisaldehyde is mixed with 10 ml glacial acetic acid, followed by 85 ml methanol and 5 ml concentrated sulphuric acid, in that order.

   The plate is sprayed with 10 ml, heated at 110 °C for 10 min, then evaluated in VIS and under UV 366 nm

   Note: The reagent has only limited stability and is no longer useable when the colour has turned to red-violet.

4. Description:

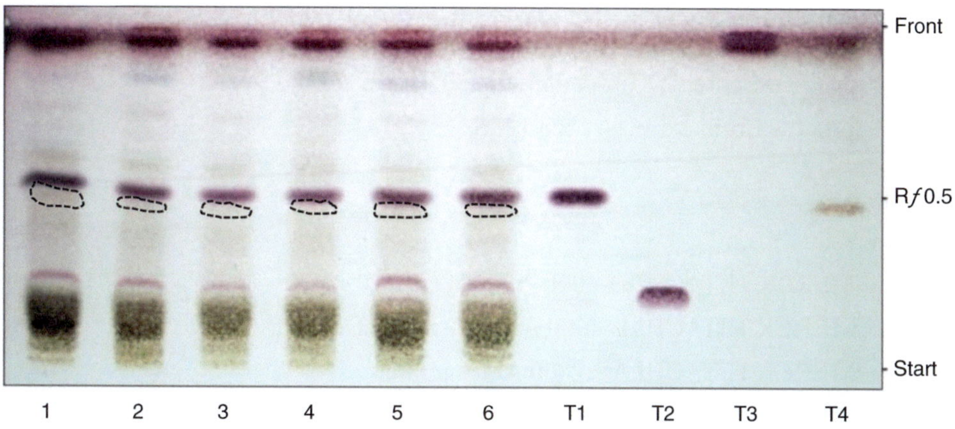

Fig. 2a Thin layer chromatogram of the methanol extracts of Fructus Corni, sprayed with Anisaldehyde – Sulphuric acid reagent (VIS)

Description of Fig. 2a:

The TLC in VIS is characterized by brown/red zones on the front at R*f* = 0.95 / 0.99 (oleanolic acid, **T3**) and any of the volatile compounds (e.g. asarone). A violet zone appears at R*f* = 0.50 (loganin, **T1**), direct underneath at R*f* = 0.49 sweroside (**T4**), a further violet zone at R*f* = 0.24 (loganic acid, **T2**) and at the low R*f*-range a broad zone between R*f* = 0.18 and R*f* = 0.07, which might contain cornusiin A-C and other gallotannins.

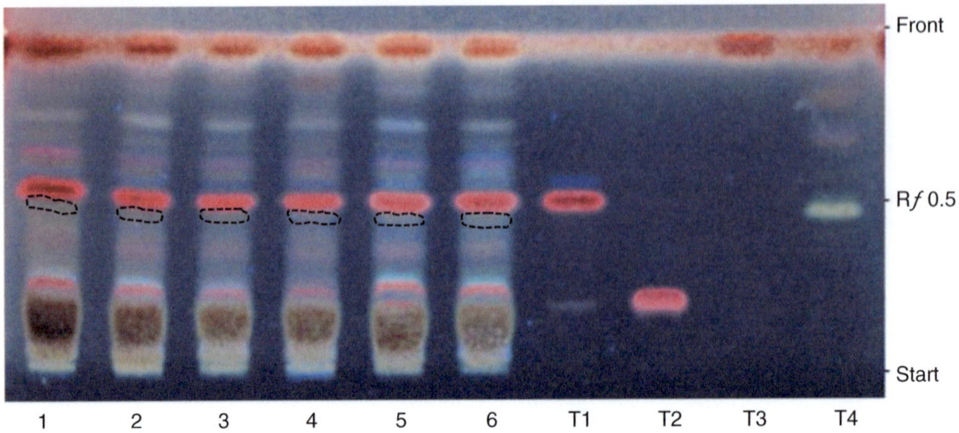

Fig. 2b Thin layer chromatogram of the methanol extracts of Fructus Corni, sprayed with Anisaldehyde – Sulphuric acid reagent (UV 366 nm)

Description of Fig. 2b:

This TLC under UV 366 nm shows the same pattern of zones but with primarily light red or weak grey coloured zones. The assignment of the main red zones is the same as in Fig. 2a, except the grey zones between R*f* = 0.5 and R*f* = 0.85. These may be assigned to the different triterpenoids. The gallotannins or galloglycosides are again centred around R*f* = 0.18.

HPLC-Fingerprint Analysis

1.	Sample preparation:	1.0 g powdered drug is ultrasonicated with 10 ml ethanol for 30 min. The extract is filtered and evaporated to dryness. The residue is dissolved in 1 ml ethanol and filtered over Millipore® filtration unit, type 0.45 µm
2.	Injection volume:	Fructus Corni extracts: each 10 µl
3.	HPLC parameter:	
	Apparatus:	MERCK HITACHI D-6000 A Interface MERCK HITACHI L-4500 A Diode Array Detector MERCK HITACHI AS-2000 Autosampler MERCK HITACHI L-6200 A Intelligent Pump
	Separation column:	LiChroCART® 250-4 LiChrospher® 100 RP-18 (5 µm), Merck
	Precolumn:	LiChroCART® 4-4 LiChrospher® 100 RP-18 (5 µm), Merck
	Solvent:	A: 0.001 % Phosphoric acid/Water (Millipore Ultra Clear UV plus® filtered) B: Acetonitrile (Fa. VWR)
	Gradient:	0 % B for 5 min, 0–100 % B in 60 min, 100 % B for 5 min, total runtime: 70 min
	Flow:	1.0 ml/min
	Detection:	210 nm

Retention times of the main peaks

Peak	Rt (min)	Compound
1	15.3	Gallic acid
2	18.6	Loganic acid
3	19.2	Not identified iridoid glycoside (morroniside?)
4	21.0	Loganin
5	19.9	Sweroside
6	21.9	Not identified
7	63.1	Oleanolic acid

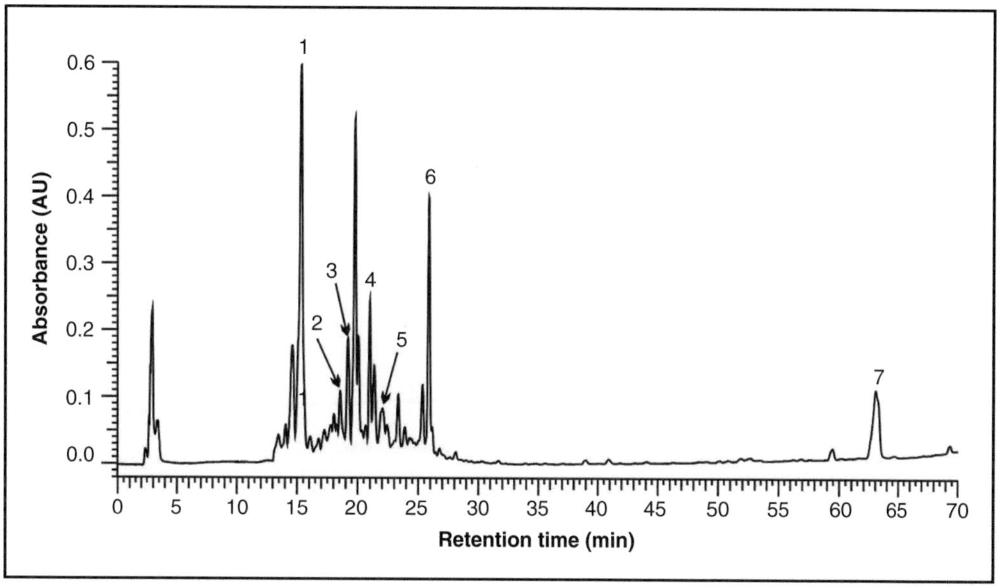

Fig. 3a HPLC-fingerprint analysis of the extract of Fructus Corni, sample 1

Description of Figs. 3a and 3b:

We see an accumulation of 6 peaks (peak **1–6**). The non-identified compound at Rt = 19 might be assignable to morroniside (**3**) according to the UV-spectrum. Peak **6** shows nearly the same UV-profile as gallic acid (**1**) and could be assigned to a more lipophilic galloylester.

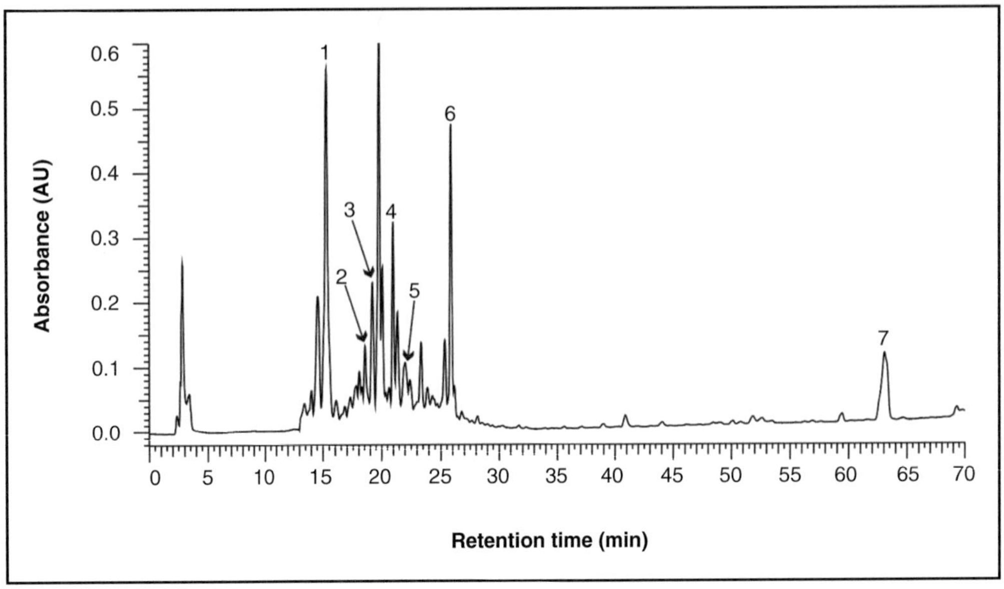

Fig. 3b HPLC-fingerprint analysis of the extract of Fructus Corni, sample 6

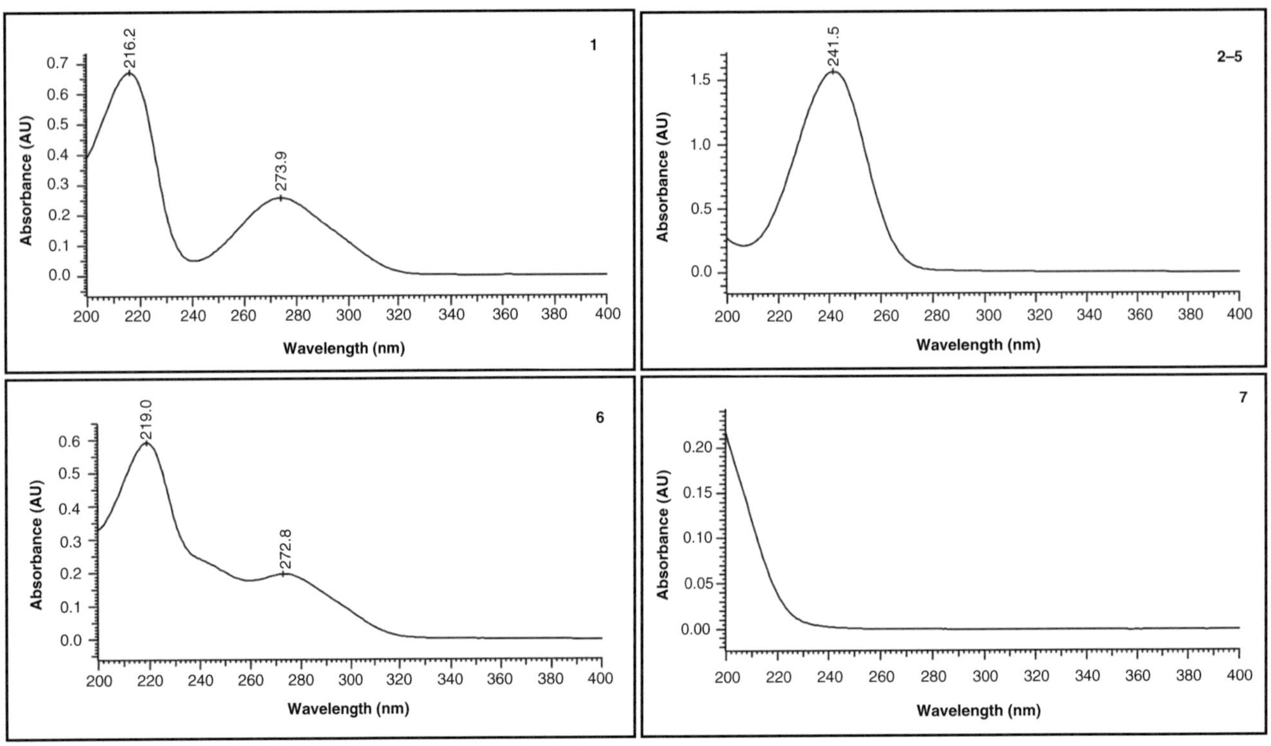

Fig. 4 On line UV-spectra of the main peaks of Fructus Corni

Note: According to the Chinese Pharmacopeia 2010 Fructus Corni contains not less than 0.60 % of loganin, calculated with reference to the dried drug [1].

Conclusion

The methanol extracts of Fructus Corni show in the chosen TLC- and HPLC-analysis the same distinct authentic chromatographic profiles.

References

1. Pharmacopoeia of the People's Republic of China. English Edition, vol. 1. People's Medical Publishing House, Beijing (2010)
2. Porkert, M.: Klinische Chinesische Pharmakologie. Verlag für Medizin Dr. Ewald Fischer, Heidelberg (1978)
3. Keys, J.D.: Chinese herbs: their botany, chemistry, and pharmacodynamics. Charles E. Tuttle Company, Rutland/Tokyo (1976)
4. Flora of China, *Cornus officinalis*, FOC 14, 215 (www.eFloras.org)
5. Paulus, E., Ding, Y.-H.: Handbuch der traditionellen chinesischen Heilpflanzen. Karl F. Haug Verlag, Heidelberg (1987)
6. Ma, W., Wang, K.J., Cheng, C.S., Yan, G.Q., Lu, W.L., Ge, J.F., Cheng, Y.X., Li, N.: Bioactive compounds from *Cornus officinalis* fruits and their effects on diabetic nephropathy. J. Ethnopharmacol. **153**(3), 840–845 (2014)
7. He, L.H.: Comparative study for α-glucosidase inhibitory effects of total iridoid glycosides in the crude products and the Wine-processed products from *Cornus officinalis*. Yakugaku Zasshi **131**(12), 1801–1805 (2011)
8. Stöger, E.A.: Arzneibuch der chinesischen Medizin. Dtsch. Apotheker-Verlag, Stuttgart (2009)
9. Hempen, C.-H., Fischer, T.: A materia medica for Chinese medicine, english edition. Elsevier GmbH, Urban & Fischer, Munich (2009)
10. Tang, W., Eisenbrand, G.: Handbook of Chinese medicinal plants: chemistry, pharmacology, toxicology, vol. 1. WILEY-VCH Verlag GmbH & Co. KgaA, Weinheim (2011)
11. Wu, Y., Wang, X., Shen, B., Kang, L., Fan, E.: Extraction, structure and bioactivities of the polysaccharides from frutcus corni. Recent Pat. Food Nutr. Agric. **5**(1), 57–61 (2013)
12. Park, C.H., Tanaka, T., Kim, J.H., Cho, E.J., Park, J.C., Shibahara, N., Yokozawa, T.: Hepato-protective effects of loganin, iridoid glycoside from Corni Fructus, against hyperglycemia-activated signaling pathway in liver of type 2 diabetic *db/db* mice. Toxicology **290**(1), 14–21 (2011)
13. Li, X., Huo, C., Wang, Q., Zhang, X., Sheng, X., Zhang, L.: Identification of new metabolites of morroniside produced by rat intestinal bacteria and HPLC-PDA analysis of metabolites in vivo. J. Pharm. Biomed. Anal. **45**(2), 268–274 (2007)
14. Liang, J., He, J., Zhu, S., Zhao, W., Zhang, Y., Ito, Y., Sun, W.: Preparative isolation and purification of iridoid glycosides from Fructus Corni by high-speed countercurrent chromatography. J. Liq. Chromatogr. Relat. Tech. **36**(8), 983–999 (2013)
15. Kim, D.K., Kwak, J.H.: A furan derivative from *Cornus officinalis*. Arch. Pharm. Res. **21**(6), 787–789 (1998)
16. Wang, M.Y., Zhao, F.M., Peng, H.Y., Lou, C.H., Li, Y., Ding, X., Yu, X.Y., Yang, G.M., Xu, D.Q., Jiang, L.H., Zhnag, X., Ye, L.H., Cai, B.C.: Investigation on the morphological protective effect of 5-hydroxymethylfurfural extracted from wine-processed Fructus corni on human L02 hepatocytes. J. Ethnopharmacol. **130**(2), 424–428 (2010)
17. Zhang, Q.C., Zhao, Y., Bian, H.M.: Antiplatelet activity of a novel formula composed of malic acid, succinic acid and citric acid from *Cornus officinalis* Fruit. Phytother. Res. **27**(12), 1894–1896 (2013)
18. Park, C.H., Tanaka, T., Cho, E.J., Park, J.C., Shibahara, N., Yokozawa, T.: Glycerol-induced renal damage improved by 7-*O*-galloyl-D-sedoheptulose treatment throough attenuating oxidative stress. Biol. Pharm. Bull. **35**(1), 34–41 (2012)
19. Lee, N.H., Seo, C.S., Lee, H.Y., Jung, D.Y., Lee, J.K., Lee, J.A., Song, K.Y., Shin, H.K., Lee, M.Y., Seo, Y.B., Kim, H., Ha, H.: Hepatoprotective and antioxidative activities of *Cornus officinalis* against acetaminophen-induced hepatotoxicity in mice. Evid. Based Complement Alternat. Med. **2012**, 804834 (2012)
20. Wang, W., Xu, J., Li, L., Wang, P., Ji, X., Ai, H., Zhang, L., Li, L.: Neuroprotective effect of morroniside on focal cerebral ischemia in rats. Brain Res. Bull. **83**(5), 196–201 (2010)
21. Jeong, E.J., Kim, T.B., Yang, H., Kang, S.Y., Kim, S.Y., Sung, S.H., Kim, Y.C.: Neuroprotective iridoid glycosides from Cornus officnalis fruits against glutamate-induced toxicity in HT22 hippocampal cells. Phytomedicine **19**(3-4), 317–321 (2012)

22. Yokozawa, T., Yamabe, N., Kim, H.Y., Kang, K.S., Hur, J.M., Park, C.H., Tanaka, T.: Protective effects of morroniside isolated from corni fructus against renal damage in streptozotocin-induced diabetic rats. Biol. Pharm. Bull. **31**(7), 1422–1428 (2008)
23. Jiang, W.L., Zhang, S.M., Tang, X.X., Liu, H.Z.: Protective roles of cornuside in acute myocardial ischemia and reperfusion injury in rats. Phytomedicine **18**(4), 266–271 (2011)
24. Miyazawa, M., Kameoka, H.: Volatile flavor components of corni fructus (*Cornus officinalis* Sieb. Et Zucc.). Agric. Biol. Chem. **53**(12), 3337–3340 (1989)
25. Kim, H.J., Kim, B.H., Kim, Y.C.: Antioxidative action of corni fructus aqueous extract on kidneys of diabetic mice. Toxicol. Res. **27**(1), 37–41 (2011)
26. Yokozawa, T., Kang, K.S., Park, C.H., Noh, J.S., Yamabe, N., Shibahara, N., Tanaka, T.: Bioactive constituents of Corni Fructus: the therapeutic use of morroniside, loganin, and 7-*O*-galloyl-D-sedoheptulose as renoprotective agents in type 2 diabetes. Drug. Discov. Ther. **4**(4), 223–234 (2010)
27. Jiang, W.L., Chen, X.G., Zhu, H.B., Hou, J., Tian, J.W.: Cornuside attenuates apoptosis and ameliorates mitochondrial energy metabolism in rat cortical neurons. Pharmacology **84**(3), 162–170 (2009)
28. Kang, D.G., Moon, M.K., Lee, A.S., Kwon, T.O., Kim, J.S., Lee, H.S.: Cornuside suppresses cytokine-induced proinflammatory and adhesion molecules in the human umbilical vein endothelial cells. Biol. Pharm. Bull. **30**(9), 1796–1799 (2007)
29. Park, C.H., Tanaka, T., Yokozawa, T.: Evaluation of 7-*O*-galloyl-D-sedoheptulose, isolated from Corni Fructus, in the adipose tissue of type 2 diabetes db/db mice. Fitoterapia **89**, 131–142 (2013)
30. Peng, Q., Wei, Z., Lau, B.H.S.: Fructus corni enhances endothelial cell antioxidant defenses. Gen. Pharmac. **31**(2), 221–225 (1998)

Fructus Kochiae – *Difuzi*

Pharmacopoeia: [1] Pharmacopoeia of the people's Republic of China, English Edition Vol. I, 2010

Official drug: [1] Belvedere Fruit is the dried ripe fruit of *Kochia scoparia* (L.) Schrad. (Fam. Chenopodiaceae=Amaranthaceae)

The plant is collected in autumn when the fruit is ripe, dried in the sun, the fruit is gathered, and removed from foreign matters.

Origin: [2, 3] Mainly in Chinese provinces such as Shandong, Jiangsu, Henan and Hebei; Asia, Europe, widely naturalized in Africa, Australia, North and South America.

Synonym: [4] *Bassia scoparia*

Description of the drug: [1] Oblate-spheroidal, five-pointed star shape, 1–3 mm in diameter, surrounded by a persistant perianth. Externally greyish-green or pale brown, with 5 membranous winglets, the centre of dorsal surface having a slightly prominent, pointed fruit stalk scar and 5–10 radial veins; membranous pericarp visible when the perianth stripped, translucent. Seeds flattened-ovoid, about 1 mm long, black. Odour, slight; taste, slightly bitter.

Medicinal use: [5–10] Used for the treatment of external skin dermatomycoses and cutaneous pruritus. It is also used as remedy of diabetes mellitus and rheumatoidal arthritis and as diuretic agent.

Effects and indications of Fructus Kochiae according to Traditional Chinese Medicine [1, 11–13]	
Taste:	Bitter, sweet
Temperature:	Cold
Channels entered:	*Orbis vesicalis, o. renalis*
Effects (functions):	To clear heat and drain dampness, and dispel wind to relieve itching.
Symptoms and indications:	Slow and painful urination, pudendal itching, vaginal discharge, rubella, eczema, itching of skin.

H. Wagner et al. (eds.), *Chromatographic Fingerprint Analysis of Herbal Medicines*, Vol. 4,
DOI 10.1007/978-3-319-32328-2_5, © Springer International Publishing Switzerland 2016

Main constituents:

-Triterpene saponins [5–10, 12, 14–19]

oleanolic acid, oleanolic acid 3-*O*-glucuronide, momordin Ic (= kochioside Ic), momordin IIc (momordin Ic 28-β-D-glucopyranosyl ester), momordin Ic 6'-methyl ester, 2'-O-β-D-glucopyranosyl-momordin Ic, 2'-O-β-D-glucopyranosyl-momordin IIc, kochianoside I-IV, scoparianosides A-C

-Flavonol glycosides [20]

Quercetin-7-*O*-β-D-glucopyranoside, quercetin-7-*O*-β-D-sophoroside, quercetin-3-*O*-β-D-apiofuranosyl-(1 → 2)-β-D-galactopyranoside, quercetin-3-*O*-β-D-apiofuranosyl-(1 → 2)-β-D-galactopyranosyl-7-*O*-β-D-glucopyranoside, quercetin-3-*O*-α-L-rhamnopyranosyl-(1 → 6)-β-D-galactopyranosyl-7-*O*-β -D-sophoroside, quercetin-3-*O*-β-D-galactopyranosyl-7-*O*-β-D-glucopyranoside

Minor constituents:

Alkaloids [12, 15, 21]
Phytoecdysteroids [15]
Fatty acids (5-hexadecenoic acid) [22]

Reported pharmacology:

-anti-fungal [12]
-anti-inflammatory [5–7, 9, 10, 17, 23]
-antiallergic [5, 6, 8, 9, 17]
-antipruritic [5, 8, 9, 17]
-inhibitory activity against dermatomycoses [13]
-hepatoprotective [5]
-analgetic effect [5]
-antinociceptice [5]
-peripheral antinociceptive [6, 7, 9, 10, 17, 23]
-antidiabetic [6]
-pro-apoptotic [10]
-gastroprotective effects [10]
-inhibits ethanol and indomethacin-induced gastric mucosal lesions [17]
-inhibits glucose uptake in the small intestine [17]
-induces HepG2cell apoptosis in a ROS-mediated PI3K and MAPK pathway-dependent manner [14, 16]
-contraction of the ileum caused by histamine or barium chloride [13]

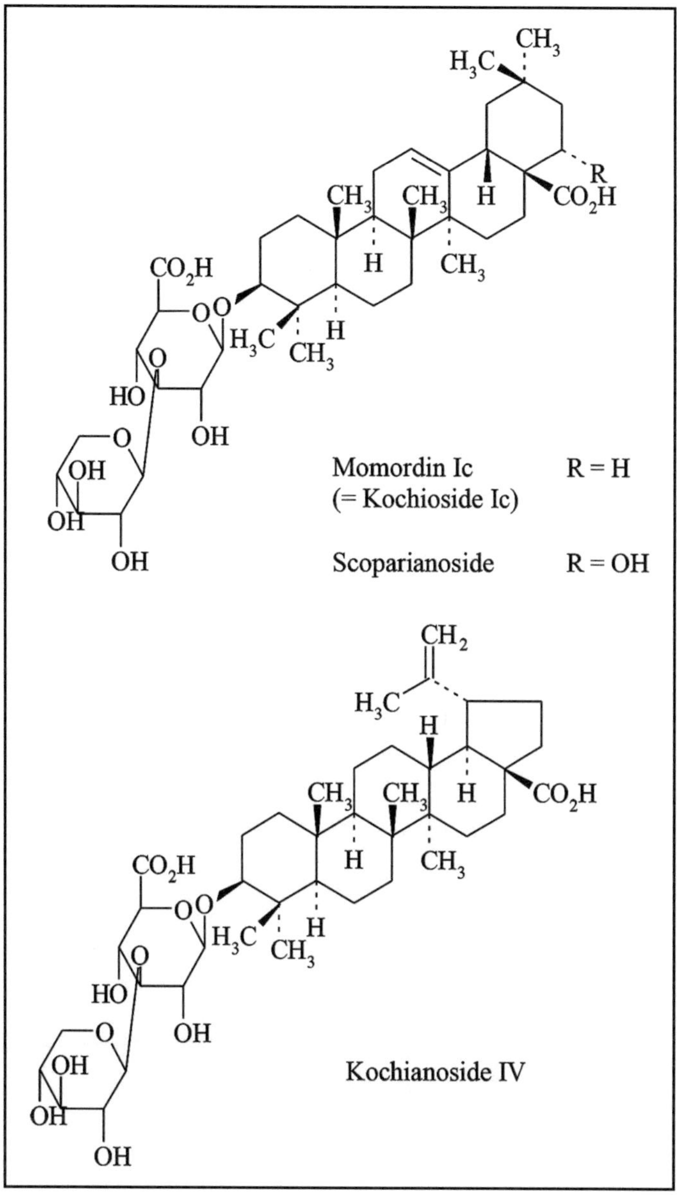

Fig. 1 Formulae of the main compounds of Fructus Kochiae [5]

TLC-Fingerprint Analysis

Drug samples	Origin
1 Fructus Kochiae / *Kochia scoparia*	Sample of commercial drug obtained from China Medica (origin: Hebei, China)
2 Fructus Kochiae / *Kochia scoparia*	Province Jiangsu, China
3 Fructus Kochiae / *Kochia scoparia*	Sample of commercial drug obtained from TCM- Clinic Bad Kötzting (Charge: K30.08.2000)
4 Fructus Kochiae / *Kochia scoparia*	Province Heilongjiang, China
5 Fructus Kochiae / *Kochia scoparia*	Province Shanxi, China

1. TLC-Fingerprint Analysis of the Triterpene Saponins [1]

Reference compounds of Fig. 2		R*f*
T1	Momordin Ic	0.64
T2	Oleanolic acid	0.98

1. Sample preparation — 1.0 g powdered drug is extracted under reflux with 10 ml ethanol (70 %) for 1 h. The extract is evaporated to dryness and the residue dissolved in 1 ml ethanol.
2. Reference compounds — 0.5 mg are dissolved in 0.5 ml methanol.
3. Separation parameter

Plate:	HPTLC Silica gel 60 F_{254}, Merck
Applied amounts:	Fructus Kochiae extracts: each 10 µl, Reference compounds: each 10 µl
Solvent system:	Chloroform + methanol + water (16 + 9 + 2)
Detection:	10 % ethanolic Sulphuric acid The plate is sprayed with 8 ml reagent and heated at 105 °C for 10 min. The plate is evaluated in VIS (Fig. 2a) and under UV 366 nm (Fig. 2b).

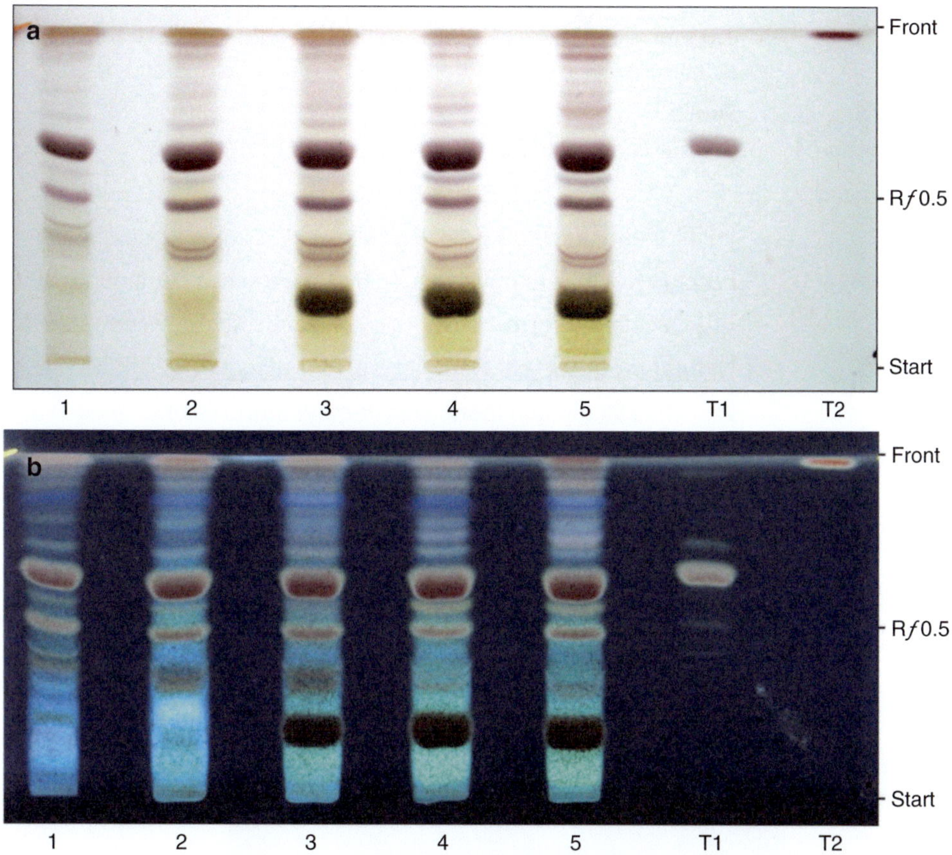

Fig. 2a/b Thin layer chromatogram of the methanol extracts of Fructus Kochiae sprayed with 10% ethanolic sulphuric acid (a=VIS, b=UV 366 nm)

4. Description of Fig. 2a, b:

2a: The VIS-TLC of the Fructus Kochiae extracts is characterised by main dark brown zones at R*f*=0.64 (Momordin Ic, **T1**) and at R*f*=0.18 one of the Kochianosides A, B or C. Further weak zones between R*f*=0.25 and 0.48 and R*f*=0.63 and 0.92 can be assigned to the other triterpenglycosides and glycoside esters. Oleanolic acid (**T2**) possesses the R*f*=0.98.
The extract samples 1 and 2 of Chinese origin differ from the other extracts samples by lacking or only weak presence of the polar triterpenoside at R*f*=0.18.

2b: Under UV 366 nm the extracts show the same triterpene saponin pattern with a green fluorescent background.

2. TLC-Fingerprint Analysis of the Flavonol Glycosides

Reference compound of Fig. 3		**R*f***
T3	Isoquercitrin	0.71

1. Sample preparation: 1.0 g powdered drug is extracted under reflux with 10 ml ethanol (70 %) for 1 h. The extract is evaporated to dryness and the residue dissolved in 1 ml ethanol.
2. Reference compound: 0.5 mg is dissolved in 0.5 ml methanol.
3. Separation parameter:

Plate:	HPTLC Silica gel 60 F_{254}, Merck
Applied amounts:	Fructus Kochiae extracts: each 10 µl, Reference compound: 10 µl
Solvent system:	Ethyl acetate + formic acid + glacial acetic acid + water (10 + 1.1 + 1.1 + 2.6)
Detection:	Natural products – Polyethylene glycol reagent (NP/PEG)

I: 1 % diphenylboric acid-*β*-ethylamino ester (= diphenylboryloxyethylamine, NP) in methanol

II: 5 % Polyethylene glycol-4000 (PEG) in ethanol (80 %)

The plate is sprayed first with solution **I** and then with solution **II**.

After 1 h the evaluation is carried out under UV 366 nm.

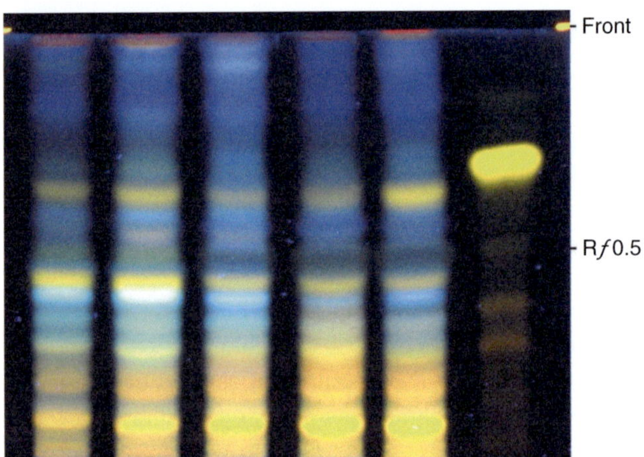

Fig. 3 Thin layer chromatogram of the methanol extracts of Fructus Kochiae sprayed with NP/PEG reagent (UV 366 nm)

4) Description of Fig. 3:

   In the deep R*f*-range appear from R*f*=0.65 to the start 9–10 yellow and orange fluorescent zones which can be assigned to the known or not yet identified flavon- and flavonol di-, tri- and tetraglycosides of quercetin or kaempferol (e.g. sophorosides, gentiobiosides etc.).

HPLC-Fingerprint Analysis

1. Sample preparation: 1.0 g powdered drug is extracted under reflux with 10 ml ethanol (70 %) for 1 h. The extract is evaporated to dryness and the residue dissolved in 1 ml ethanol. The extract is filtered over Chromafil® filtration unit, type 0-20 µm/25 mm.

2. Injection volume: Fructus Kochiae: each 10 µl

3. HPLC parameter:

Apparatus:	MERCK HITACHI D-6000 A Interface
	MERCK HITACHI L-4500 A Diode Array Detector
	MERCK HITACHI AS-2000 Autosampler
	MERCK HITACHI L-6200 A Intelligent Pump
Separation column:	LiChroCART® 125-4 LiChrospher® 100 RP-18 (5 µm), Merck
Precolumn:	LiChroCART® 4-4 LiChrospher® 100 RP-18 (5 µm), Merck
Solvent system:	A: 0.001 % phosphoric acid/water (Millipore Ultra Clear UV plus® filtered)
	B: acetonitrile (VWR)
Gradient:	0 % B for 5 min,
	0–30 % B in 25 min,
	30–100 % B in 30 min,
	100 % B for 10 min
	total run time: 70 min
Flow:	1.0 ml/min
Detection:	215 nm

Retention times of the main peaks:

Peak	Rt (min)	Compound
A	19.0–23.0	Flavon-/flavonol glycosides
1	43.1	Momordin Ic
2	55.5	Oleanolic acid

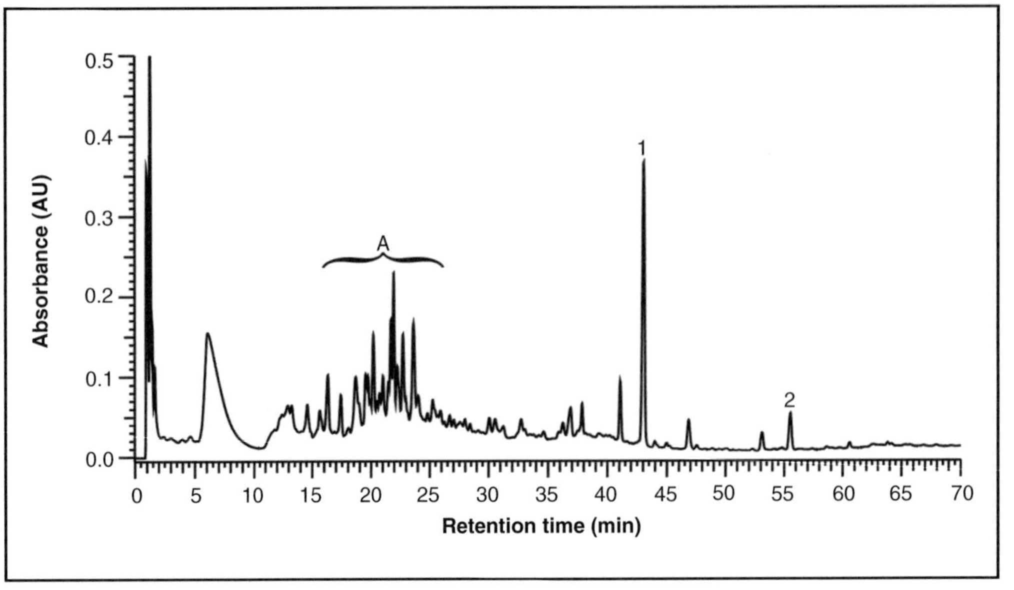

Fig. 4a HPLC-fingerprint analysis of the ethanol extract of Fructus Kochiae, sample 3

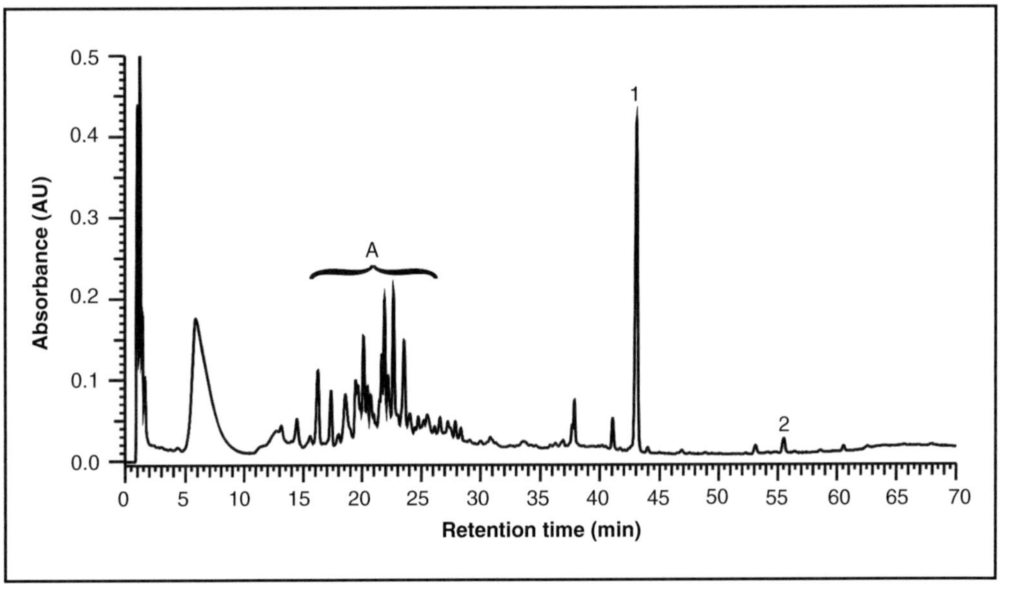

Fig. 4b HPLC-fingerprint analysis of the ethanol extract of Fructus Kochiae, sample 4

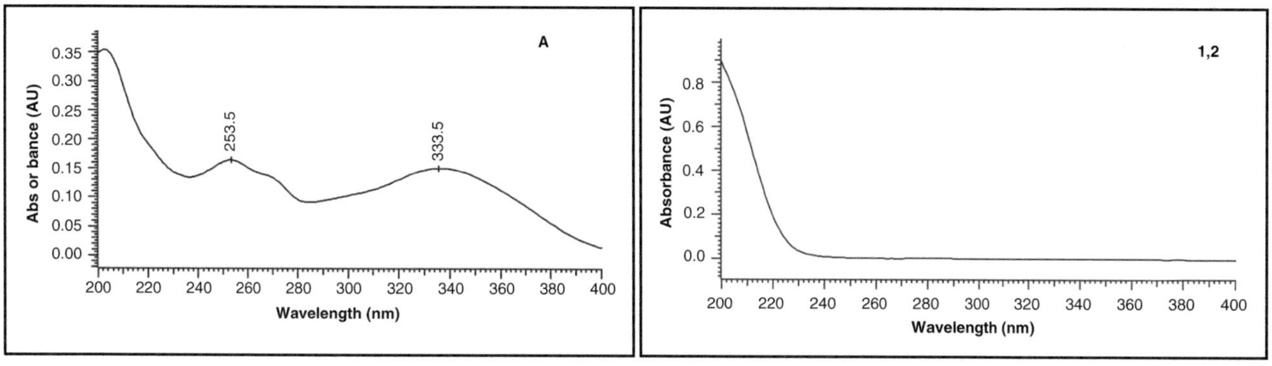

Fig. 5 On line UV-spectra of the reference compounds of Fructus Kochiae

4) Description of the HPLC-Figures

The ethanol extracts of Fructus Kochiae extracts provide in the Rt-range **A** of Fig. 4a, b the accumulation of 9–12 flavon(ol) di-, tri- and tetraglycosides as evidenced by the characteristic UV-spectra with the two maxima at 250–260 nm and 330–340 nm. The triterpene saponins appear with the main peaks at Rt=38.0, 41.0 and 43 (**1**, Momordin Ic) and Rt=55 (**2**, Oleanolic acid). They possess the characteristic UV end-absorptions at 200 nm (Fig. 5).

Note: According to the Chinese Pharmacopeia 2010 Fructus Kochiae contains not less than 1.8 % of momordin Ic, calculated with reference to the dried drug [1].

Conclusion

Fructus Kochiae extracts can be easily authenticated by their characteristic TLC-fingerprints and HPLC-graphs.

References

1. Pharmacopoeia of the people's republic of China, English edition, vol. I. People's Medical Publishing House, Beijing (2010)
2. Zhang, Z.Z.: An illustrated Chinese materia medica in Hong Kong. School of Chinese Medicines/Hong Kong Baptist University, Hong Kong (2004)
3. Flora of China, *Kochia scoparia*, FOC 5, 385 (www.eFloras.org)
4. Whitney, H.M., Sayanova, O., Pickett, J.A., Napier, J.A.: Isolation and expression pattern of two putative acyl-ACP desaturase cDNAs from *Bassia scoparia*. J. Exp. Bot. **55**(397), 787–789 (2004)
5. Tang, W., Eisenbrand, G.: Handbook of Chinese medicinal plants – chemistry, pharmacology, toxicology, vol. 1. WILEY-VCH Verlag GmbH & Co, KgaA, Weinheim (2011)
6. Choi, J., Lee, K.T., Jung, H.J., Park, H.S., Park, H.J.: Anti-rheumatoid arthritis effect of the *Kochia scoparia* fruits and activity comparison of momordin Ic, its prosapogenin and sapogenin. Arch. Pharm. Res. **25**(3), 336–342 (2002)
7. Shin, K.M., Kim, Y.H., Park, W.S., Kang, I., Ha, J., Choi, J.W., Park, H.J., Lee, K.T.: Inhibition of methanol extract from the fruits of *Kochia scoparia* on lipopolysaccharide-induced nitric oxide, prostaglandin E_2, and tumor necrosis factor-α production from murine macrophage RAW 264.7 cells. Biol. Pharm. Bull. **27**(4), 538–543 (2004)
8. Matsuda, H., Dai, Y., Ido, Y., Yoshikawa, M., Kubo, M.: Studies on Kochiae Fructus IV. Anti-allergic effects of 70% ethanol extract and its component, momordin Ic from dried fruits of Kochia scoparia L. Biol. Pharm. Bull. **20**(11), 1165–1170 (1997)
9. Matsuda, H., Dai, Y., Ido, Y., Murakami, T., Matsuda, H., Yoshikawa, M., Kubo, M.: Studies on Kochiae Fructus V. Antipruritic effects of oleanolic acid glycosides and the structure-requirement. Biol. Pharm. Bull. **21**(11), 1231–1233 (1998)

10. Choi, Y.Y., Kim, M.H., Lee JY, J.Y., Hong, J., Kim, S.H., Yang, W.M.: Topical application of *Kochia scoparia* inhibits the development of contact dermatitis in mice. J. Ethnopharmacol. **154**(2), 380–385 (2014)
11. Porkert, M.: Klinische Chinesische Pharmakologie, Verlag für Medizin Dr. Ewald Fischer, Heidelberg (1978)
12. Hempen, C.H., Fischer, T.: A materia medica for Chinese medicine. Elsevier GmbH, Urban & Fischer, Munich (2009)
13. Wen, Y., Chen, Y., Cui, Z., Li, J., Wang, Z.: Triterpenoid glycosides from the fruits of *Kochia scoparia*. Planta Med. **61**(5), 450–452 (1995)
14. Wang, J., Yuan, L., Xiao, H., Wang, C., Ciao, C., Wang, Y., Liu, X.: A novel mechanism for momordin Ic-induced HepG2 apoptosis: involvement of PI3K- and MAPK-dependent PPARγ activation. Food Funct. **5**(5), 859–868 (2014)
15. Yoshikawa, M., Shimada, H., Morikawa, T., Yoshizumi, S., Matsumura, N., Murakami, T., Matsuda, H., Hori, K., Yamahara, J.: Medicinal foodstuffs VII. On the saponin constituents with glucose and alcohol absoprtion-inhibitory activity from a food garnish "Tonburi", the fruit of Japanese Kochia scoparia (L.) SCHRAD.: Structures of scoparianosides A, B, and C. Chem. Pharm. Bull. **45**(8), 1300–1305 (1997)
16. Wang, J., Yuan, L., Xiao, H., Xiao, C., Wang, Y., Liu, X.: Momordin Ic induces HepG2 cell apoptosis through MAPK and PI3K/Akt-mediated mitochondrial pathways. Apoptosis **18**(6), 751–765 (2013)
17. Han, L.K., Nose, R., Li, W., Gong, X.J., Zheng, Y.N., Yoshikawa, M., Koike, K., Nikaido, T., Okuda, H., Kimura, Y.: Reduction of fat storage in mice fed a high-fat diet long term by treatment with saponins prepared from *Kochia scoparia* fruit. Phytother. Res. **20**(10), 877–882 (2006)
18. Matsuda, H., Li, Y., Yoshikawa, M.: Roles of capsaicin-sensitive sensory nerves, endogenous nitric oxide, sulfhydryls, and prostaglandins in gastroprotection by momordin Ic, an oleanolic acid oligoglycoside, on ethanol-induced gastric mucosal lesions in rats. Life Sci. **65**(2), PL27–PL32 (1999)
19. Balsevich, J.J., Bishop, G.G., Deibert, L.K.: Use of digitoxin and digoxin as internal standards in HPLC analysis of triterpene saponin-containing extracts. Phytochem. Anal. **20**(1), 38–49 (2009)
20. Xu, Y.H., Huang, H., Zhnag, N., Kong, D.Y., Hua, M.L.: Studies on the flavone glycosides from Fructus Kochiae. J. Asian Nat. Prod. Res. **16**(2), 141–147 (2014)
21. Rankins Jr., D.L., Smith, G.S.: Nutritional and toxicological evaluations of Kochia hay (*Kochia scoparia*) fed to lambs. J. Anim. Sci. **69**(7), 2925–2931 (1991)
22. Whitney, H., Sayanova, O., Lewis, M.J., Pickett, J., Napier, J.A.: Isolation of two putative acyl-acyl carrier protein desaturase enzymes from *Kochia scoparia*. Biochem. Soc. Trans. **28**(6), 623–624 (2000)
23. Matsuda, H., Dai, Y., Ido, Y., Ko, S., Yoshikawa, M., Kubo, M.: Studies on Kochiae Fructus III. Antinociceptive and antiinflammatory effects of 70% ethanol extract and its component, momordin Ic from dried fruits of *Kochia scoparia* L. Biol. Pharm. Bull. **20**(10), 1086–1091 (1997)

Fructus Psoraleae – *Buguzhi*

Pharmacopoeia: [1]	Pharmacopoeia of the People's Republic of China, English Edition Vol. I, 2010
Official drug: [1]	Malaytea Scurfpea Fruit is the dried ripe fruit of *Psoralea corylifolia* L. (Fam. Leguminosae) The infructescence is collected in autumn when the fruit is ripe, dried in the sun, the fruit is gathered, and then removed from foreign matters.
Synonyms: [2, 3]	*Cullen corylifolium, Cullen corylifolia, Lotodes corylifolia, Psoralea patersoniae, Trifolium unifolium*
Origin: [3–6]	India, Southern Africa and China (Henan, Anhui, Guangdong, Shaanxi, Shanxi, Jiangxi, Sichuan, Yunnan and Guizhou), Vietnam and Burma
Description of the drug: [1]	Reniform, slightly flattened, 3–5 mm long, 2–4 mm wide, about 1.5 mm thick. Externally black, blackish-brown or greyish-brown, with finely reticulate wrinkles. Apex obtuse-rounded, with a small prominence and the depressed end showing a scar of fruit stalk. Texture hard. Pericarp thin, uneasily stripped from seed; seed 1, cotyledons 2, yellowish-white, oily. Odour, aromatic; taste, pungent and slightly bitter.
Medicinal use: [7]	Used together with UV light to treat psoriasis, vitiligo, leukoderma, leprosy and mycosis fungoides. The herb can also be used to treat alopecia and menorrhagia.
Toxicity: [7]	The herb contains furanocoumarins which may also have a phototoxic effect. In high doses teratogenic effects have been found in animal studies.

Effects and indications of Fructus Psoraleae according to Traditional Chinese Medicine [1, 3–23]

Taste:	Pungent and bitter
Temperature:	Warm
Channels entered:	*Orbis renalis, o. lienalis*
Effects (functions):	To warm the kidney and assist yang, absorb qi to relieve panting, warm the spleen to check diarrhea
Symptoms and indications:	Deficiency of kidney yang, impotence, menstruation disorders and seminal emission, enuresis and frequent urination, cold pain in the lower back and knees, pianting caused by kidney deficiency, diarrhea before dawn, asthma. Treatment of bone diseases Topical application: skin diseases such as vitiligo, psoriasis, leukoderma, leprosy and alopecia

H. Wagner et al. (eds.), *Chromatographic Fingerprint Analysis of Herbal Medicines*, Vol. 4,
DOI 10.1007/978-3-319-32328-2_6, © Springer International Publishing Switzerland 2016

Main constituents:

- (Furano)coumarins [2–4, 6–10, 12–25]

Psoralen (ficusin), isopsoralen (angelicin), psoralidin

- Flavonoids [2–4, 6, 8, 9, 12–16, 18–20, 22, 24]

Bavachalcone, isobavachalcone (corylifolinin), bavachin, isobavachin, neobavaisoflavone, broussochalcone, bavachinin, corylin, corylinin, psoralenol, corilyfol A,B,C,E, corylifolin, bavacoumestan B, corylinal, 6-prenylnaringenin, daidzein, genistein, biochanin A

- Meroterpenoids [3, 8, 9, 12–15, 18, 19, 22, 24]

Δ^1,3-hydroxybakuchiol; Δ^3,2-hydroxybakuchiol; 12,13-dihydroxybakuchiol; 12,13-epoxybakuchiol; 13-hydroxyisobakuchiol, bakuchiol, 3-hydroxybakuchiol

- phenolics [8, 9, 13]

Protocatechualdehyde, 3-hydroxybenzaldehyde, *p*-hydroxy-benzaldehyde, hydroquinone

- benzofuran glycosides [3, 12, 15]

Corylifonol, isocorylifonol, psoralenoside, isopsoralenoside

Minor constituents:

- linoleic acid, linolenic acid [8]

- essential oil (caryophyllene; α-humulene; (+)-aromadendrene; naphthalene; 1,2,3,4-tetrahydro-1,6-dimethyl-4-(1-methyl)-,(1S-cis); (-)-caryophyllene oxide; trans-caryophyllene; methyl hexadecanoate; phenol; 4-(3,7-dimethyl-3-ethenylocta-1,6-dienyl); limonene; α-elemene; γ-elemene; β-caryophelenoxide; 4-terpineol; linalool; geranylacetate [3, 11, 16]

- uracil, sophoracoumestan A [2]

Reported pharmacology:

- antibacterial/antiviral [2, 3, 12, 14–18, 20, 22]
- dna/topoisomerase ii polymerase inhibitory activity [2, 8, 9, 14]
- antitumor [3, 11, 13–19, 22, 24]
- antihyperglycemic [3, 13, 19]
- antidepressant [3, 19, 20]
- antioxidative [3, 6, 8, 9, 11–14, 16, 18, 20, 22]
- antiplatelet [8, 13, 14, 16, 18, 22]
- hepatoprotective [9, 13, 20]
- antimicrobial [9, 11, 18, 19]
- anti-inflammatory [9, 12, 16, 19, 22]
- cytotoxic [9, 21, 24]
- antidiabetic [9, 20]
- phytoestrogenic [10, 14, 15, 17, 18]
- antiallergic [11]
- insecticidal [11]
- anticancer [12, 20, 21]
- antidermatophytic/antifungal [13, 14, 19]
- skin-photosensitizing [16, 22]
- antipsoriatic [16, 22]
- hemostatic [17]
- immunoenhancing [17, 18, 20, 21]
- anti-aging [17]
- neuroprotective [18]

TLC-Fingerprint Analysis

Drug samples		**Origin**
1	Fructus Psoraleae / *Psoralea corylifolia*	Sample of commercial drug obtained from TCM-Clinic Bad Kötzting (Charge: K30.08.2000)
2	Fructus Psoraleae / *Psoralea corylifolia*	Sample of commercial drug obtained from TCM-Clinic Bad Kötzting (K08.10.2004)

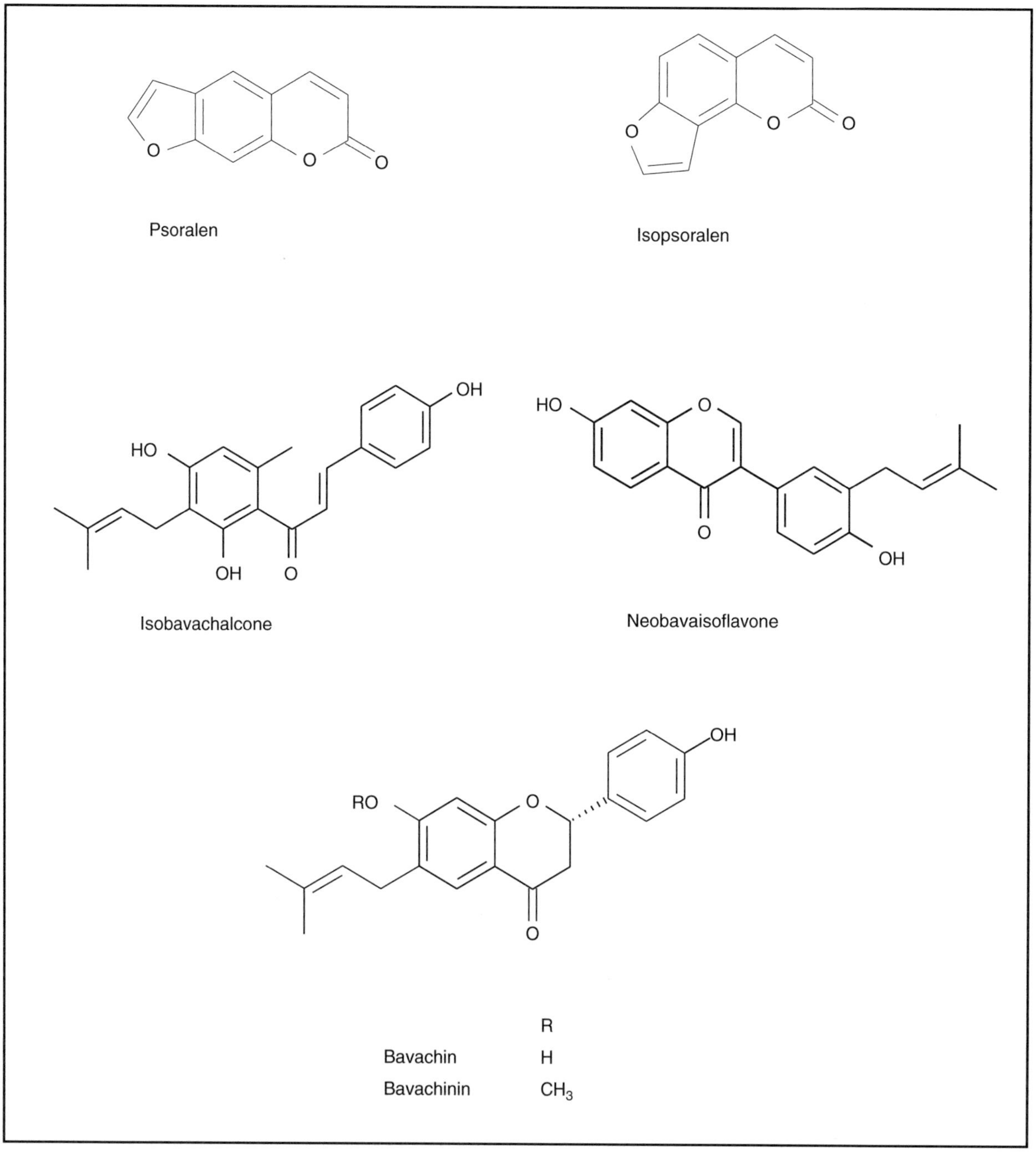

Fig. 1 Formulae of the main compounds of Fructus Psoraleae [13]

Drug samples		Origin
3	Fructus Psoraleae / *Psoralea corylifolia*	Sample of commercial drug obtained from HerbaSinica (origin: Gansu)
4	Fructus Psoraleae / *Psoralea corylifolia*	Sample of commercial drug obtained from Munich pharmacy (origin: Jiangjin, Chongqing)
5	Fructus Psoraleae / *Psoralea corylifolia*	Sample of commercial drug obtained from TCM-Clinic Bad Kötzting (A28.06.2002)

Reference compounds of Fig. 2a–d		R*f*
T1	Psoralen	0.65
T2	Isopsoralen	0.72
T3	Bavachinin	0.68
T4	Isobavachalcone	0.58

1. Extraction: 1.0 g powdered drug is extracted with 10 ml methanol under reflux for 30 min. The extract is filtered, evaporated to dryness and the residue dissolved in 1 ml methanol.

2. Reference compounds: 1.0 mg is dissolved in 1.0 ml methanol

3. Separation parameters:

Plate: HPTLC Silica gel 60 F_{254}, Merck

Applied amounts: Fructus Psoraleae extracts: each 8 µl

Reference compounds: each 10 µl

Solvent system: Petroleum ether (60–90 °C) + ethyl acetate + glacial acetic acid (10 + 10 + 0.4)

Detection: 1. Potassium hydroxide reagent (KOH)

The plate is sprayed with 8 ml 10% ethanolic KOH and evaluated in VIS (**Fig. 2a**) and under UV 366 nm (**Fig. 2b**).

2. Anisaldehyde – Sulphuric acid reagent

0.5 ml anisaldehyde is mixed with 10 ml glacial acetic acid, followed by 85 ml methanol and 5 ml concentrated sulphuric acid, in that order.

The plate is sprayed with 10 ml, heated at 110 °C for 10 min and evaluated after 30 min in VIS (**Fig. 2c**).

Note: The reagent has only limited stability and is no longer useable when the colour has turned to red-violet.

4. Description:

   In **Fig. 2a** the MeOH extract samples of Fructus Psoraleae 1–5 provide in the TLC (sprayed with KOH) three yellow/orange zones in the R*f*-range R*f*=0.5 up to R*f*=0.75 which could be identified as Psoralen (**T1**), Isopsoralen (**T2**), Bavachinin (**T3**) and Isobavachalcone (**T4**). In the deeper R*f*-range (R*f*=0.25–0.45) further pink zones of minor concentration could be detected.

   In **Fig. 2b** Psoralen and Isopsoralen appear under UV 366 nm as blue-white fluorescent zones, partly overlapped by Bavachinin and Isobavachalcone. All other compounds in the deeper R*f*-range are also detectable as blue-white fluorescent zones.

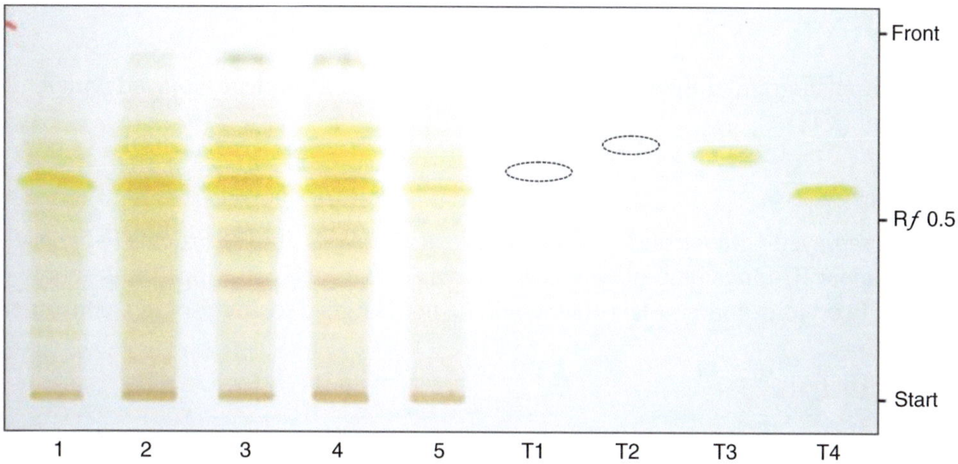

Fig. 2a Thin layer chromatogram of the methanol extracts of Fructus Psoraleae, sprayed with 10 % ethanolic KOH (VIS)

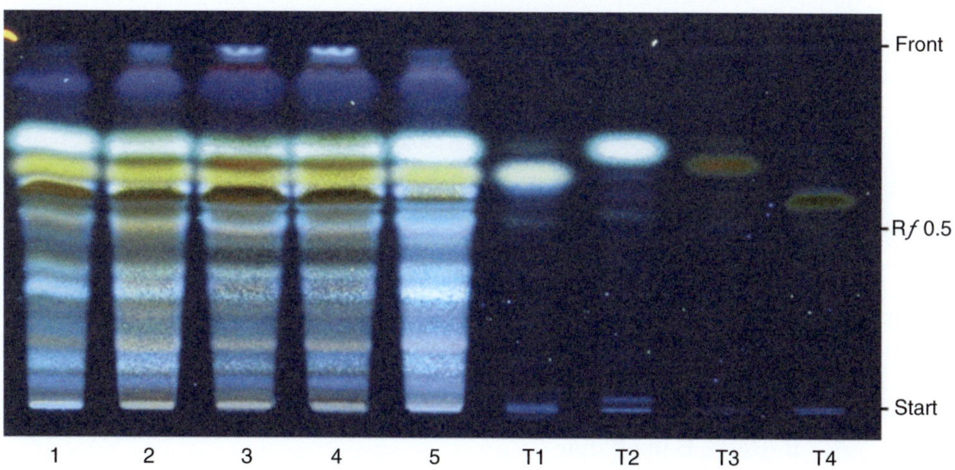

Fig. 2b Thin layer chromatogram of the methanol extracts of Fructus Psoraleae, sprayed with 10 % ethanolic KOH (UV 366 nm)

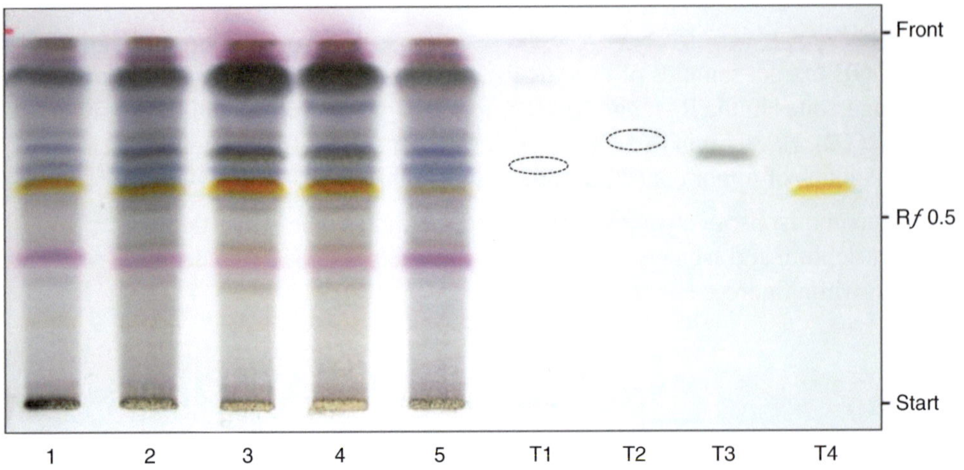

Fig. 2c Thin layer chromatogram of the methanol extracts of Fructus Psoraleae, sprayed with Anisaldehyde – Sulphuric acid reagent (VIS)

Description of Fig. 2c:

The TLC of **Fig. 2c** is sprayed with anisaldehyde-sulphuric acid. In VIS all five Fructus Psoraleae extracts show in the upper R*f*-range blue-grey zones, whereas isobavachalcone appears as yellow-orange zone (**T4**, R*f*=0.58). Two violet zones at R*f*=0.40 could be not assigned to any psoralen constituent.

HPLC-Fingerprint Analysis

1. Extraction: 1.0 g powdered drug is extracted with 10 ml methanol under reflux for 30 min. The extract is filtered, evaporated to dryness and the residue is dissolved in 1 ml methanol. The extract is filtered over Chromafil®, Type 0.20 µm.

2. Injection volume: Fructus Psoraleae extracts: each 5 µl

3. HPLC parameter:

Apparatus:	MERCK HITACHI D-6000 A Interface
	MERCK HITACHI L-4500 A Diode Array Detector
	MERCK HITACHI AS-2000 Autosampler
	MERCK HITACHI L-6200 A Intelligent Pump
Separation column:	LiChroCART® 250-4 LiChrospher® 100 RP-18 (5 µm), Merc
Precolumn:	LiChroCART® 4-4 LiChrospher® 100 RP-18 (5 µm), Merck
Solvent system:	A: 0.1 % phosphoric acid/water (Millipore Ultra Clear UV plus® filtered)
	B: acetonitrile (VWR)
Gradient:	0 % B for 5 min,
	0–100 % B in 60 min,
	100 % B for 5 min
	total runtime: 70 min
Flow:	1.0 ml/min
Detection:	210 nm

Retention times of the main peaks

Peak	Rt (min)	Compound
1	22.7	Psoralenoside[a]
2	23.0	Isopsoralenoside[a]
3	35.3	Psoralen
4	35.9	Isopsoralen[a]
5	41.2	Not identified
6	42.0	Not identified
7	43.4	Not identified
8	43.9	Not identified
9	46.0	Bavachromene[a]
10	47.9	Isobavachalcone
11	48.9	Bavachinin
12	50.2	Not identified
13	52.9	Not identified
14	57.4	Bakuchiol[a]

[a]According to references: [16, 25]

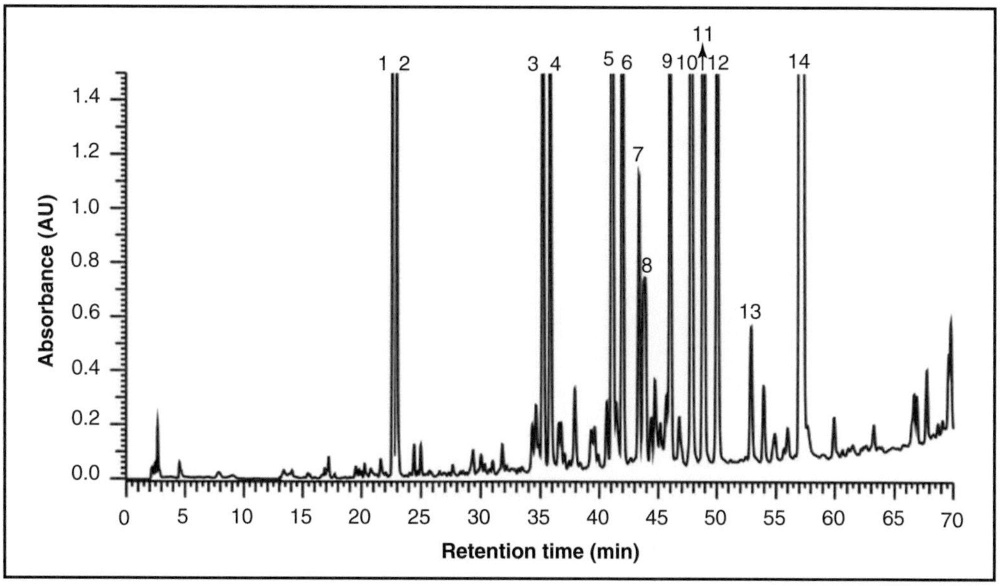

Fig. 3a HPLC-fingerprint analysis of the methanol extract of Fructus Psoraleae, sample 3

4. Description of the HPLC-Figures

All extract samples show a very homogenous peak profile consisting of 16 detectable compounds inclusive all characteristic psoralens and prenylated flavones, flavanones and chalcones as shown in Fig. 1 and the corresponding TLC Figures 2a–c. The concentration of the detected constituents is also comparable with those of the TLC (e.g. peaks **9** and **10**; bavachromene and isobavachalcone).

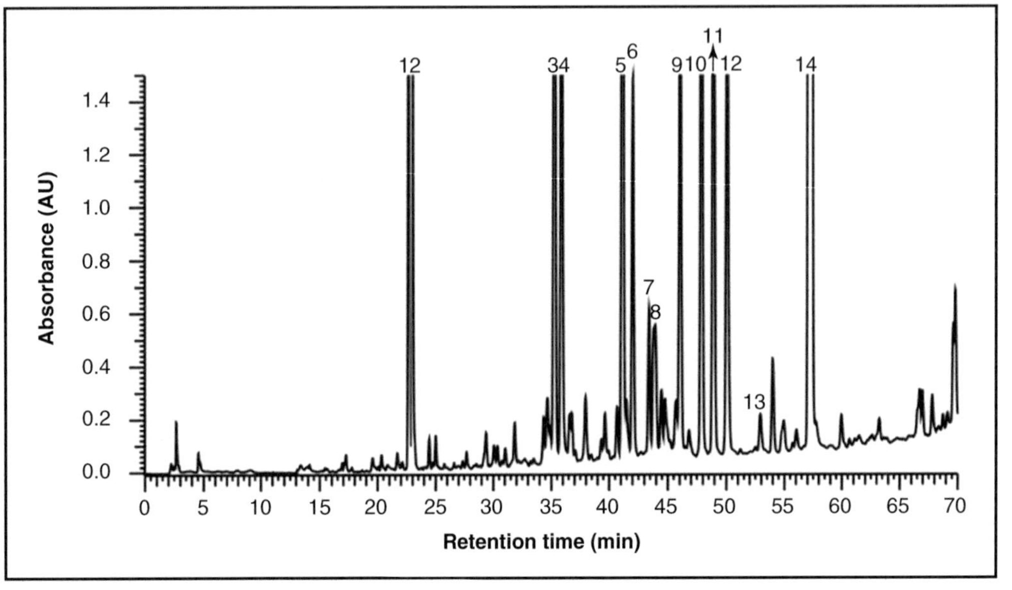

Fig. 3b HPLC-fingerprint analysis of the methanol extract of Fructus Psoraleae, sample 4

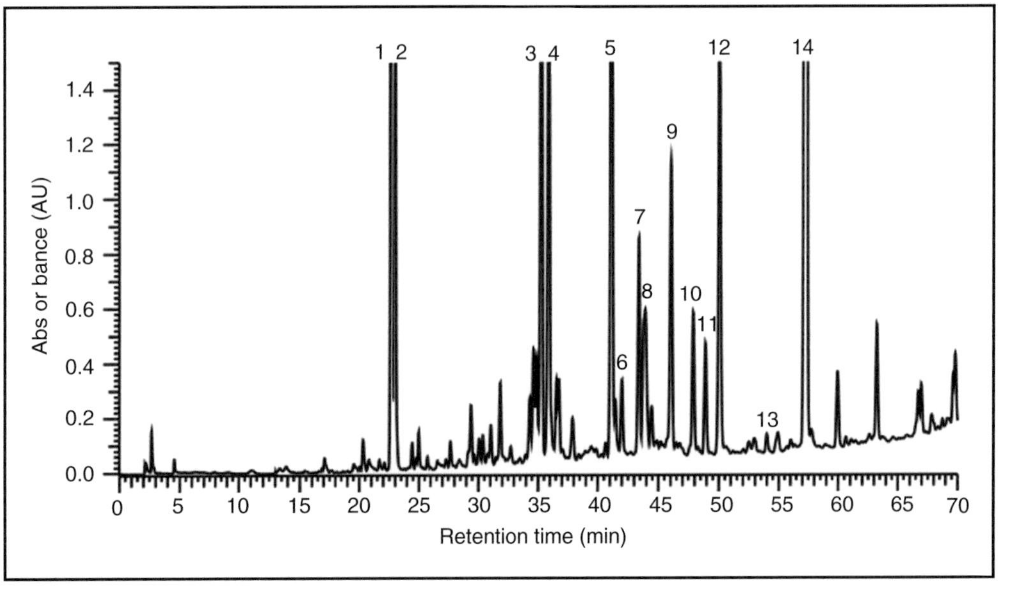

Fig. 3c HPLC-fingerprint analysis of the methanol extract of Fructus Psoraleae, sample 5

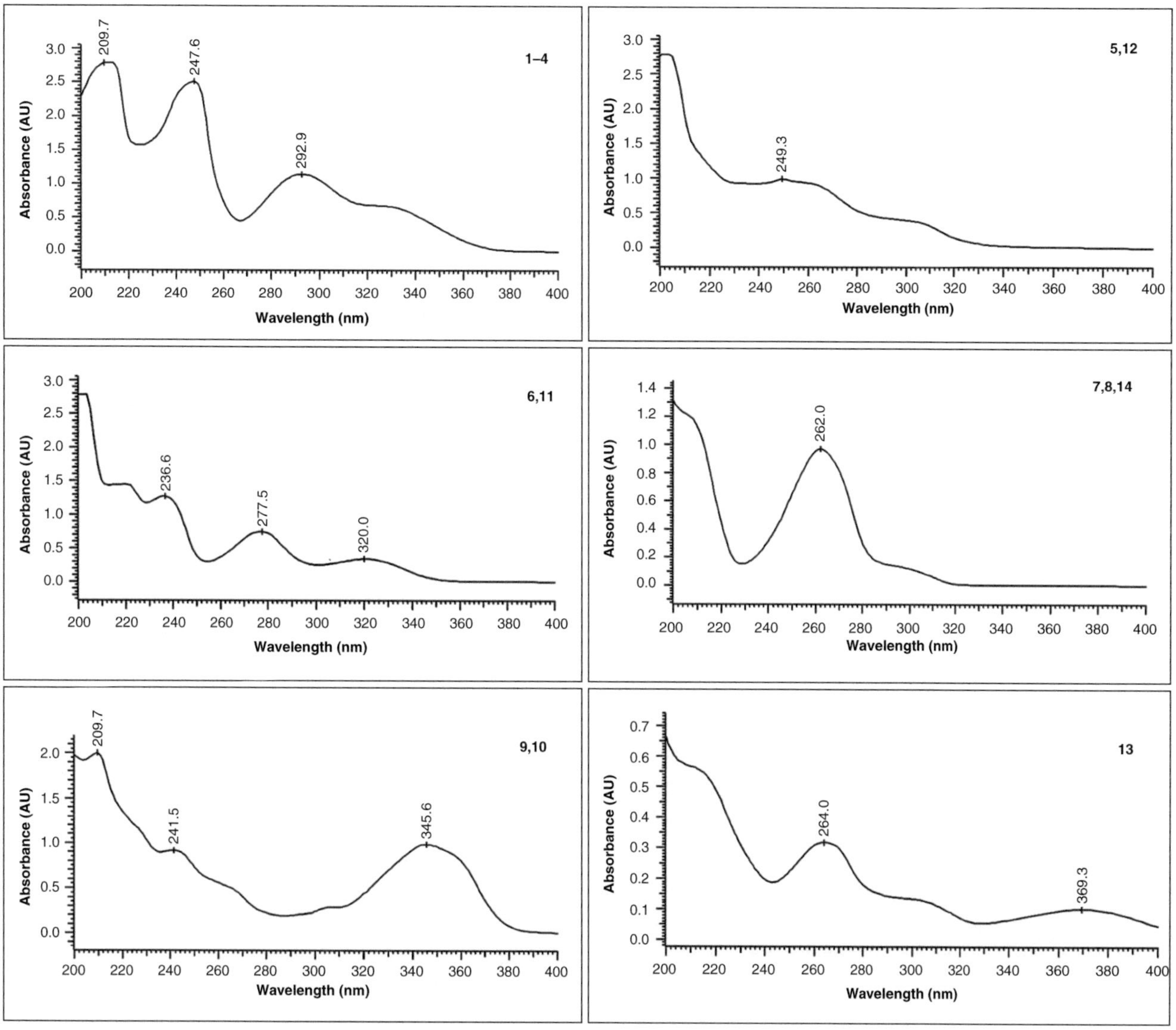

Fig. 4 On line UV-spectra of the main peaks of Fructus Psoraleae extracts

Note: According to the Chinese Pharmacopeia 2010 Fructus Psoraleae contains not less than 0.7 % of the total amount of psoralen and isopsoralen, calculated with reference to the dried drug. [1]

Conclusion

The authentication of Fructus Psoraleae is easy to carry out by TLC as well as HPLC.

References

1. Pharmacopoeia of the People's Republic of China, English Edition, vol. I. People's Medical Publishing House, Beijing (2010)
2. Chopra, B., Dhingra, A.K., Dhar, K.L.: *Psoralea corylifolia* L. (Buguchi) – folklore to modern evidence: review. Fitoterapia **90**, 44–56 (2013)
3. Khushboo, P.S., Jadhav, V.M., Kadam, V.J., Sathe, N.S.: Psoralea corylifolia Linn. – "Kushtanashini". Pharmacogn. Rev. **4**(7), 69–76 (2010)
4. Paulus, E., Ding, Y.H.: Handbuch der traditionellen chinesischen Heilpflanzen, Karl F. Haug Verlag, Heidelberg (1987)
5. Wong, R.W., Rabie, A.B.: Systemic effect of Fructus Psoraleae extract on bone in mice. Phytother. Res. **24**(10), 1578–1580 (2010)
6. Zhao, L.H., Wu, M.H., Xiang, B.R.: Analysis of *Psoralea corylifolia* L. Fruits in different regions. Chem. Pharm. Bull. **53**(8), 1054–1057 (2005)
7. Hempen, C.-H., Fischer, T.: A materia medica for Chinese medicine. Elsevier GmbH, Urban & Fischer, Munich (2009)
8. Matsuda, H., Sugimoto, S., Morikawa, T., Matsuhira, K., Mizuguchi, E., Nakamura, S., Yoshikawa, M.: Bioactive constituents from Chinese natural medicines XX: inhibitors of antigen-induced degranulation in RBL-2H3 cells from the seeds of *Psoralea corylifolia*. Chem. Pharm. Bull. **55**(1), 106–110 (2007)
9. Matsuda, H., Kiyohara, S., Sugimoto, S., Ahndo, S., Nakamura, S., Yoshikawa, M.: Bioactive constituents from Chinese natural medicines XXXIII: inhibitors from the seeds of *Psoralea corylifolia* on production of nitric oxide in lipopolysaccaride-activated macrophages. Biol. Pharm. Bull. **32**(1), 147–149 (2009)
10. Wang, J., Jiang, Z., Ji, J., Li, Y., Chen, M., Wang, Y., Zhang, Y., Tai, T., Wang, T., Zhang, L.: Evaluation of hepatotoxicity and cholestasis in rats treated with EtOH extract of Fructus Psoraleae. J. Ethnopharmacol. **144**(1), 73–81 (2012)
11. Dua, V.K., Kumar, A., Pandey, A.C., Kumar, S.: Insecticidal and genotoxic activity of *Psoralea corylifolia* Linn. (Fabaceae) against *Culex quinquefasciatus* Say, 1823. Parasit. Vectors **6**, 30 (2013)
12. Szliszka, E., Skaba, D., Czuba, Z.P., Krol, W.: Inhibition of inflammatory mediators by Neobavaisoflavone in activated RAW264.7 macrophages. Molecules **16**(5), 3701–3712 (2011)
13. Won, T.H., Song, I.H., Kim, K.H., Yang, W.Y., Lee, S.K., Oh, D.C., Oh, W.K., Oh, K.B., Shin, J.: Bioactive metabolites from the fruits of *Psoralea corylifolia*. J. Nat. Prod. **78**(4), 666–673 (2015)
14. Yin, S., Fan, C.Q., Wang, Y., Dong, L., Yue, J.M.: Antibacterial prenylflavone derivatives from *Psoralea corylifolia*, and their structure-activity relationship study. Bioorg. Med. Chem. **12**(16), 4387–4392 (2004)
15. Wang, Y.F., Liu, Y.N., Xiong, W., Yan, D.M., Zhu, Y., Gao, X.M., Xu, Y.T., Qi, A.D.: A UPLC-MS/MS method for in vivo and in vitro pharmacokinetic studies of psoralenoside, isopsoralenoside, psoralen and isopsoralen from *Psoralea corylifolia* extract. J. Ethnopharmacol. **151**(1), 609–617 (2014)
16. Tan, G., Yang, T., Miao, H., Chen, H., Chai, Y., Wu, H.: Characterization of compounds in Psoralea corylifolia using high-performance liquid chromatography diode array detection time-of-flight mass spectrometry and quadrupole ion trap mass spectrometry. J. Chromatogr. Sci. **53**(9), 1455–1462 (2015)
17. Xu, M., Tian, X.Y., Leung, K.S., Lee, K.C., Chow, T.C., Deng, B., Yiu, C.M., Chow, C.I., Zhao, Z.Z., Yang, Z.J., Fun, F.W., Xiao, T.T., Xu, W., Deng, P.X.: Embryotoxicity of *Psoralea corylifolia* L.: in vivo and in vitro studies. Birth Defects Res. B Dev. Reprod. Toxicol. **95**(6), 386–394 (2012)
18. Lee, K.M., Kim, J.M., Baik, E.J., Ryu, J.H., Lee, S.H.: Isobavachalcone attenuates lipopolysaccharide-induced ICAM-1 expression in brain endothelial cells through blockade of toll-like receptor 4 signaling pathways. Eur. J. Pharmacol. **754**, 11–18 (2015)
19. Kim, K.A., Shim, S.H., Ahn, H.R., Jung, S.H.: Protective effects of the compounds isolated from the seed of *Psoralea corylifolia* on oxidative stress-induced retinal damage. Toxicol. Appl. Pharmacol. **269**(2), 109–120 (2013)
20. Wu, C., Sun, Z., Ye, Y., Han, X., Song, X., Liu, S.: Psoralen inhibits bone metastasis of breast cancer in mice. Fitoterapia **91**, 205–210 (2013)

21. Wang Y, Hong C, Zhou C, Xu D, Qu HB.: Screening Antitumor Compounds Psoralen and Isopsoralen from *Psoralea corylifolia* L. Seeds. Evid. Based Complement Alternat. Med. Article ID 363052 (2011)

22. Xu, M.J., Wu, B., Ding, T., Chu, J.H., Li, C.Y., Zhang, J., Wu, T., Wu, J., Liu, S.J., Liu, S.L., Ju, W.Z., Li, P.: Simultaneous characterization of prenylated flavonoids and isoflavonoids in *Psoralea corylifolia* L. by liquid chromatography with diode-array detection and quadrupole time-of-flight mass spectrometry. Rapid Commun. Mass Spectrom. **26**(19), 2343–2358 (2012)

23. Tang, W., Eisenbrand, G.: Handbook of Chinese medicinal plants – chemistry, pharmacology, toxicology, vol. 2. Wiley-VCH Verlag GmbH & Co KgaA, Weinheim (2011)

24. Latha, P.G., Nayar, M.N.: Cytotoxic fatty acid glycerides from the seeds of *Psoralea corylifolia*. Anc. Sci. Life **19**(1–2), 36–38 (1999)

25. Qiao, C.F., Han, Q.B., Song, J.Z., Mo, S.F., Kong, L.D., Kung, H.F., Xu, H.X.: Chemical fingerprint and quantitative analysis of Fructus Psoraleae by high-performance liquid chromatography. J. Sep. Sci. **30**(6), 813–818 (2007)

Fructus Trichosanthis – *Gualou*
Semen Trichosanthis – *Chaogualouzi*

Pharmacopoeia: [1]	Pharmacopoeia of the People's Republic of China, English Edition Vol. I, 2010
Official drug: [1]	Fructus: Snakegourd Fruit is the dried ripe fruit of *Trichosanthes kirilowii* Maxim. or *Trichosanthes rosthornii* Harms (Fam. Cucurbitaceae) The drug is collected with fruit stalk in autumn, and dried in ventilated shade. Semen: Snakegourd Seed is the dried ripe seed of *Trichosanthes krilowii* Maxim. or *Trichosanthes rosthornii* Harms (Fam. Cucurbitaceae). The ripe fruit is picked in outumn and dissected. The seed is collected, washed clean, and dried in the sun.
Origin: [2]	Mainly in chinese provinces such as Anhui, Sichuan and Henan.
Description of the drug: [1]	Fructus: Subspherical or broadly ellipsoidal, 7–15 cm long, 6–10 cm in diameter. The outer surface orange- red or orange-yellow, shrunken or relatively smooth, Apex with a rounded style scar, base slightly acute, with remains of fruit stalk. Varying in weight. Texture fragile, easily split, the inner surface yellowish-white, with reddish yellow reticulation sarcocarp orange-yellow, viscous, adhering to many seeds into a mass. Seeds of *Trichosanthes kirilowii*: Flattened- elliptical, 12–15 mm long, 6–10 mm wide, about 3.5 mm thick. Externally pale brown to brown, smooth with a circle or furrow along the edge. Apex relatively acute, with a hilum, base obtuse or relatively narrow. Testa hard; tegmen membraneous, greyish-green, cotyledons 2, yellowish-white, oily. Odour, slight; Taste, weak. Seeds of *Trichosanthes rosthornii*: Relatively large and flattened, 15-19 mm long, 8-10 mm wide, about 2.5 mm thick. Externally brown, the circular furrow distinct and rather near the inner par. Apex even and truncate.
Pretreatment of the raw drug: [1]	*Fructus*: Fruit stalk and soil are eliminated, the fruit is flattened and cut into slivers or pieces. *Semen*: Foreign matter and blighted seeds are eliminated, the seeds are washed clean, and dried in the sun and broken into pieces before use.
Medicinal use: [3, 4]	The fruits are used as antipyretic, expectorant, and to treat Angina pectoris. The seeds possess a laxative activity for treatment of constipations.

H. Wagner et al. (eds.), *Chromatographic Fingerprint Analysis of Herbal Medicines*, Vol. 4,
DOI 10.1007/978-3-319-32328-2_7, © Springer International Publishing Switzerland 2016

Effects and indications of Fructus and Semen Trichonsathis according to Traditional Chinese Medicine [1, 4]	
Taste:	*Fructus*: slightly bitter and sweet *Seeds*: sweet
Temperature:	*Fructus*: cold *Seeds*: cold
Channels entered:	*Orbis pulmonalis, Orbis stomachi, Orbis intestini crassi*
Effects (functions):	To clear heat and flush away phlegm, soothe the chest and dissipate binds, moisten dryness and lubricate intestine
Symptoms and indications:	*Fructus*: Phlegm- heat cough, oppression in the chest and hypochondriac pain *Semen*: Dryness cough with greasy phlegm, constipation caused by intestinal dryness

Main constituents: [5–14]

Fructus* and *seeds:

- **Triterpenoids**:

karounidiol [D:C-Friedo-oleana-7,9 (11)-diene-3 alpha,29-diol] and 7-oxodihydrokarounidiol [7-oxo-D:C-friedo-olean-8-ene-3 alpha,29-diol], 7-oxodihydrokarounidiol [7-oxo-D:C-friedo-olean-8-ene-3 alpha,29-diol], 7-oxodihydrokarounidiol [7-oxo-D:C-friedo-olean-8-ene-3 alpha, 29-diol],

bryonolic acid (3 beta-hydroxy-D:C-friedo-olean-8-en-29-oic acid) and bryononic acid (3-oxo-D:C-friedo-olean-8-en-29-oic acid),

- **Phytosterols**:

Stigmastane-3 β,6α-diol, poriferastane-3β,6α-diol, stigmast-5-ene-3β,4β-diol, poriferast-5-ene-3β,4β-diol, and poriferasta- 5,25-diene-3β,4β-diol

- **Lignans**:

Hanultarin [(-)-1-O-feruloylsecoisolariciresinol], (-)-secoisolariciresinol.1, 4-O-diferuloylsecoisolariciresinol, (-)-pinoresinol and 4-ketopinoresinol

-**Proteins**:

α and β-kirilowin, trichokirin, karasurin

-**Lipids**:

triglycerides, glycolipids and phospholipids contain the unsaturated fatty acids linoleic and oleic acid and the saturated palmitic acid, trichosanic acid (punicic acid)

- **Saccharides**:

mono- and oligosaccharides, monosaccharides: rhamnose, galactose and glucose, oligosaccharides: a mixture of tri- and tetrasaccharides

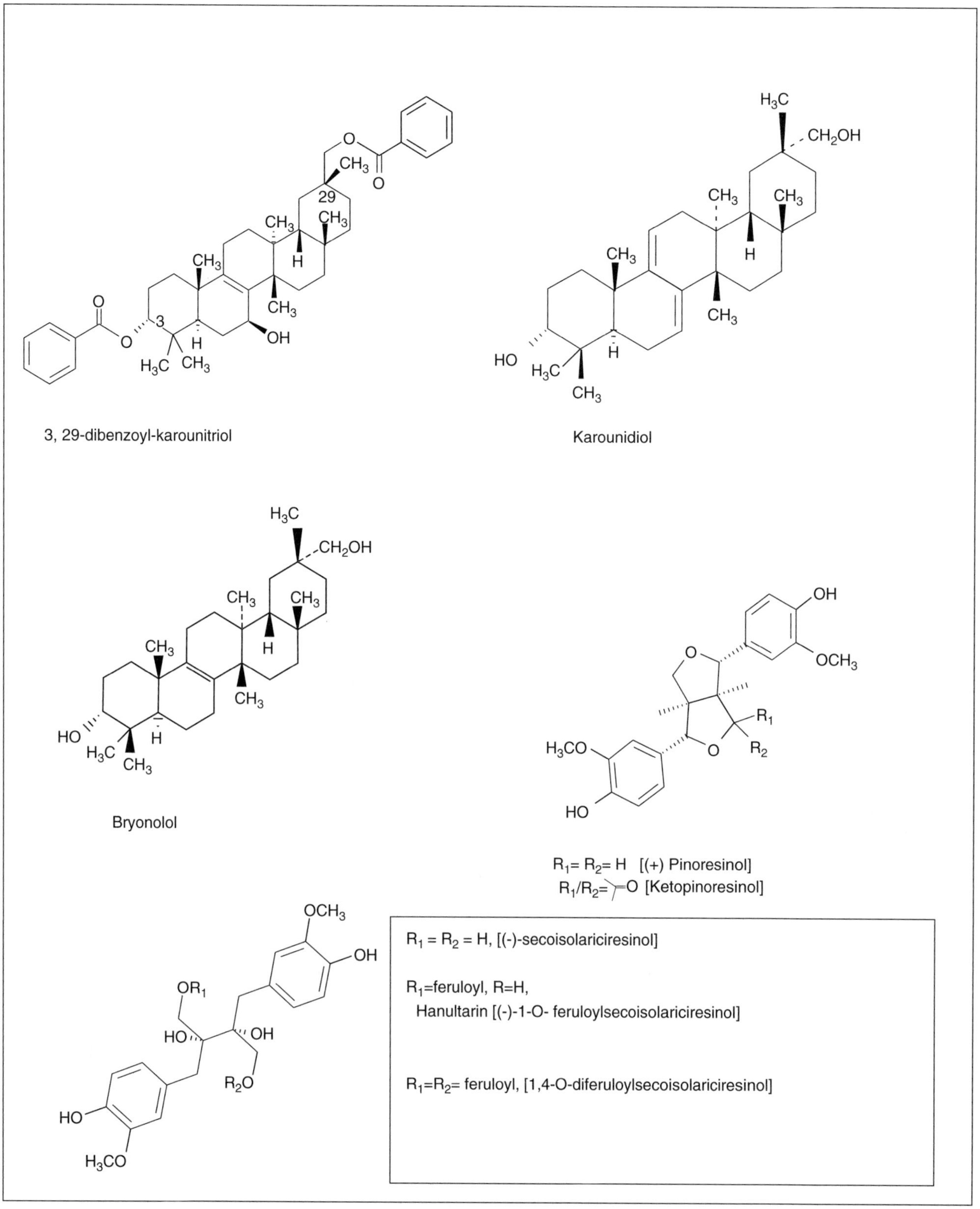

Fig. 1 Formulae of the main constituents of Semen and Fructus Trichosanthis

Reported pharmacological activities of *Fructus und Semen*: [3, 7, 9, 11, 15, 16]

- expectorant
- febrifugal
- inhibitory activity on protein synthesis in cell-free rabbit reticulocyte lysate system
- anti-tumor promoting effects (karounidiol)
- anti-inflammatory effect and analgesic effect (Triterpens)
- analgesic effect
- laxative effect

TLC Fingerprint Analysis

	Drug samples	Origin
1	Fructus Trichosanthis / *Trichosanthes kirilowii*	Sample of commercial drug, obtained from HerbaSinica (origin: Anhui)
1*	Fructus Trichosanthis / *Trichosanthes kirilowii* **(seeds are removed)**	Sample of commercial drug, obtained from HerbaSinica (origin: Anhui)
2	Fructus Trichosanthis / *Trichosanthes kirilowii*	Sample of commercial drug, obtained from China Medica
2*	Fructus Trichosanthis / *Trichosanthes kirilowii* **(seeds are removed)**	Sample of commercial drug, obtained from China Medica
3	Fructus Trichosanthis / *Trichosanthes kirilowii*	Provence Hebei (China)
4	Fructus Trichosanthis / *Trichosanthes rosthornii*	Province Shanxi (China)
5	Fructus Trichosanthis / *Trichosanthes kirilowii* Maxim	Sample of commercial drug, obtained from TCM-hospital Bad Kötzting (Charge: 19613112014)
6	Fructus Trichosanthis / *Trichosanthes kirilowii* Maxim	Sample of commercial drug, obtained from TCM-hospital Bad Kötzting (Charge: 19601072004)
7	**Semen Trichosanthis / *Trichosanthes kirilowii* Maxim**	**Sample of commercial drug, obtained from Firm China Medica**

1. **TLC-fingerprint analysis of Triterpenoids and Lignans:**

Reference compounds of Fig. 2a		R*f*
T 1	Stigmasterin	0.77
T 2	(+)– Pinoresinol	0.25

1.	Extraction:	2 g powdered drug are extracted with 20 ml 80 % ethanol under reflux for 1 h. The extract is cooled, filtered and evaporated to dryness. The residue is dissolved in 1 ml ethanol and filtered over Chromafil® filtration unit, type 0-20 µm/25 mm.
2.	Reference compounds:	Each 1 mg is dissolved in 0.5 ml ethanol

3. Separation parameters:

Plate:	HPTLC Silica gel 60 F_{254}, Merck
Applied amounts:	Semen / Fructus Trichosanthis extracts: 20 µl each Reference compounds: 10 µl each
Solvent system:	cyclohexane + ethyl acetate + methanol (6 + 2 + 1)
Detection:	Anisaldehyde- sulphuric acid reagent: 0.5 ml anisaldehyde is mixed with 10 ml glacial acetic acid, followed by 85 ml methanol and 5 ml sulphuric acid in that order. The plate is sprayed with 10 ml and then heated at 110 °C for 7–10 min. Evaluation is carried out under visible light.

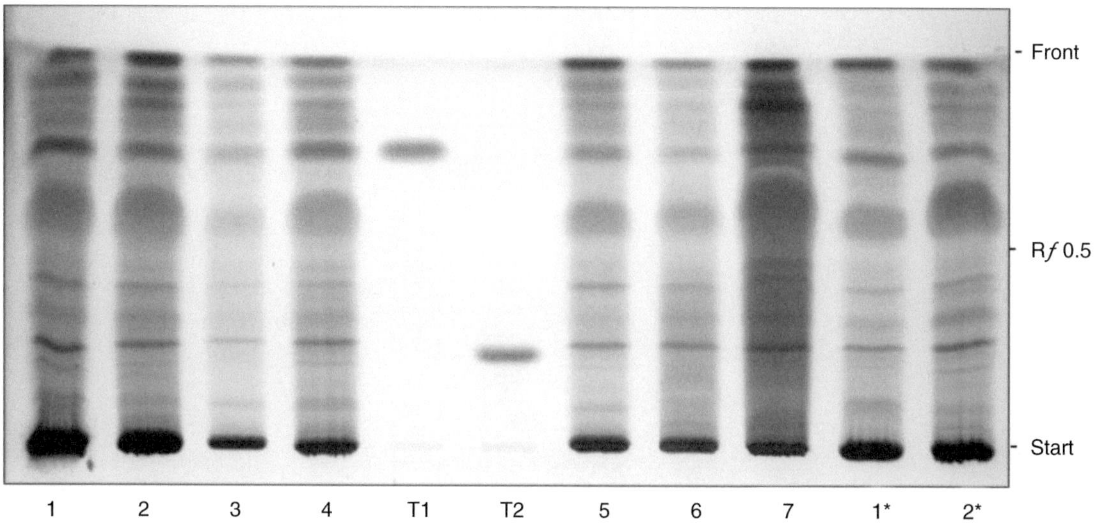

Fig. 2a Thin layer chromatogram of the 80 % ethanol extracts of **Semen- Fructus** Trichosanthis sprayed with Anisaldehyde – sulphuric acid reagent, detected under VIS

4. Description of Fig. 2a:

The Fructus Trichosanthis extracts 1–6 are characterized by about 12 more or less brown zones distributed over the whole R*f*-range with one zone at R*f* 0.77 (=**T1** Stigmasterin) and at R*f* 0.25 (=**T2** (+) Pinoresinol). The triterpenoids of the karoundiol type may be assignable in the R*f*- range 0.25 up to the TLC solvent front. The Semen Trichosanthis extract 7 differs from those of Fructus extracts by stronger brown coloured zone in the middle and higher R*f*-range. This shows that the Semen drug of *Trichosanthes kirilowii* is the dominant drug part. The Fructus Trichosanthis after removement of all seeds is shown in extracts 1* and 2*.

2. **TLC-fingerprint analysis of lignans:**

Reference compounds of Fig. 2b		**R*f***
T2	(+)– Pinoresinol	0.75

1. Extraction: The same extract used for TLC-fingerprint analysis (1)

   Reference compound: 10 µl

   Solvent system: Chloroform + methanol (85 + 15)

   Detection: Potassium hexacyanoferrate-iron-III-chloride reagent (PF)

   1 % aqueous potassium hexacyanoferrate mixed with an equal volume of 5 % aqueous iron-III- chloride.

   The plate is sprayed with 10 ml solution and evaluated in Vis.

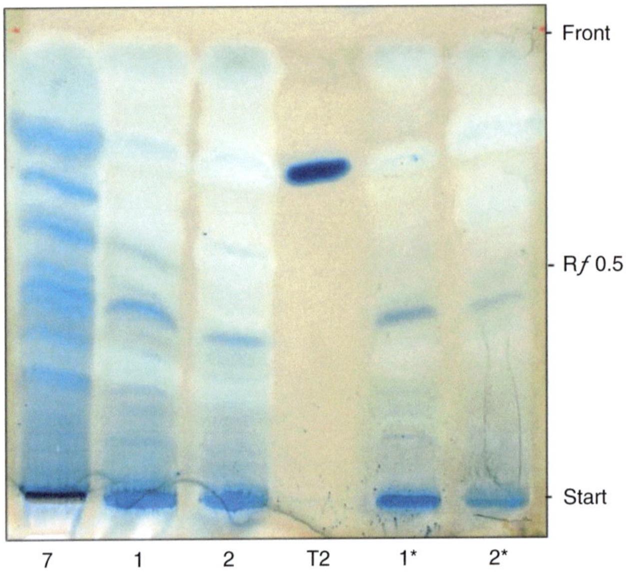

Fig. 2b Thin layer chromatogram of the 80 % ethanol extracts of **Semen** and **Fructus** Trichosanthis Trichosanthis sprayed with potassium hexacyanoferrate-iron-III-chloride reagent (PF) (VIS)

TLC- Discription of **Fig.** 2b:

The TLC of **Semen Trichosanthis** extract sample 7 sprayed with Potassium hexacyanoferrate-iron-III-chloride reagent (PF), provides in VIS the strong blue coloured zones of pinoresinol and many other blue zones from R*f*=0.25 up to the Pinoresinol which might be assigned to other lignans derivatives.

HPLC-Fingerprint Analysis

1. Sample preparation: The same extract used for TLC-fingerprint analysis (1)
2. Injection volume: Semen and Fructus Trichosanthis extracts 30 µl each
3. HPLC parameter:

Apparatus:	MERCK HITACHI D-6000 A Interface MERCK HITACHI L-4500 A Diode Array Detector MERCK HITACHI AS-2000 Autosampler MERCK HITACHI L-6200 A Intelligent Pump
Separation column:	LiChroCART® 250 -4 LiChrospher® 100 RP-18 (5 µm), Merck
Precolumn:	LiChroCART® 4-4 LiChrospher® 100 RP-18 (5 µm), Merck
Solvent System:	A: 0.001 % aq. H_3PO_4: (Millipore Ultra Clear UV plus® filtered) B: Acetonitril (VWR)
Gradient:	0–42 % B in 25 min 42–90 % B in 25 min 90–100 % B in 5 min 100 % B for 15 min total run time: 70 min
Flow:	1 ml/min
Detection:	210, 230 and 254 nm

Retention times of the main peaks recorded at 210, 230 and 254 nm

Peak	Rt (min)	Compound
1	16.3	Not identified
2	18.6	Not identified
3	20.1	Lignan (e.g. Hanultarin)
4	23.2	Ketopinoresinol?
5	24.4	Lignan (e.g 1,4-O- diferuloylsecoisolariciresinol
6	26.6	Pinoresinol
7	32.1	Not identified, (-)-secoisolariciresinol?
8	37.8	Triterpene (e.g karounidiol)
9	43.5	3, 29 Dibenzoyl-karounitriol?
10,11, 12	51.1–53.8	Triterpenoid
13	57.1	Triterpenoid
14	59.0	Triterpenoid
15	65.3	Stigmasterin
A, B (only in Fructus)	54.0, 55.6	Sterols

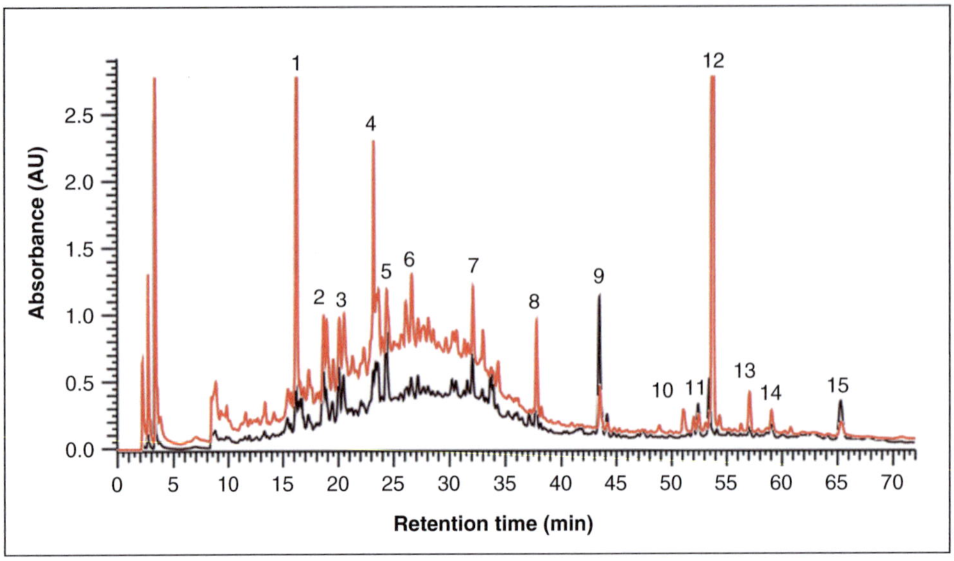

Fig. 3a HPLC-fingerprint analysis of the 80% ethanol extract of **Semen** Trichosanthis (Sample 7) detected at (210 nm-red) and (230 nm-black)

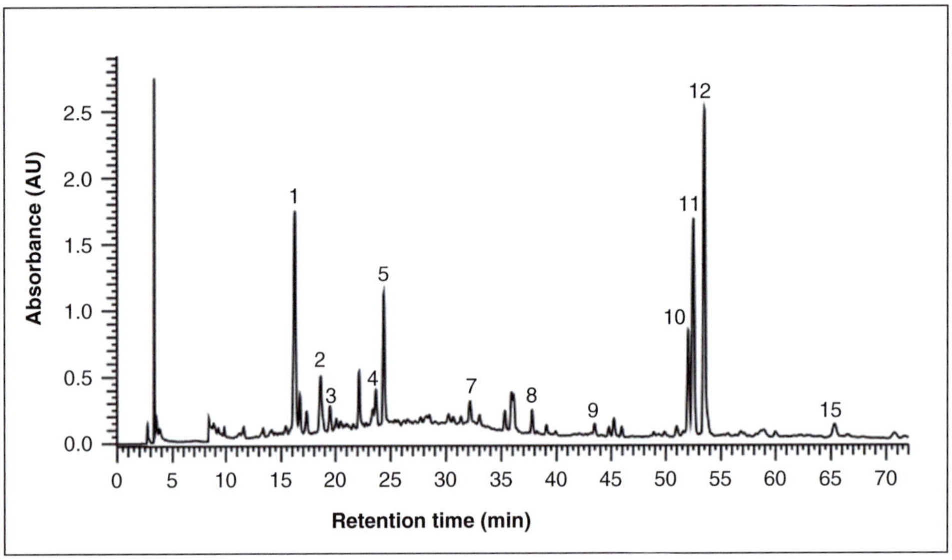

Fig. 3b HPLC-fingerprint analysis of the 80% ethanol extract of **Semen** Trichosanthis (sample 7) detected at **254** nm

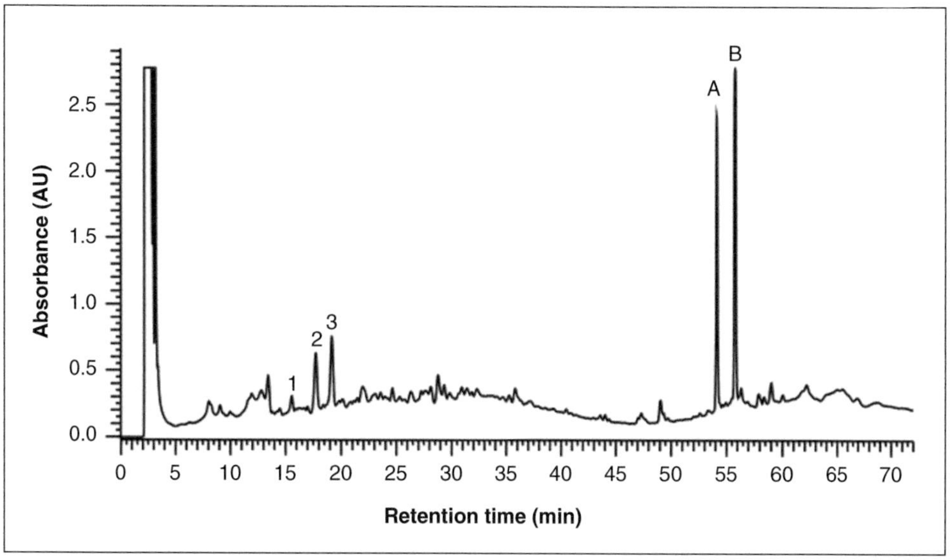

Fig. 3c HPLC-fingerprint analysis of the 80 % ethanol extract of **Fructus** Trichosanthis (sample 2) detected at **210** nm

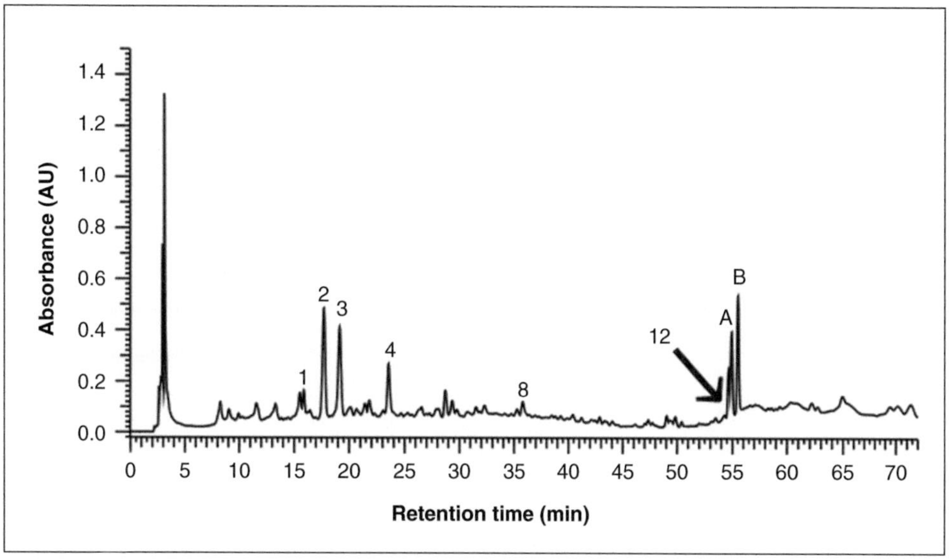

Fig. 3d HPLC-fingerprint analysis of the 80 % ethanol extract of **Fructus** Trichosanthis (sample 2) detected at **254** nm

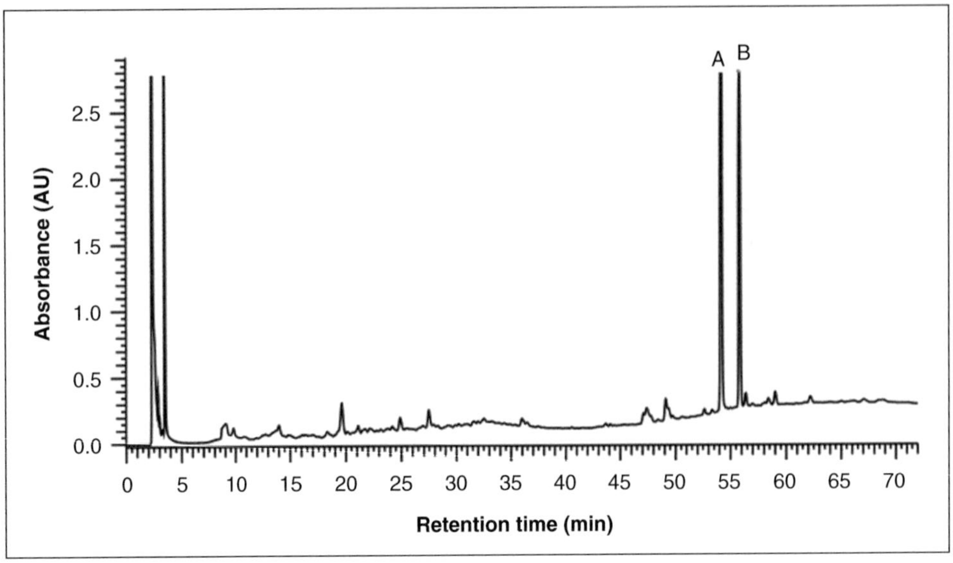

Fig. 3e HPLC-fingerprint analysis of the 80 % ethanol extract of **Fructus** Trichosanthis sample 2* (with semen removed), detected at **210** nm

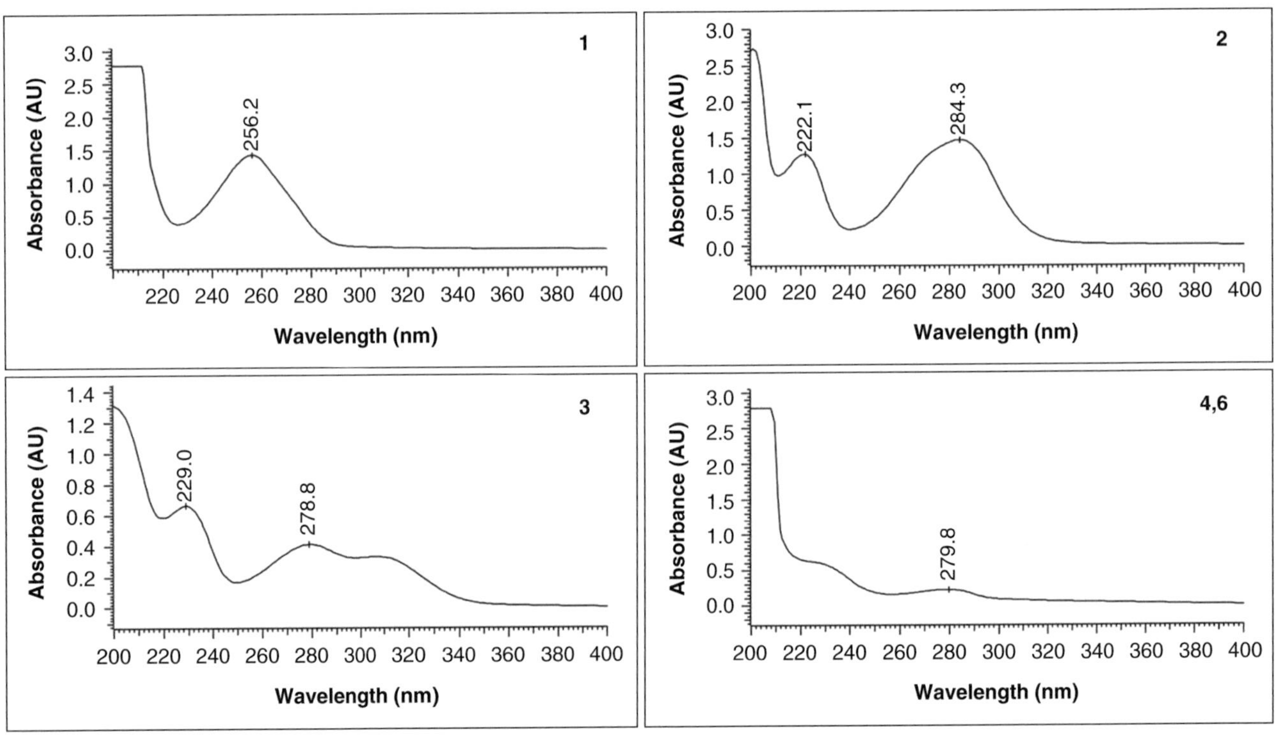

Fig. 4 On line UV-spectra of the main peaks of **Semen** and **Fructus** Trichosanthis

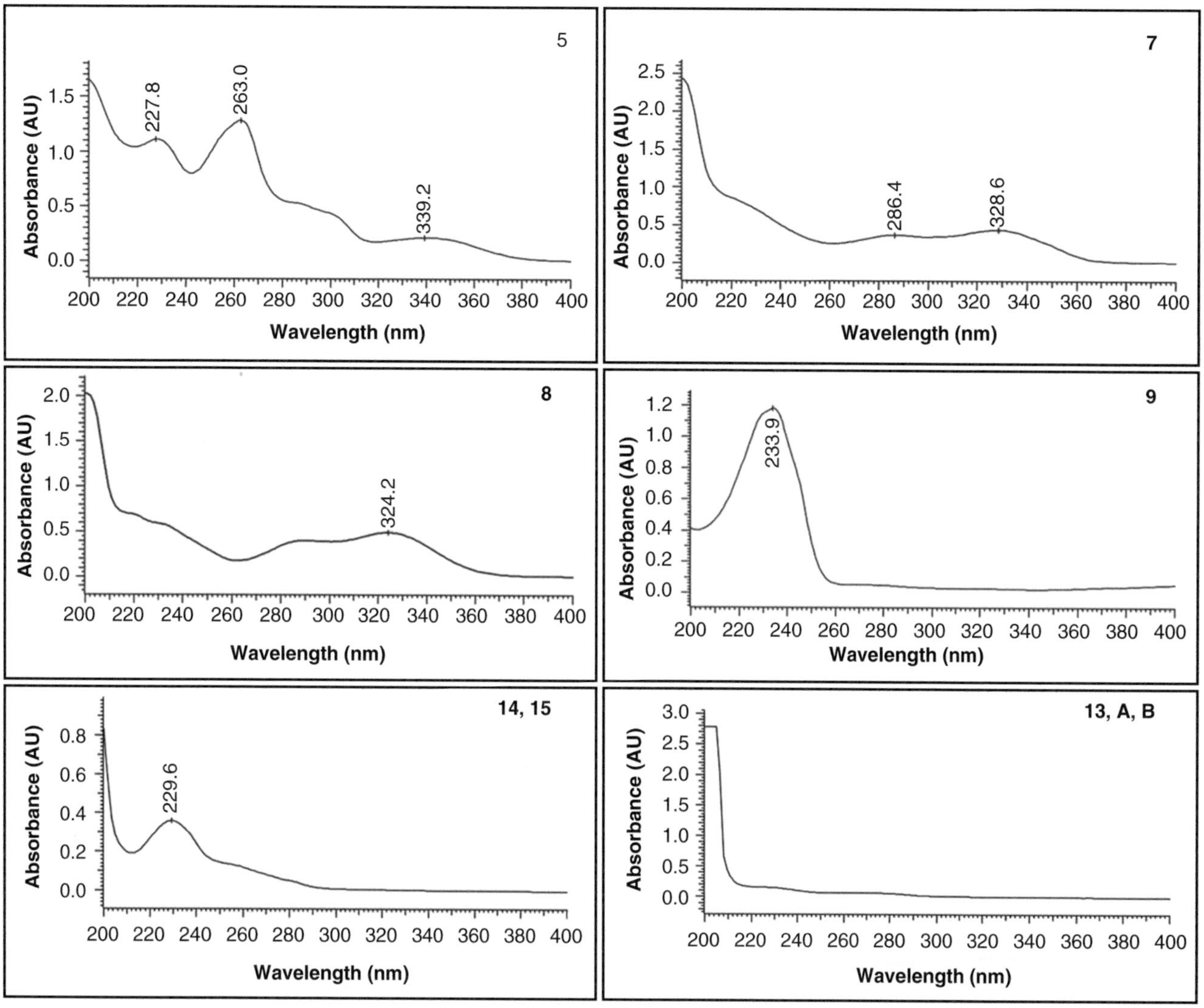

Fig. 4 (continued)

4. Description of the HPLC-Figures

Description of **semen** extract sample 7 at different UV-wavelength (210, 230, and 254 nm): the peak profile of the Semen extract samples are characterized by a high concentration of peak 1–8 in the Rt – range from Rt=15 till 40, which show an assembly of lignans. In the second Rt- range from Rt 40.0–55.0 appear the various triterpenoids and phytosterols (e.g. stigmasterin's compounds).

The HPLC-fingerprint of the **Fructus** samples shows only a small amount of the lignans. The two main peaks **A/B** are triterpenoids. This peak profile is very similar in the one **Fructus** Trichosanthis after removement of all seeds.

Conclusion

From the different TLC and HPLC- fingerprints **Semen** and **Fructus** Trichosanthis can be easily distinguished from each other by the dominance of lignans in Semen Trichosanthis (see Fig. 3a and b).

References

1. Pharmacopoeia of the people's republic of China, English edition, vol. I. People's Medical Publishing House, Beijing (2010)
2. Paulus, E, Ding, Y-H: Handbuch der traditionellen chinesischen Heilpflanzen, Karl F. Haug Verlag, Heidelberg (1987)
3. Tang, W, Eisenbrand, G: Handbuch of chinese medicinal plants. Wiley-VCH Verlag, Wheinheim (2011)
4. Hempen, C-H, Fischer, T: Leitfaden chinesische phytotherapie 2. Auflage Urban & Fischer, Munich (2007)
5. Tang, W, Eisenbrand, G: Chinese drugs of plant origin chemistry, pharmacology, and use tradionaland modern medicine. Springer, Berlin/Heidelberg (1992)
6. Yuan, GF, Yuan, JQ, Li, D: Punicic acid from *Trichosanthes kirilowii* seed oil is rapidly metabolized to conjugated linoleic acid in rats. J. Med. Food **12**(2), 416–422 (2009)
7. Mooon, SS, Rahman, AA, Kim, JY, Kee, SH: Hanultarin, a cytotoxic lignan as an inhibitor of actin cytoskeleton polymerization from the seeds of T*richosanthes kirilowii*. Bioorg. Med. Chem. **16**(15), 7264–7269 (2008)
8. Li, F, Yang, XX, Hu, WG, Xia, HC, Li, Z, Zhang, ZC: Purification and characterization of trichokirin-S1, a novel ribosome-inactivating peptide from seeds of *Trichosanthes kirilowii*. Sheng Wu Hua Xue Yu Sheng Wu Wu Li Xue Bao (Shanghai) **35**(9), 841–846 (2003)
9. Akihisa, T, Tokuda, H, Ichiishi, E, Mukainaka, T, Toriumi, M, Ukiya, M, Yasukawa, K, Nishino, H: Anti-tumor promoting effects of multiflorane-type triterpenoids and cytotoxic activity of karounidiol against human cancer cell lines. Cancer Lett. **173**(1), 9–14 (2001)
10. Huang, Y, He, P, Bader, KP, Radunz, A, Schmid, GH: Seeds of *Trichosanthes kirilowii*, an energy- rich diet. Z. Naturforsch. C **55**(3–4), 189–194 (2000)
11. Akihisa, T, Yasukawa, K, Kimura, Y, Takido, M, Kokke, WC, Tamura, T: Five D:C-friedo-oleanane triterpenes from the seeds of *Trichosanthes kirilowii* Maxim. and their anti-inflammatory effects. Bio. Pharm. Bull. **42**(5), 1101–1105 (1994)
12. Akihisa, T, Kokke, WC, Tamura, T, Nambara, T: 7-Oxodihydrokarounidiol [7-oxo-D:C-friedo-olean-8-ene-3 alpha,29-diol], a novel triterpene from *Trichosanthes kirilowii*. Chem. Pharm. Bull (Tokyo) **40**(5), 1199–1202 (1992)
13. Casellas, P, Dussossoy, D, Falasca, AI, Barbieri, L, Guillemot, JC, Ferrara, P, Bolognesi, A, Cenini, P, Stirpe, F: Trichokirin, a ribosome-inactivating protein from the seeds of *Trichosanthes kirilowii* Maximowicz. Purification, partial characterization and use for preparation of immunotoxins. Eur. J. Biochem. **176**(3), 581–588 (1988)
14. Chao, Z, He, B: Studies on chemical constituents from fruits of *Trichosanthes kirilowii* Maxim. Zhongguo Zhong Yao Za Zhi Chao **24**(10), 612–613 (1999)
15. Ozaki, Y, Xing, L, Satake, M: Anti-inflammatory effect of *Trichosanthes kirilowii* Maxim, and its effective parts. Bio Pharm Bull **19**(8), 1046–1048 (1996)
16. Yasukawa, K, Akihisa, T, Tamura, T, Takido, M: Inhibitory effect of karounidiol on 12-O-tetradecanoylphorbol-13-acetate-induced tumor promotion. Biol. Pharm. Bull. **17**(3), 460–462 (1994)

Fructus Viticis – *Manjingzi*

Pharmacopoeia: [1] Pharmacopoeia of the People's Republic of China, English Edition Vol. I, 2010

Official drug: [1] Shrub Chastetree Fruit is the dried ripe fruit of *Vitex trifolia* L. var. *simplicifolia* Cham. or *Vitex trifolia* L. (Fam. Verbenaceae).

The drug is collected in autumn when ripe, removed from foreign matter, and dried in the sun.

Origin: [2–5] Mainly in Chinese provinces such as Fujian, Guangdong, Guangxi, Jiangsu, Liaoning, Taiwan, Yunnan, Hebei, Henan, Shaanxi, Shanxi, Anhui, Shandong, Jiangxi and Zheijiang; also in Hong Kong, Australia and Pacific Islands.

Description of the drug: [1] Speroidal, 4–6 mm in diameter. Externally greyish-black or blackish-brown, covered with greyish-white frost-like hairs, bearing four longitudinal shallow furrows; apex slightly concave, with greyish-white persistent calyx and short fruit stalk at base. Calyx is 1/3–2/3 length of the fruit, 5-toothed, two of them relatively deep and covered with densely pubescent. Texture light and hard, uneasily broken. Transverse section showing 4 loculi, each with 1 seed. Odour, characteristic and aromatic; taste, weak and slightly pungent.

Processing: [1] *Fructus Viticis* (*stir-baked*)

The clean Fructus Viticis have to be stir-baked gently as described under the method for simple stir-baking (Appendix II D). Break into pieces before use.

Medicinal use: [6] Due to the similar constituents with Fructus Agni casti it can be used for the treatment of premenstrual symptoms, usually menstruation tempo anomalies and mastodynia.

Effects and indications of Fructus Viticis according to Traditional Chinese Medicine [1–3, 5, 7, 8]	
Taste:	Pungent and bitter
Temperature:	Mild cold
Channels entered:	*Orbis vesicalis, o. hepaticus, o. stomachi*
Effects (functions):	To disperse wind-heat, clear and soothe head
Symptoms and indications:	Common cold caused by wind-heat, cough, headache, gum swelling and pain, red eyes and hyperdacryosis, dim and blurred vision, diziness and vertigo

H. Wagner et al. (eds.), *Chromatographic Fingerprint Analysis of Herbal Medicines*, Vol. 4,
DOI 10.1007/978-3-319-32328-2_8, © Springer International Publishing Switzerland 2016

Main constituents:

- **Flavones** [9]

Vitexicarpin (= casticin)

- **Diterpenoids** [5, 8, 10, 11]

Vitextrifolin A-H, vitexifolin E,F; vitexilactone, vitexilactone B, previtexilactone, deacetylvitexilactone, rotundifuran, viteagnusin I, negundol, dihydrosolidagenone, 6-acetoxy-9-hydroxy-13 (14)-labden-16,15-olide, abietatriene 3β-ol

Minor constituents:

- Fatty acids (stearic acid, palmitic acid, palmitoleic acid, linoleic acid, myristic acid, oleic acid), sterols (β-sitosterol, β-sitosterol-3-O-glucoside) [9]

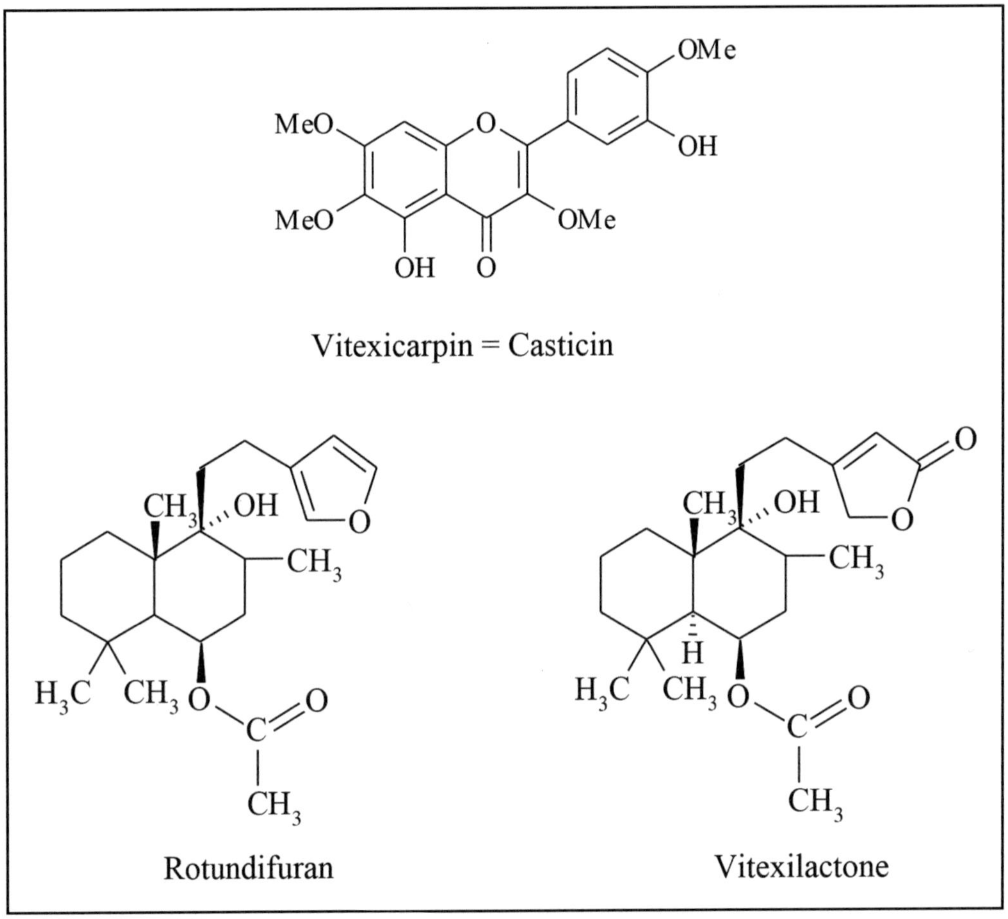

Fig. 1 Formulae of the main compounds of Fructus Viticis [9, 11]

Reported pharmacology: - antipyretic [7, 9]
- antibiotic [7]
- analgesic [7]
- sedating [7]
- anti-oxidative [5]

TLC Fingerprint Analysis

Drug samples		Origin
1	Fructus Viticis / *Vitex trifolia* var. *simplicifolia*	Sample of commercial drug obtained from HerbaSinica (origin: Shandong)
2	Fructus Viticis / without botanical assignment	Sample of commercial drug obtained from TCM- Clinic Bad Kötzting (Charge: 24711102003)
3	Fructus Viticis / without botanical assignment	Sample of commercial drug obtained from China Medica
4	Fructus Viticis / without botanical assignment	Province Guangxi (China)
5	**Fructus Agni casti tot.** / *Vitex agnus-castus*[a]	Sample of commercial drug obtained from Caelo

[a]For comparison

Reference compound		R*f*
T	Casticin	0.18

1. Extraction: 1.0 g powdered drug is ultrasonicated with 10 ml ethanol (90 %) for 1 h. The extract is filtered, evaporated to dryness and the residue is dissolved in 1 ml ethanol.

2. Reference compound: 1.0 mg is dissolved in 1.0 ml methanol

3. Separation parameters:

Plate:	HPTLC Silica gel 60 F_{254}, Merck
Applied amounts:	Fructus Viticis extracts: each 10 µl, Reference compound: 10 µl
Solvent system:	Toluene + ethyl acetate (9 + 2)
Detection:	10 % ethanolic sulphuric acid The plate is sprayed with the reagent, heated at 110 °C for 10 minutes and evaluated in VIS

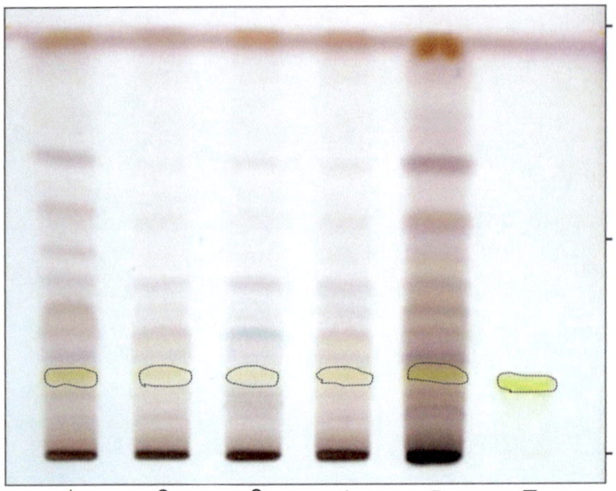

Fig. 2 Thin layer chromatogram of the ethanol extracts of Fructus Viticis, sprayed with 10 % ethanolic sulphuric acid (VIS)

4. Description:

   The five *Vitex* extract samples are characterized by a fairly equal zone profile with 8–10 brown-violet zones from the TLC-solvent front down to the solvent start with one dominant zone in sample 1 and 5 at R*f*=0.68 (rotundifuran) and the yellow zone of casticin (**T**) at R*f*=0.18. The *Vitex* extract sample 5 possess the highest content of casticin (see also HPLC-Fig. 3c).

HPLC-Fingerprint Analysis

1. Extraction: 1.0 g powdered drug is ultrasonicated with 10 ml ethanol (90 %) for 1 h. The extract is filtered, evaporated to dryness and the residue is dissolved in 1 ml ethanol. The extract is filtered over Chromafil®, Type 0.20 µm.

2. Injection volume: Fructus Viticis extracts: each 10.0 µl

3. HPLC parameter:

Apparatus:	MERCK HITACHI D-6000 A Interface
	MERCK HITACHI L-4500 A Diode Array Detector
	MERCK HITACHI AS-2000 Autosampler
	MERCK HITACHI L-6200 A Intelligent Pump
Separation column:	LiChroCART® 250-4 LiChrospher® 100 RP-18 (5 µm), Merck
Precolumn:	LiChroCART® 4-4 LiChrospher® 100 RP-18 (5 µm), Merck
Solvent system:	A: 0.001 % phosphoric acid/water (Millipore Ultra Clear UV plus® filtered)
	B: acetonitrile (VWR)
Gradient:	0 % B for 5 min,
	0–90 % B in 50 min,
	total runtime: 50 min
Flow:	1.0 ml/min
Detection:	210 nm

Retention times of the main peaks

Peak	Rt (min)	Compound
1	16.8	Not identified Diterpenoid
2	19.7	Not identified Diterpenoid
3	22.4	Not identified Diterpenoid
4	27.4	Not identified Diterpenoid
5	37.8	Casticin
6	43.7	Not identified

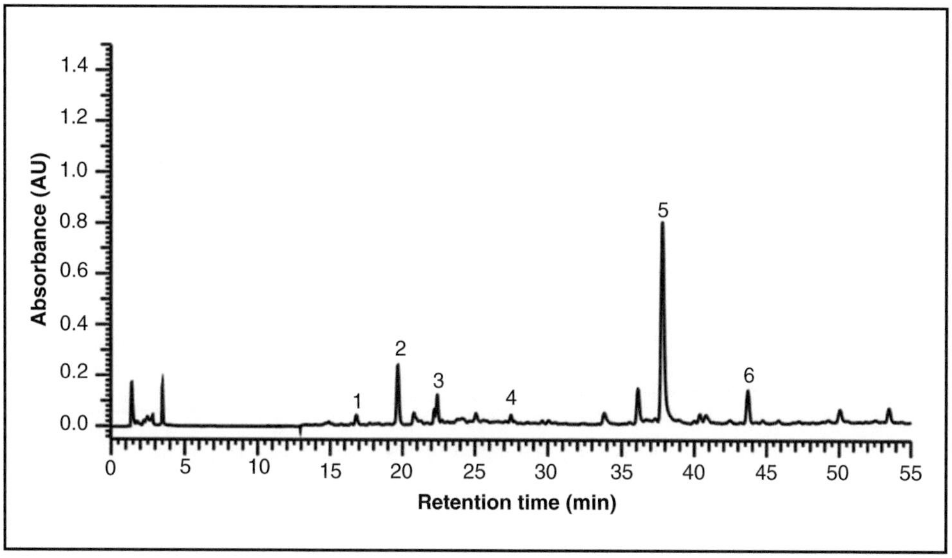

Fig. 3a HPLC-fingerprint analysis of the methanol extract of Fructus Viticis, sample 1

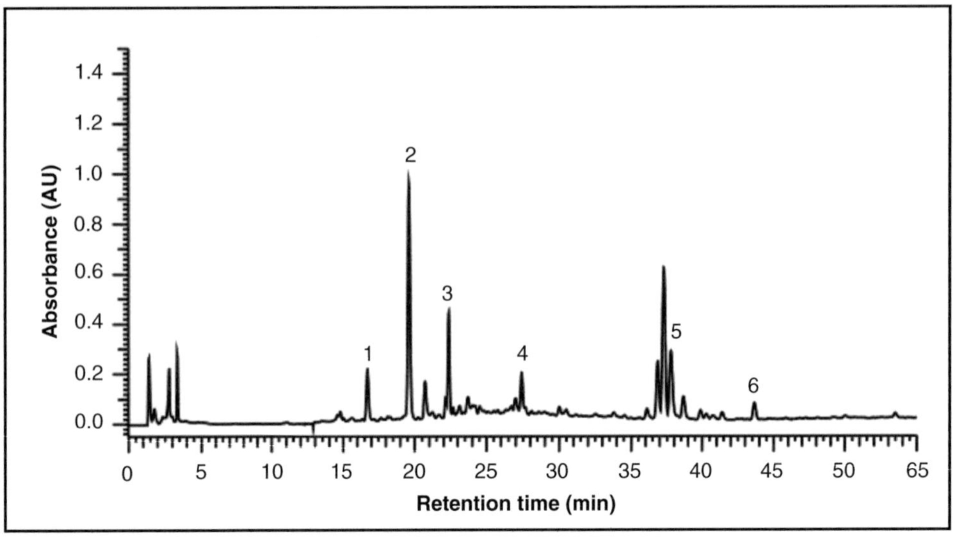

Fig. 3b HPLC-fingerprint analysis of the methanol extract of Fructus Viticis, sample 3

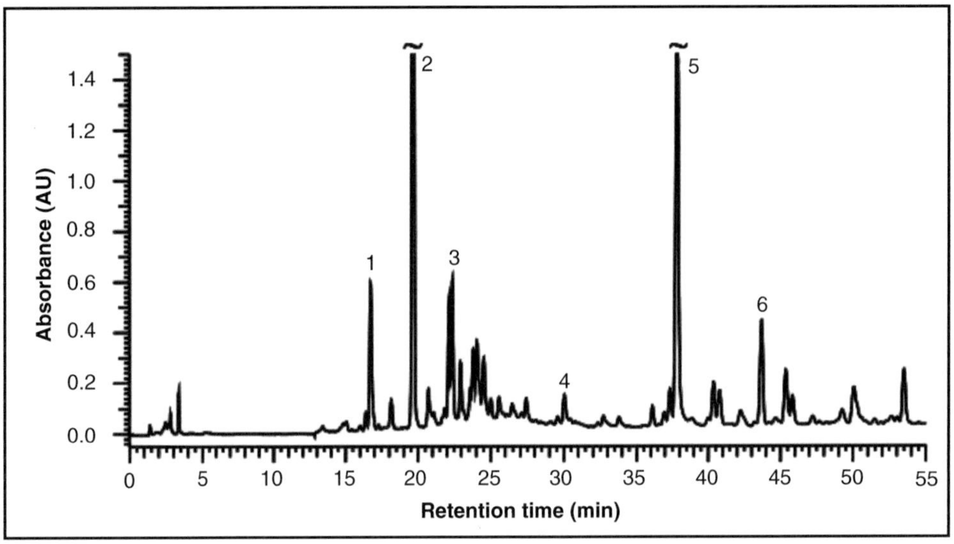

Fig. 3c HPLC-fingerprint analysis of the methanol extract of **Fructus Agni casti tot.**, sample 5

4. Description of the HPLC-Figures

   The HPLC-graphs show analogue to the TLC a very homogenous pattern of 6 peaks with the dominant peaks **2** and **5** (casticin).

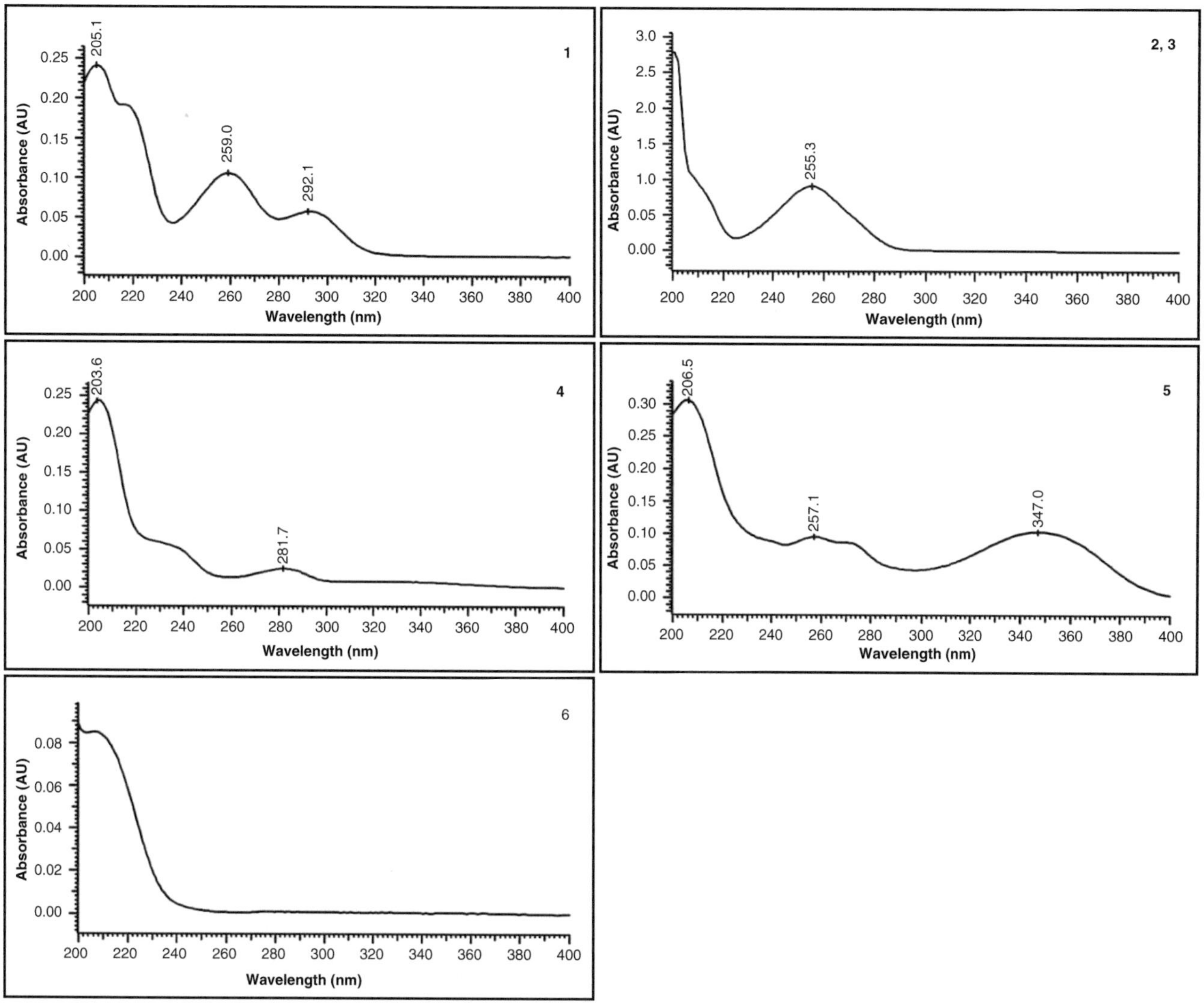

Fig. 4 On line UV-spectra of the main peaks of Fructus Viticis extracts

Note: According to the Chinese Pharmacopeia 2010 Fructus Viticis contains not less than 0.03 % of vitexicarpin, calculated with reference to the dried drug [1].

Conclusion

Although the exact botanical assignment of some Vitex samples is not available the presence of the characteristic compound of Vitex species is sufficient for authentication.

References

1. Pharmacopoeia of the people's republic of China, English edition, vol. I. People's Medical Publishing House, Beijing (2010)
2. Zhang, ZZ: An illustrated Chinese materia medica in Hong Kong. School of Chinese Medicine/Hong Kong Baptist University, Hong Kong (2004)
3. Paulus, E, Ding, Y-H: Handbuch der traditionellen chinesischen Heilpflanzen, Karl F. Haug Verlag, Heidelberg (1987)
4. Flora of China, *Vitex trifolia*, FOC 17, 29 (www.eFloras.org)
5. Ono, M, Sawamura, H, Ito, Y, Mizuki, K, Nohara, T: Diterpenoids from the fruits of *Vitex trifolia*. Phytochemistry **55**(8), 873–877 (2000)
6. Wagner, H, Vollmer, A, Bechthold, A: Pharmazeutische Biologie 2, Biogene Arzneistoffe und Grundlagen von Gentechnik und Immunologie. Wissenschaftliche Verlagshesellschaft mbh, Stuttgart (2007)
7. Hempen, C-H, Fischer, T: A materia medica for Chinese medicine, english edition. Elsevier GmbH, Urban & Fischer, Munich (2009)
8. Zheng, CJ, Zhu, JY, Yu, W, Ma, XQ, Rahman, K, Qin, LP: Labdane-type diterpenoids from the fruits of *Vitex trifolia*. J. Nat. Prod. **76**(2), 287–291 (2013)
9. Tang, W, Eisenbrand, G: Handbook of Chinese medicinal plants – chemistry, pharmacology, toxicology, vol. 2. WILEY-VCH Verlag GmbH & Co KgaA, Weinheim (2011)
10. Ono, M, Ito, Y, Nohara, T: Four new halimane-type diterpenes, vitetrifolins D-G, from the fruit of *Vitex trifolia*. Chem. Pharm. Bull. **49**(9), 1220–1222 (2001)
11. Kiuchi, F, Matsuo, K, Ito, M, Qui, TK, Honda, G: New norditerpenoids with trypanocidial activtiy from *Vitex trifolia*. Chem. Pharm. Bull. **52**(12), 1492–1494 (2004)

Fructus Xanthii – *Cang'erzi*

Pharmacopoeia: [1]	Pharmacopoeia of the People's Republic of China, English Edition Vol. I, 2010
Official drug: [1]	Siberian Cocklebur Fruit is the dried ripe fruit with involucre of *Xanthium sibiricum* Patr. (Fam. Asteraceae). The drug is collected in autumn when ripe, dried and removed from stalk, leaf and other foreign matter.
Synonym: [2–6]	*Xanthium strumarium* L.
Origin: [7]	In most regions of China, including Hong Kong.
Description of the drug: [1]	Fusiform or ovoid, 1–1.5 cm long, 0.4–0.7 cm in diameter. Externally yellowish-brown or yellowish-green, with hooked spines throughout. Summit with 2 relatively thick spines, separated or linked up, base with a fruit stalk scar. Texture hard and flexible, the centre of transverse section showing a septum and 2 loculi, each having an achene. Achene slightly fusiform, relatively even at one side, apex with a protruding remains of style, pericarp thin, greyish-black, with longitudinal wrinkles. Testa membranous, pale grey, cotyledons 2, oily. Odour, slight; taste, slightly bitter.
Medicinal use: [2, 3, 5, 8, 9]	Application for headaches, tinnitus, toothache, malaria and mumps. Treatment of chronic sinusitis, ozena, topical application for chronic rhinitis and intramuscular administration for back pain.
Toxicity: [8, 10]	Overdose can lead to nausea, vomiting, diarrhea, abdominal pain. Boiling or stir-baking reduces toxicity.

Effects and indications of Fructus Xanthii according to Traditional Chinese Medicine [1, 4–6, 8–16]

Taste:	Pungent and bitter, sweet
Temperature:	Warm
Channels entered:	*Orbis pulmonalis, o. hepaticus*
Effects (functions):	To dispers wind-cold, relieve the stuffy nose, dispel wind-dampness
Symptoms and indications:	Wind-cold headache due to rheumatism and skin pruritus, nasal congestion, nasal discharge, allergic rhinitis, sinusitis, urticarial, arthritis, itching caused by rubella, spasm caused by fixed impediment

H. Wagner et al. (eds.), *Chromatographic Fingerprint Analysis of Herbal Medicines*, Vol. 4,
DOI 10.1007/978-3-319-32328-2_9, © Springer International Publishing Switzerland 2016

Main constituents:

- **Phenol carboxylic acids** [3–7, 9, 16]

  Chlorogenic acid; neochlorogenic acid (=5-caffeoylquinic acid); cryptochlorogenic acid (=4-caffeoylquinic acid); caffeic acid; 1,4-dicaffeoylquinic acid; 1-caffeoylquinic acid; cynarin (=1,5-dicaffeoylquinic acid); 1,3,5-tri-O-caffeoylquinic acid; isochlorogenic acid A (=3,5-dicaffeoylquinic acid); isochlorogenic acid B (=4,5-dicaffeoylquinic acid); isochlorogenic acid C (=3,4-dicaffeoylquinic acid); chlorogenate, ferulic acid

- **N- and S-containing glycosides (some derived from monoterpenoids)** [4, 13–15, 17]

  -3β-norpinan-2-one-3-O-β-D-apiofuranosyl-(1 → 6)- β-D-glucopyranoside; (6Z)-3-hydroxymethyl-7-methylocta-1,6-dien-3-ol-8-O- β-D-glucopyranoside; (6E)-3-hydroxymethyl-7-methylocta-1,6-dien-3-ol-8-O-β-D-glucopyranoside; 7-[(β-D-apiofuranosyl-(1 → 6)-β-D-glucopyranosyl) oxymethy]-8,8-dimethyl-4,8-dihydro-benzo [1,4] thiazine-3,5-dione

  -xanthiside, xanthiazinone, xanthiazone, xanthienopyran, 5-hydroxypyrrolidin-2-one, adenosine, (+) sibiricumin A, (-)-sibiricumin A

Minor constituents:

Fatty oils, alkaloids, essential oils, diterpenes, saponins, flavones, sesquiterpene lactones, thiazinediones, amino acids, vitamin C, linoleic acid, oleic acid, palmitic acid, stearic acid, β,γ,ε-sitosterin, lecithin, cephaeline, xanthostrumarin, ceryl alcohol, resin [3, 6–8, 10, 14, 16]

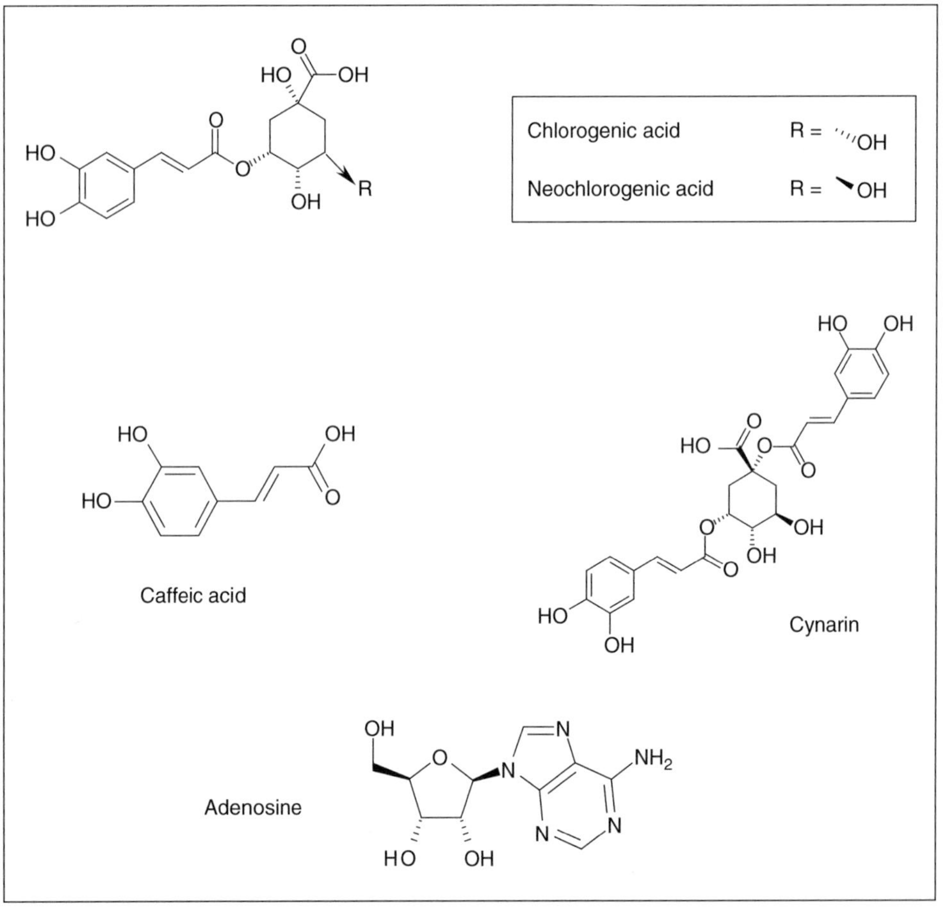

Fig. 1 Formulae of the main compounds of Fructus Xanthii [13, 14]

Reported pharmacology: - anti-inflammatory [5, 14–16]
- anti-allergic [16]
- antibiotic [10]
- decrease of serum glucose [10]
- analgesic activity [5, 16]

TLC Fingerprint Analysis

Drug samples	Origin
1 Fructus Xanthii (praep.) / *Xanthium sibiricum*	Sample of commercial drug obtained from HerbaSinica (origin: Jilin, China)
2 Fructus Xanthii (praep.) / *Xanthium sibiricum*	Sample of commercial drug obtained from China Medica (origin: Anhui, China)
3 Fructus Xanthii / *Xanthium sibiricum*	Province Neimeng, China
4 Fructus Xanthii / *Xanthium sibiricum*	Province Dongbei, China
5 Fructus Xanthii / *Xanthium sibiricum*	Sample of commercial drug obtained from TCM-Clinic Bad Kötzting (Charge: K30.08.2000)

1. TLC-fingerprint analysis of phenol carboxylic acid: [18]

Reference compounds of Fig. 2		R*f*
T1	Chlorogenic acid	0.37
T2	Caffeic acid	0.94
T3	Hyperosid	0.46

1. Extraction: 1.0 g powdered drug is extracted with 10 ml methanol under reflux for 20 min. The extract is filtered, evaporated to dryness and the residue is dissolved in 1.0 ml methanol.

2. Reference compounds: 1.0 mg is dissolved in 1.0 ml methanol

3. Separation parameters:

Plate:	HPTLC Silica gel 60 F_{254}, Merck
Applied amounts:	Fructus Xanthii extracts: each 10 µl, Reference compounds: each 10 µl
Solvent system:	Ethyl acetate + formic acid + water + toluene (20 + 2 + 2 + 1)
Detection:	Natural products – Polyethylene glycol reagent (NP/PEG)

I: 1 % diphenylboric acid-*β*-ethylamino ester (= diphenylboryloxyethylamine, NP) in methanol

II: 5 % Polyethylene glycol-4000 (PEG) in ethanol (80 %)

The plate is sprayed first with solution **I** and then with solution **II**. After 1 h the evaluation is carried out under UV 366 nm.

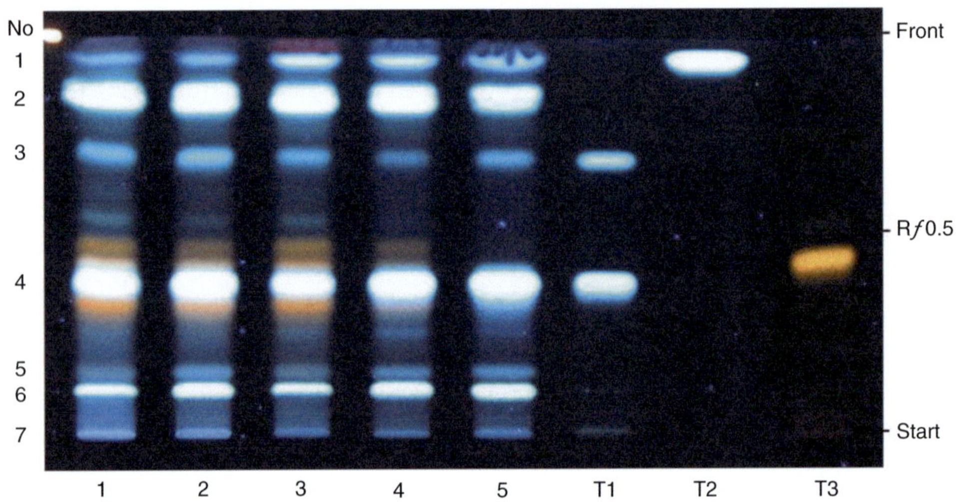

Fig. 2 Thin layer chromatogram of the methanol extracts of Fructus Xanthii, sprayed with NP/PEG (UV 366 nm)

4. Description:

   The TLC sprayed with NP/PEG and evaluated under UV 366 nm is characterized by 6–7 main white/blue fluorescent zones from the TLC-front down to the solvent start.

No.	Compound	R*f*
1	Caffeic acid (**T2**)	0.94
2	Isochlorogenic acids	0.83
3	Dicaffeoylquinic acids (e.g. Cynarin)	0.69
4	Chlorogenic acid (**T1**)	0.37
5	5 or 4-monocaffeoylquinic	0.15
6	acids (crypto- or	0.11
7	neochlorogenic acid)	start

Flavonol glycosides were detectable in the extract samples 1–4 as orange fluorescent zones below and above of chlorogenic acid (R*f*=0.46 (**T3**=Hyperoside); 0.44; 0.31).

2. TLC-fingerprint analysis in a more lipophilic solvent system for the detection of terpenoids of essential oils and fatty acids: [18]

Reference compound of Fig. 3a/b		R*f*
T4	Linoleic acid	0.18

1. Extraction: 1.0 g powdered drug is extracted with 10 ml methanol under reflux for 20 min. The extract is filtered, evaporated to dryness and the residue is dissolved in 1.0 ml methanol.

2. Reference compound: 1.0 mg is dissolved in 1.0 ml methanol

3. Separation parameters:

   Plate: HPTLC Silica gel 60 F_{254}, Merck

   Applied amounts: Fructus Xanthii extracts: each 10 µl, Reference compound: 10 µl

   Solvent system: Toluene + ethyl acetate (9 + 1)

   Detection: 10% ethanolic sulphuric acid

   The plate is sprayed with 8 ml reagent and heated at 110 °C for 10 min. The plate is evaluated in VIS (Fig. 3a) and under UV 366 nm (Fig. 3b).

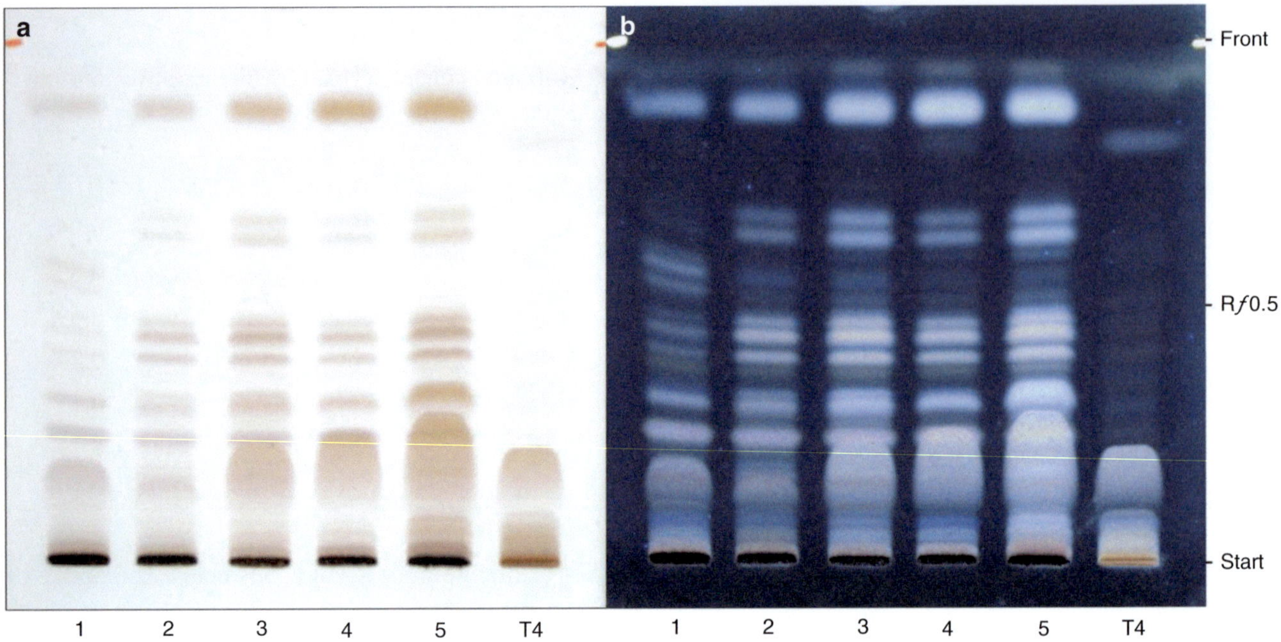

Fig. 3a/b Thin layer chromatogram of the methanol extracts of Fructus Xanthii, sprayed with 10 % ethanolic sulphuric acid (**a** = VIS, **b** = UV 366 nm)

4. Description:

   In Fig. 3a all five extracts show in the low and middle R*f*-range a very homogeneous pattern of light brown coloured zones which could be assigned to mono- and sesquiterpenoids apart of linoleic acid (**T4**, R*f*=0.18).

   In Fig. 3b the same zones appear with light blue fluorescence.

Note: Both figures with their characteristic zone profiles provide an additional definite authentication for Fructus Xanthii.

3. TLC-fingerprint analysis of the nucleoside Adenosine: [18]

Reference compound of Fig. 4a/b		R*f*
T5	Adenosine	0.66

1. Extraction: 1.0 g powdered drug is extracted with 10 ml methanol under reflux for 20 min. The extract is filtered, evaporated to dryness and the residue is dissolved in 1.0 ml methanol.

2. Reference compound: 1.0 mg is dissolved in 1.0 ml methanol

3. Separation parameters:

   Plate: HPTLC Silica gel 60 F_{254}, Merck

   Applied amounts: Fructus Xanthii extracts: each 10 µl Reference compound: 10 µl

   Solvent system: *n*-butanol + ethanol + glacial acetic acid + water (12 + 4 + 4 + 6)

   Direct evaluation: UV 254 nm (**Fig. 4a**)

   Spray reagent: Potassium permanganate / sodium hydroxide solution

   0.5 g potassium permanganate is dissolved in aqueous sodium hydroxide solution (1 mol).

   The plate is sprayed with 8 ml reagent and evaluated in VIS (**Fig. 4b**).

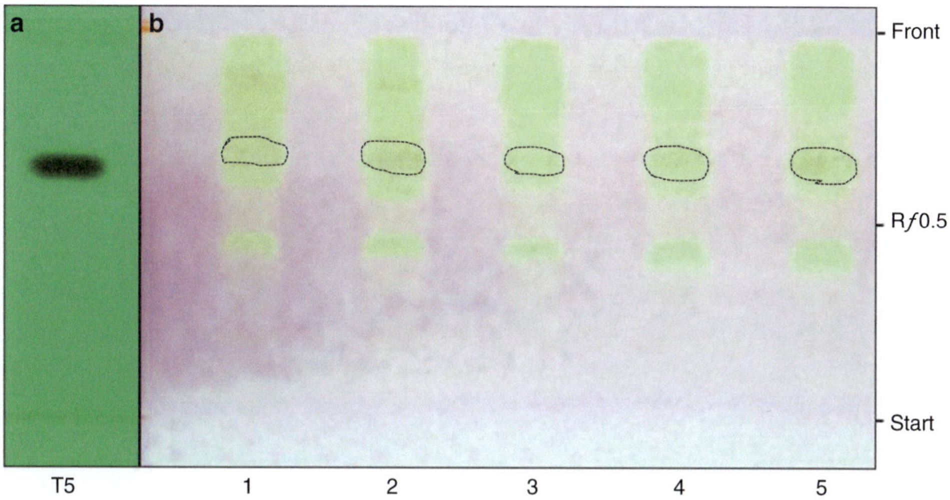

Fig. 4a/b Thin layer chromatogram of the methanol extracts of Fructus Xanthii, sprayed with $KMnO_4$/NaOH solution (**a**) UV 254 nm, (**b**) VIS

4. Description:

   Adenosine (**T5**) shows in Fig. 4a a black zone at R*f*=0.66 under a green fluorescent background.

   In Fig. 4b the extract samples 1–5 provide light green zones from the solvent front down to R*f*=0.45. The green zone at R*f*=0.66 can be assigned to Adenosine.

HPLC-Fingerprint Analysis

1. Extraction: 1.0 g powdered drug is extracted with 10 ml methanol under reflux for 20 min. The extract is filtered, evaporated to dryness and the residue is dissolved in 1.0 ml methanol. The extract is filtered over Chromafil®, Type 0.20 µm.

2. Injection volume: Fructus Xanthii extracts: each 15 µl

3. HPLC parameter:

Apparatus:	MERCK HITACHI D-6000 A Interface
	MERCK HITACHI L-4500 A Diode Array Detector
	MERCK HITACHI AS-2000 Autosampler
	MERCK HITACHI L-6200 A Intelligent Pump
Separation column:	LiChroCART® 250-4 LiChrospher® 100 RP-18 (5 µm), Merck
Precolumn:	LiChroCART® 4-4 LiChrospher® 100 RP-18 (5 µm), Merck
Solvent system:	A: 0.1 % phosphoric acid/water (Millipore Ultra Clear UV plus® filtered)
	B: acetonitrile (VWR)
Gradient:	10–22 % B in 15 min,
	22 % B for 5 min,
	22–40 % B in 15 min,
	total runtime: 35 min
Flow:	1.0 ml/min
Detection:	210 nm

Retention times of the main peaks

Peak	Rt (min)	Compound
1	2.3	Not identified
2	3.0	Adenosine
X	5.5	Not identified
3	9.6	Chlorogenic acid
4	10.6	Not identified
5	15.3	Not identified
6	19.3	Unknown flavonoid
7	21.8	Unknown chlorogenic
8	23.5	acid isomers /
9	28.6	dicaffeoylquinic acids

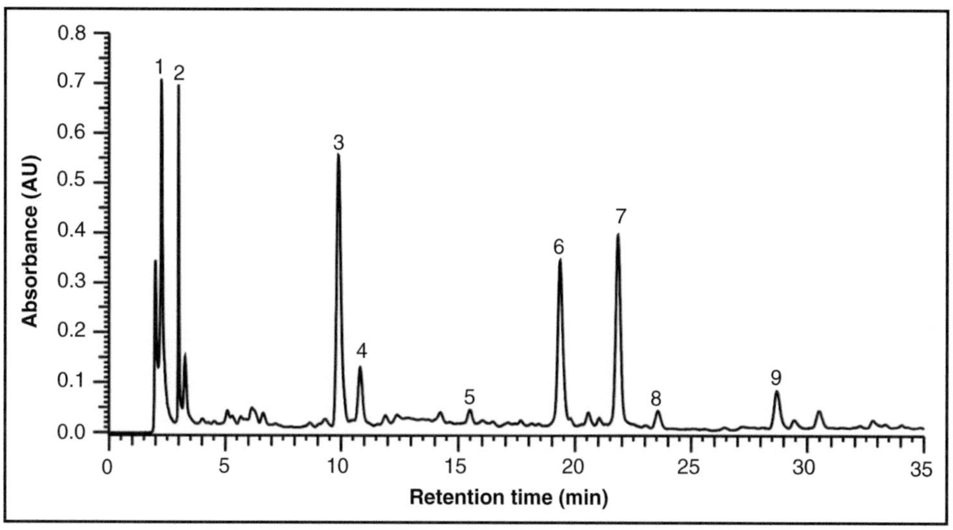

Fig. 5a HPLC-fingerprint analysis of the methanol extract of Fructus Xanthii, sample 1

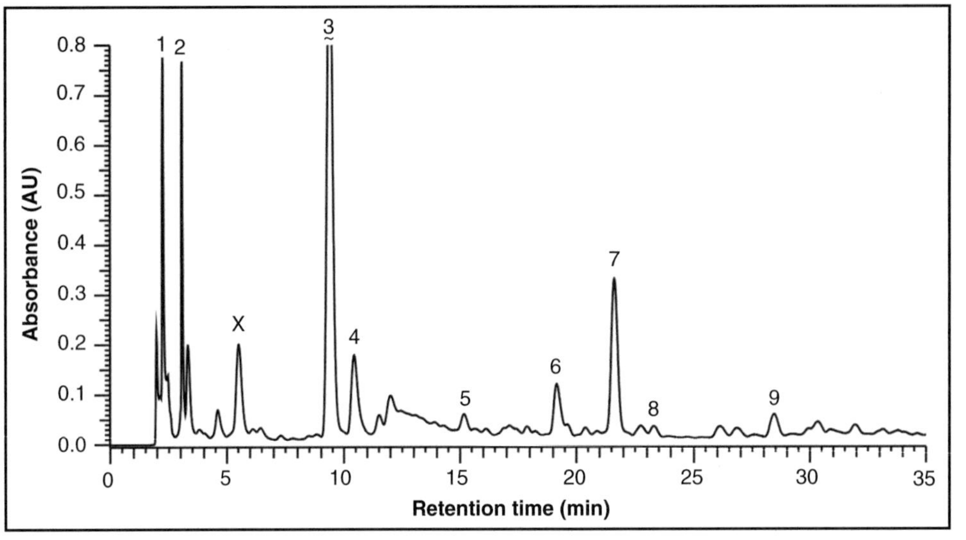

Fig. 5b HPLC-fingerprint analysis of the methanol extract of Fructus Xanthii, sample 4

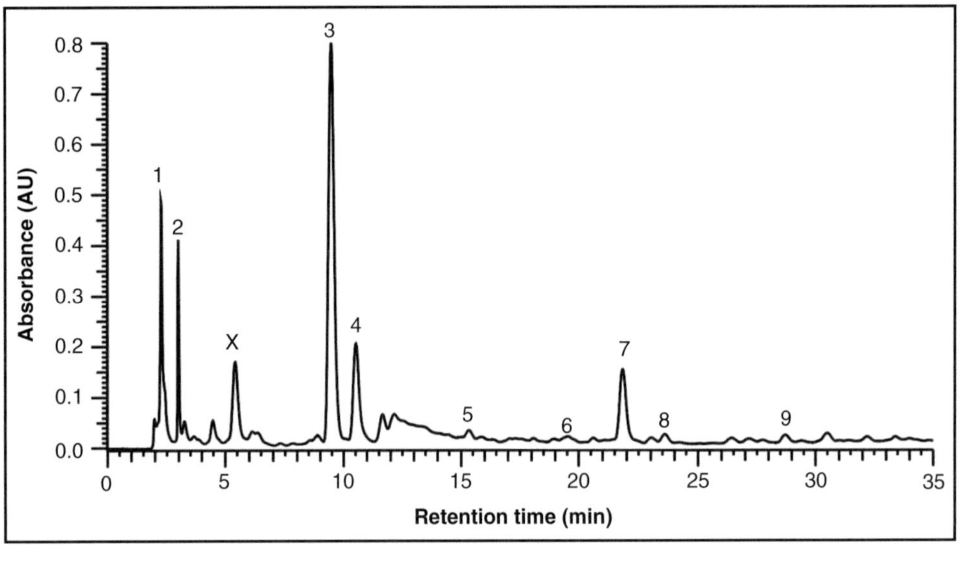

Fig. 5c HPLC-fingerprint analysis of the methanol extract of Fructus Xanthii, sample 5

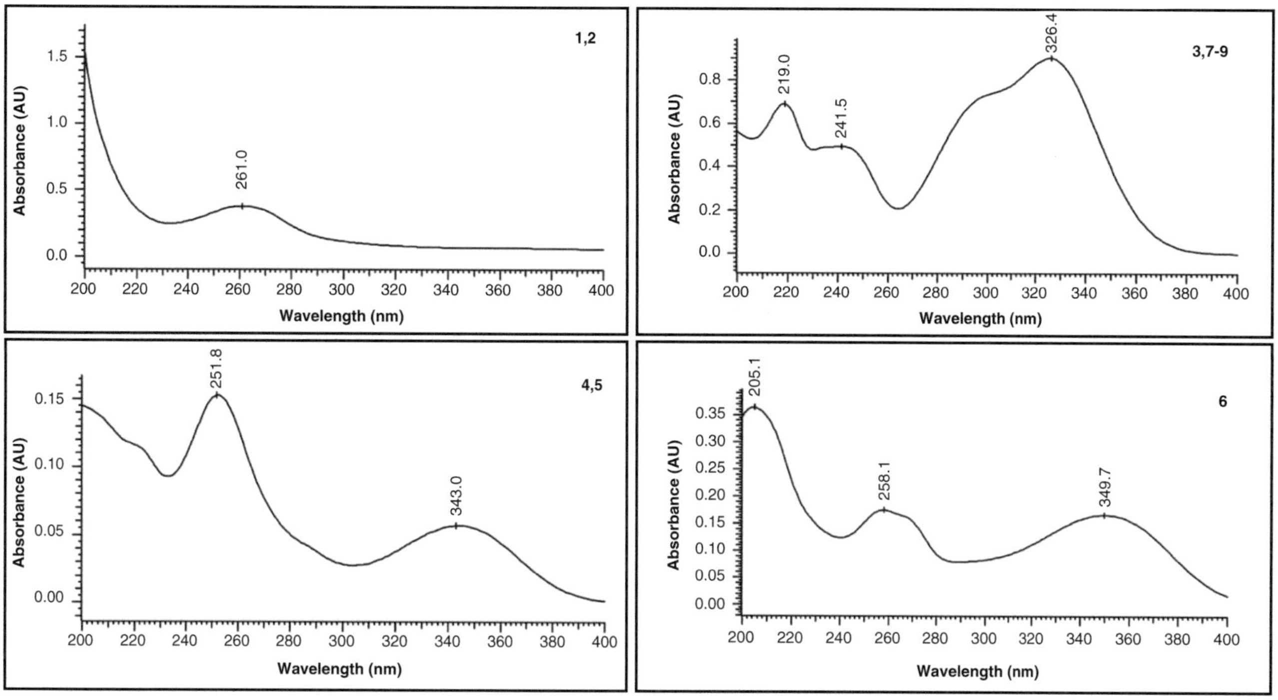

Fig. 6 On line UV-spectra of the main peaks of Fructus Xanthii extracts

4. Description of the HPLC-Fig. 5a, b, c

   Peak **1** and **2** can be assigned to the nucleosides Guanin (**1**) and Adenosin (**2**), identified with reference compound). Peak **3** is chlorogenic acid as the only monocaffeoylquinic acid. The peak numbers **4–6** with the characteristic double maxima at 257 and 348 nm can be assigned to flavonoids. The peak **7–9** possess a very similar UV-spectrum as peak **3** (chlorogenic acid) and can be identified as dicaffeoylquinic acids (see also TLC-Fig. 2).

Conclusion

The TLC of Figs. 2 and 3a/b are sufficient for authentication of Fructus Xanthii but can be definitely assessed be the additional detection of the nucleosides adenosine and guanosin. The HPLC confirms the TLC authentication by a very similar HPLC-peak profile.

References

1. Pharmacopoeia of the people's republic of China, English edition, vol. I. People's Medical Publishing House, Beijing (2010)
2. Zhao, Y, Yang, H, Zheng, YB, Wong, YO, Leung, PC: The effects of Fructus Xanthii extract on cytokine release from human mast cell line (HMC-1) and peripheral blood mononuclear cells. Immunopharmacol. Immunotoxicol. **30**(3), 543–552 (2008)
3. Hong, SH, Jeong, HJ, Kim, HM: Inhibitory effects of Xanthii fructus extract on mast cell-mediated allergic reaction in murine model. J. Ethnopharmacol. **88**(2–3), 229–234 (2003)
4. Han, T, Li, HL, Hu, Y, Zhang, QY, Huang, BK, Zheng, HC, Rahman, K, Qin, LP: Phenolic acids in Fructus Xanthii and determination of contents of total phenolic acids in different species and populations of Xanthium in China. J. Chin. Integr. Med. **4**(2), 194–198 (2006)
5. Huang, MH, Wang, BS, Chiu, CS, Amagaya, S, Hsieh, WT, Huang, SS, Shie, PH, Huang, GJ: Antioxidant, antinociceptive, and anti-inflammatory activities of Xanthii Fructus extract. J. Ethnopharmacol. **135**(2), 545–552 (2011)
6. Han, T, Hu, Y, Zhou, SY, Li, HL, Zhang, QY, Zhang, H, Huang, BK, Rahman, K, Zheng, HC, Qin, LP: Correlation between the genetic diversity and variation of total phenolic acids contents in Fructus Xanthii from different populations in China. Biomed. Chromatogr. **22**(5), 478–486 (2008)
7. Zhang, ZZ: An illustrated Chinese materia medica in Hong Kong. School of Chinese Medicines/Hong Kong Baptist University, Hong Kong (2004)
8. Paulus, E, Ding, Y-H: Handbuch der traditionellen chinesischen Heilpflanzen, Karl F. Haug Verlag, Heidelberg (1987)
9. Chen, B, Ma, L, Wang, X, Shen, Y, Jia, X: Simultaneous determination of 5 phenolic acids in fried Fructus Xanthii from different production sites and its dispensing granules by using ultra-pressure liquid chromatography. Pharmacogn. Mag. **9**(34), 103–108 (2013)
10. Hempen, C-H, Fischer, T: A materia medica for Chinese medicine, english edition. Elsevier GmbH, Urban & Fischer, Munich (2009)
11. Geng, J., Huang, W., Ren, T., Ma, X.: Materia Medica der Chinesischen Arzneimitteltherapie (Bd. 2: Praxis der Chinesischen Arzneimitteltherapie), Verlag für Ganzheitliche Medizin Dr. Erich Wühr GmbH, Kötzting/Bayer. Wald (1993)
12. Porkert, M: Klinische Chinesische Pharmakologie, Verlag für Medizin Dr. Ewald Fischer, Heidelberg (1978)
13. Lu, F, Cao, M, Wu, B, Li, XZ, Liu, HY, Chen, DZ, Liu, SM: Urinary metabonomics study on toxicity biomarker discovery in rats treated with Xanthii Fructus. J. Ethnopharmacol. **149**(1), 311–320 (2013)
14. Jiang, H, Yang, L, Liu, C, Hou, H, Wang, Q, Wang, Z, Yang, B, Kuang, H: Four new glycosides from the fruit of *Xanthium sibiricum* Patr. Molecules **18**(10), 12464–12473 (2013)
15. Shi, Y, Liu, Y, Li, Y, Li, L, Qu, J, Ma, S, Yu, S: Chiral resolution and absolute configuration of a pair of rare racemic spirodienone sesquineolignans from *Xanthium sibiricum*. Org. Lett. **16**(20), 5406–5409 (2014)
16. Peng, W, Ming, QL, Han, P, Zhnag, QY, Jiang, YP, Zheng, CJ, Han, T, Qin, LP: Anti-allergic rhinitis effect of caffeoylxanthiazonoside isolated from fruits of *Xanthium strumarium* L. in rodent animals. Phytomedicine **21**(6), 824–829 (2014)
17. Dai, YH, Cui, Z, Li, JL, Wang, D: A new thiaziedione from the fruits of *Xanthium sibiricum*. J. Asian Nat. Prod. Res. **10**(3-4), 343–347 (2008)
18. Wagner, H., Bauer, R., Melchart, D., Xiao, P.G., Staudinger, A.: Chromatographic fingerprint analysis of herbal medicines, vol. I+II/III. Springer, Wien-New York (2011/2013)

Herba Artemisiae annuae – *Qinghao*
Folium Artemisiae argyi – *Aiye*

Pharmacopoeia: [1] Pharmacopoeia of the People's Republic of China, English Edition Vol. I, 2010

Official drugs: [1]

Herba Artemisiae annuae

Sweet Wormwood Herb is the dried aerial part of *Artemisia annua* L. (Fam. Asteraceae).

The drug is collected in summer at flowering, removed from older stems, and dried in the shade.

Folium Artemisiae argyi

Argy Wormwood Leaf is the dried leaf of *Artemisia argyi* Lévl. et Vant. (Fam. Asteraceae).

The drug is collected in summer before flowering, removed from foreign matter, and dried in the sun.

Origin: [2, 3, 8, 10]

Herba Artemisiae annuae

China, northern Vietnam, Siberia and India. It is also found in Iran, Afghanistan, Australia, Italy, Malaysia, Romania, Spain, Turkey, Hungary and the United States.

Folium Artemisiae argyi

China, Korea, Mongolia and Russia.

Description of the drugs: [1]

Herba Artemisiae annuae

Stems cylindrical, frequently branched at the upper part, 30–80 cm long, 2–6 mm in diameter; externally yellowish-green or brownish-yellow, with longitudinal ridges; texture slightly hard, easily broken, fracture medullated in the centre. Leaves alternate, dark green or brownish-green, rolled and crumpled, easily broken, when whole, 3-pinnatipartite, the segments and smaller segments oblong or long-elliptical, pubescent on both surfaces. Odour, characteristically aromatic; taste, slightly bitter.

Folium Artemisiae argyi

Mostly crumpled and broken, short petioled. When whole, ovate-elliptical, pinnatipartite, segments elliptical-laceolate, margin irregularly dentate; upper surface greyish-green or dark yellowish-green, sparsely pubescent and glandular-punctate, lower surface densely greyish-white tomentose. Texture soft. Odour, delicately aromatic; taste, bitter.

Medicinal use: [6–9] Herba Artemisiae annuae is mainly used in combination with mefloquin for the treatment of malaria. Folium Artemisiae argyi is mainly used for the treatment of menstrual disorders.

Toxicology of Folium Artemisiae argyi: [4] High dosages may result in side-effects such as dry mouth, nausea, vomiting, gastric problems, diarrhoea and dizziness. Extremely high doses may provoke bleeding during pregnancy and cause abortion.

H. Wagner et al. (eds.), *Chromatographic Fingerprint Analysis of Herbal Medicines*, Vol. 4,
DOI 10.1007/978-3-319-32328-2_10, © Springer International Publishing Switzerland 2016

Effects and indications of Herba Artemisiae annuae according to Traditional Chinese Medicine [1–4, 6–9, 11, 13, 14]	
Taste:	Bitter and pungent
Temperature:	Cold
Channels entered:	*Orbis hepaticus*, *o. felleus*
Effects (functions):	To clear deficiency heat, relieve streaming bone, release summerheat, interrupt malaria, abate jaundice
Symptoms and indications:	Warm pathogen damaging yin, fever at night and cool in the morning, yin deficiency fever, steaming bone and consumptive fever, fever induced by summer pathogen malaria, dampness-heat jaundice

Effects and indications of Folium Artemisiae argyi according to Traditional Chinese Medicine [1, 4]	
Taste:	Pungent and bitter
Temperature:	Warm
Channels entered:	*Orbis hepaticus*, *o. renalis*, *o. lienalis*
Effects (functions):	To warm the meridian stanch bleeding, dissipate cold and relieve pain
Symptoms and indications:	Hematemesis, epistaxis, flooding and spotting, profuse menstruation, vaginal bleeding during pregnancy, cold pain in the lower abdomen, menstrual irregularities caused by cold, infertility caused by uterine coldness; topical application: itching of skin

Main constituents of Herba Artemisiae annuae:

- **Sesquiterpenes** [2, 5, 7, 8, 10, 12–15]

Artemisinin I-IV (qinghaosu), deoxyartemisinin, artemisinic acid, dehydro-artemisinic acid, artemisilactone, artemisinol, epoxyarteannuinic, arteannuin B

- **Flavonoids** [2, 6, 7, 10, 15]

Hyperoside, isoquercitrin, luteolin, apigenin, peduletin, artemetin, astragalin, casticin, chrysosplenol D, kaempferol, patuletin, quercetin, rutin, chlorogenic acid, caffeic acid, vitexin

Minor constituents: - Coumarins (aesculetin, coumarin, isofraxidin, scopoletin, scopolin, tomentin), essential oils (pinene, 1,8-cineol, borneol, phenol, cuminic aldehyde, artemisia ketone, linalool, p-cymene, thujone, camphor), steroids, phenolics (syringaldehyde), terpenes (costunolide, stigmasterol, friedelin) [2, 3, 6, 7, 10, 13, 14]

Main constituents of Folium Artemisiae argyi:

- **Sesquiterpenes and –lactones** [5, 19, 21]

Spathulenol, juniper camphor, chamazulene, β-caryophyllene, β-caryophyllene alcohol, caryophyllene oxide, chrysartemin A + B, α-cedrene, elemol, clovandiol, eudesmanolides

- **Triterpenes** [5, 18, 21, 25]

Glutinone, fernenone, lupenone, simiarenol, α-amyrin acetate, β-amyrin acetate, ethyl palmitate, ethyl oleate, ethyl linoleate, lupenyl acetate, 24-methylene-cycloartanone, trans-phenyllitaconic, gult-5-en-3β-yl acetate, dammara-20,24-dien-3β-yl acetate, cycloartenyl acetate, cycloart-23-en-3β,25-diol, cycloart-23-en-3β,25-diol monoacetate, arteminolides

- **Flavonoids / caffeoylquinic acid** [5, 18, 20, 21, 24–26]

Eupatilin, jacoesidin, apigenin, ladanein, hispidulin, chryseriol, naringenin, 5-hydroxy-6,7,3',4',-tetramethoxyflavone, isochlorogenic acid A, 5,6-di-hydroxy-7,3',4'-trimethoxyflavone, 5,6,4'-trihydroxy-7,3'-dimethoxyflavone, 5,7,3'-trihydroxy-6,4',5'-trimethoxyflavone, nepetin

- **Essential Oil** [5, 17, 18, 19, 22, 23, 25]

•Monoterpenes: β-pinene, 3-carene, camphene, τ-terpinene, 1R-α-pinene, myrcene, 1,8-cineole (eucalyptol), carveol, α-phellandrene, camphor, isoborneol, carvone

•Terpenalcohols: borneol, longiborneol, linalool, 4-terpineol, menthol, nerolidol, dihydrocarveol, α-terpineol

•Sesquiterpenes: humulene, β-cubebene, α-gurjunene, (Z)-β-farnesene, δ-cadinene, α-bergamotene

•Terpenesters: bergamiol, bornyl acetate, cedryl acetate, nerol acetate, geranyl acetate, geranyl isovalerate

•Ketones: menthone, 3-octanone

•Aromatic compound: o-cymene

Minor constituents: - Coumarins (scopoletin, isoscopoletin, trans-*O*-coumaric acid), glycosides, polyacetylenes, steroids/sterols, sesquiterpene, ketones (moxartenolide, moxartenone), fatty acids (13-oxo-9 (Z),11 (E)-octadecadienoic acid; 13-oxo-9 (E),11 (E)-octadecadienoic acid; 9-oxo-10 (E),12 (E)-octadecadienoic acid) [5, 10, 15, 18, 21]

Reported pharmacology:

Herba Artemisiae annuae

- anti-parasitic / antimalarial [3a, 6, 8–10, 12, 15]
- antipyretic [3, 6]
- antibiotic [4]
- antibacterial [4, 7, 9, 10]
- anticancer [6, 12, 16]
- cytotoxic [6, 8]
- antitumor [7, 9, 10]
- anti-inflammatory [3a, 6, 7, 8, 10]
- antifungal [6, 9]

Folium Artemisiae argyi

- antiasthmatic effect [5]
- antimutagenic [5, 23]
- inhibition of proliferation of human ccrf-cem leukemia cells [5]
- gastroprotective effect [5]
- activity against pruritus [5]
- antiinflammatory [3b, 5]
- analgesic [5, 10]
- anxiolytic [5]
- antidepressant [5]
- sedative [5]
- antifungal [19, 23]
- antihistaminic [19, 23]
- antitumor [23]

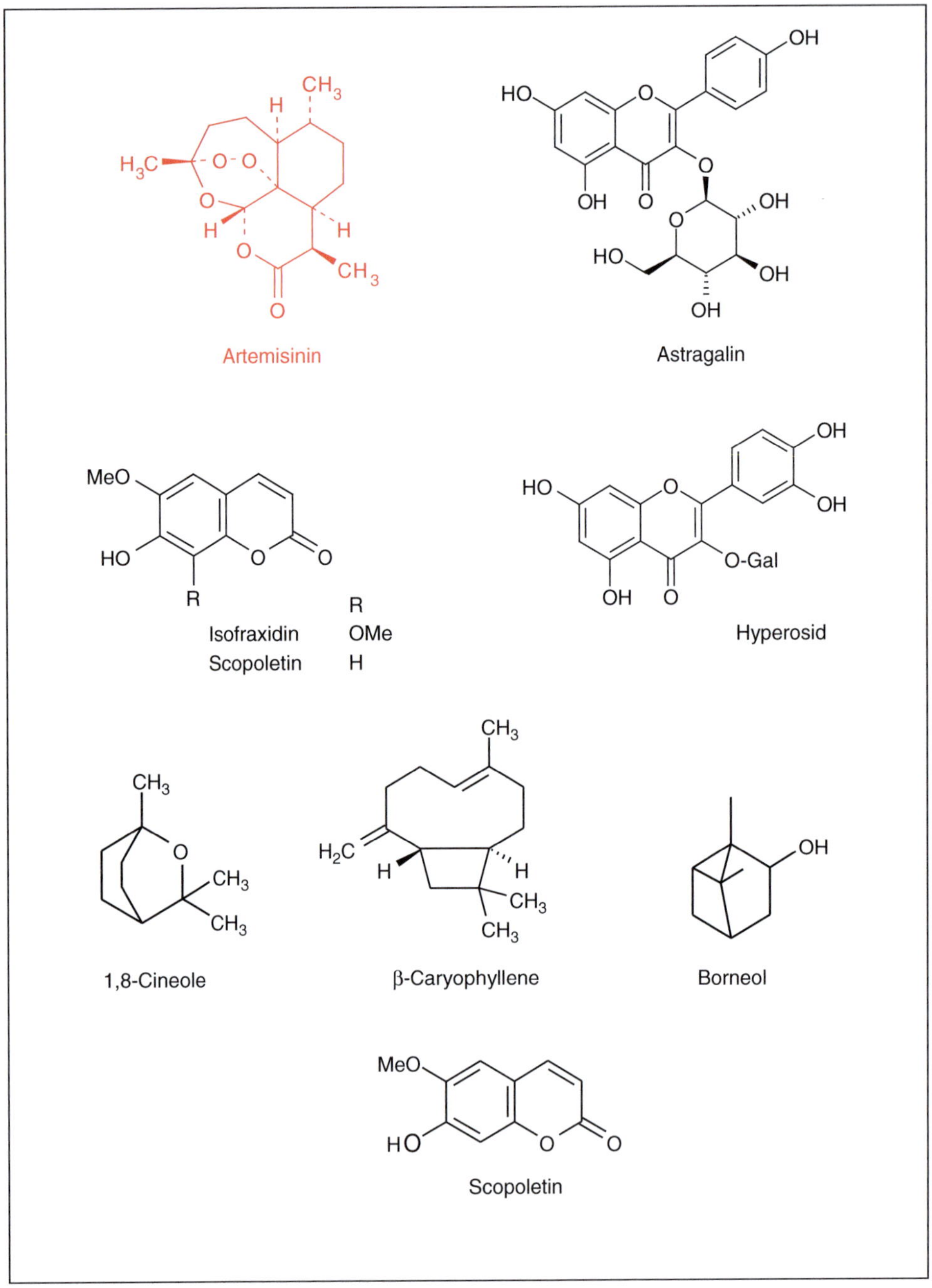

Fig. 1 Formulae of the main compounds of Herba Artemisiae annuae and Folium Artemisiae argyi [5]

TLC-Fingerprint Analysis

	Drug samples	Origin
1	Herba Artemisiae annuae / *Artemisia annua*	Sample of commercial drug obtained from HerbaSinica (origin: Neimeng, China)
2	Herba Artemisiae annuae / *Artemisia annua*	Sample of commercial drug obtained from Caelo (origin: Sichuan, China)
3	Herba Artemisiae annuae / *Artemisia annua*	Province Henan, China
4	Folium Artemisiae argyi / *Artemisia argyi*	Sample of commercial drug obtained from China Medica (origin: Jiange, Sichuan, China)
5	Folium Artemisiae argyi / *Artemisia argyi*	Sample of commercial drug obtained from TCM-Clinic Bad Kötzting (Charge: 19112072004)
6	Folium Artemisiae argyi / *Artemisia argyi*	Province Henan, China
7	**Herba Artemisiae scopariae** / *Artemisia scoparia* or *A. capillaris*[a]	Sample of commercial drug obtained from China Medica (origin: unknown)

[a]For comparison

1. TLC-fingerprint analysis of Sesquiterpenes

Reference compound of Fig. 2		R*f*
T1	Artemisinin	0.38

1. Extraction: 1.0 g powdered drug is extracted with 20 ml methanol under reflux for 30 min. The extract is filtered, evaporated to dryness and the residue is dissolved in 1 ml methanol.
2. Reference compound: 1.0 mg is dissolved in 1.0 ml methanol
3. Separation parameters:

Plate: HPTLC Silica gel 60 F_{254}, Merck

Applied amounts: Herba Artemisiae annuae extracts: each 10 µl
Folium Artemisiae argyi extracts: each 30 µl
Herba Artemisiae scopariae extract: 30 µl
Reference compound: 10 µl

Solvent system: Hexane + ethyl acetate + glacial acetic acid (16 + 4 + 0.2)

Spray reagent: 1 ml of diluted sulphuric acid (50 % v/v) is mixed with 10 ml of *p*-hydroxybenzaldehyde in methanol (2 % w/v).

The plate is sprayed with the solution and heated for 10 min at 110 °C. After 30 min the plate is evaluated in VIS.

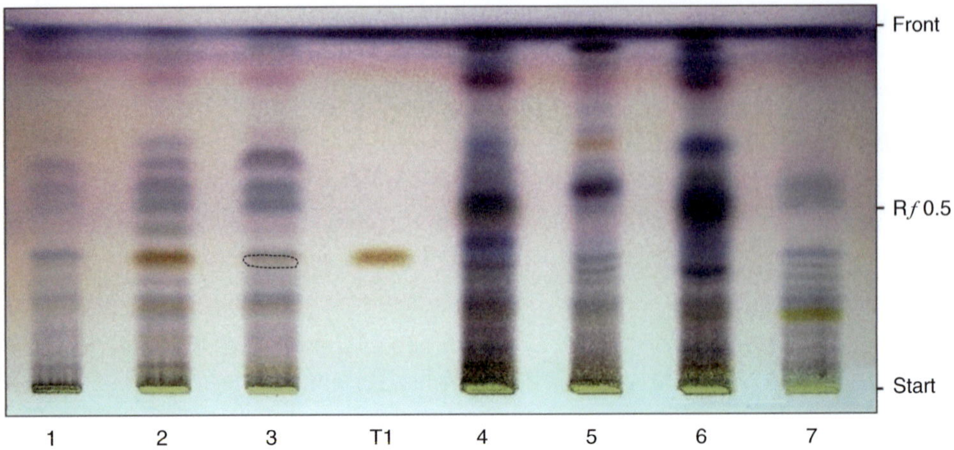

Fig. 2 Thin layer chromatogram of the different Herba and Folium Artemisiae methanol extracts, sprayed with *p*-hydroxybenzaldehyde / sulphuric acid (VIS)

4. Description:

   The Herba A. annuae extracts **1–3** show a very homogenous profile of grey and violet coloured zones in the deep, middle and upper R*f*-range with artemisinin (**T1**) as dominant orange brown zone in sample **2** and a weaker concentration in sample **3**. The Folium A. argyi extracts **4–6** differ from the Herba extracts by much stronger grey/blue/black coloured zones in the corresponding R*f*-ranges. It can be supposed that the deep blue coloured zones originate from essential oil compounds such as linalool, borneol or terpineol. Artemisinin is not detectable in these extracts.

   The Herba A. scopariae extract **7** shows a very similar TLC-profile as those of extracts **1–3** but also without artemisinin.

2. TLC-fingerprint analysis of Flavonoids, Coumarins and Phenolcarboxylic acids [29]

Reference compounds of Fig. 3		R*f*
T2	Isofraxidin	0.94
T3	Hyperoside	0.69
T4	Astragalin	0.28
T5	Scopoletin	0.98
T6	Chlorogenic acid	0.59

1. Extraction: 1.0 g powdered drug is extracted with 20 ml methanol under reflux for 30 min. The extract is filtered, evaporated to dryness and the residue is dissolved in 1 ml methanol.
2. Reference compounds: 1.0 mg is dissolved in 1.0 ml methanol

3. Separation parameters:

Plate:	HPTLC Silica gel 60 F_{254}, Merck
Applied amounts:	Herba Artemisiae annuae extracts: each 10 µl Folium Artemisiae argyi extracts: each 30 µl Herba Artemisiae scopariae extract: 30 µl Reference compound: 10 µl
Solvent system:	Ethyl acetate + formic acid + glacial acetic acid + water (20 + 2.2 + 2.2 + 5.2)
Spray reagent:	Natural products – Polyethylene glycol reagent (NP/PEG) **I**: 1 % diphenylboric acid-β-ethylamino ester (= diphenylboryloxyethylamine, NP) in methanol **II**: 5 % Polyethylene glycol-4000 (PEG) in ethanol (80 %) The plate is sprayed first with solution **I** and then with solution **II**. After 30 min the evaluation is carried out under UV 366 nm.

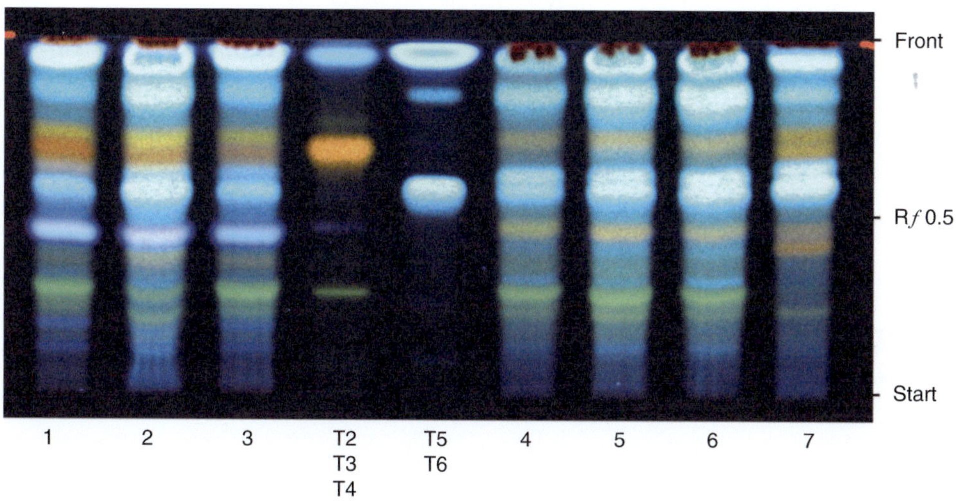

Fig. 3 Thin layer chromatogram of the different Herba and Folium Artemisiae methanol extracts, sprayed with NP/PEG (UV 366 nm)

4. Description:

All Herba and Folium Artemisiae extracts **1–7** are characterized by a very homogenous TLC-profile with blue, green and orange fluorescent zones with scopoletin (**T5**) and chlorogenic acid (**T6**) as dominant zones in all samples. Hyperoside (**T3**) is detectable in samples **1–3** and **7**. The blue-violet zone at R*f*=0.45 in samples **1–3** may be assigned to scopolin.

HPLC-Fingerprint Analysis [27, 28]

1.1. Extraction:	1.0 g powdered drug is extracted with 20 ml methanol under reflux for 30 min. The extract is filtered, evaporated to dryness and the residue is dissolved in 1 ml methanol. The extract is filtered over Chromafil®, Type 0.20 µm.
1.2. Sample derivatisation:	0.1 ml sample (extract or artemisinin) is transferred into a 1 ml measuring flask. 0.4 ml of 0.2 % (m/v) NaOH solution is added in the flask, and then left to react at 50 °C for 30 min. After cooling during 10 min, 0.1 ml of ethanol is added. Finally the flask is filled with acetic acid 0.2 N. All solutions are filtered over Chromafil®, Type 0.20 µm before injection.
2. Injection volume:	Herba Artemisiae annuae extracts: each 20 µl, Folium Artemisiae argyi exracts: each 40 µl, Herba Artemisiae scopariae extract: 40 µl
3. HPLC parameter:	
Apparatus:	MERCK HITACHI D-6000 A Interface MERCK HITACHI L-4500 A Diode Array Detector MERCK HITACHI AS-2000 Autosampler MERCK HITACHI L-6200 A Intelligent Pump
Separation column:	LiChroCART® 250-4 LiChrospher® 100 RP-18 (5 µm), Merck
Precolumn:	LiChroCART® 4-4 LiChrospher® 100 RP-18 (5 µm), Merck
Solvent system:	A: phosphate buffer 5 mM, pH 6.3 (Millipore Ultra Clear UV plus® filtered) B: methanol (VWR)
Gradient:	0–20 % B in 20 min, 20–40 % B in 35 min, 40–100 % B in 17 min, Total runtime: 72 min
Flow:	1.0 ml/min
Detection:	260 nm

Retention times of the main peaks

Peak	Rt (min)	Compound
1	13.8	Chlorogenic acid
2	14.9	Not identified
3	16.6	Not identified
4	20.3	Not identified chlorogenic acid isomers
5	26.4	Not identified chlorogenic acid isomers
6	32.0	Not identified chlorogenic acid isomers
7	38.3	Not identified
8	42.0	Isofraxidin or Scopoletin
9	57.0	Artemisinin
10	70.4	Not identified

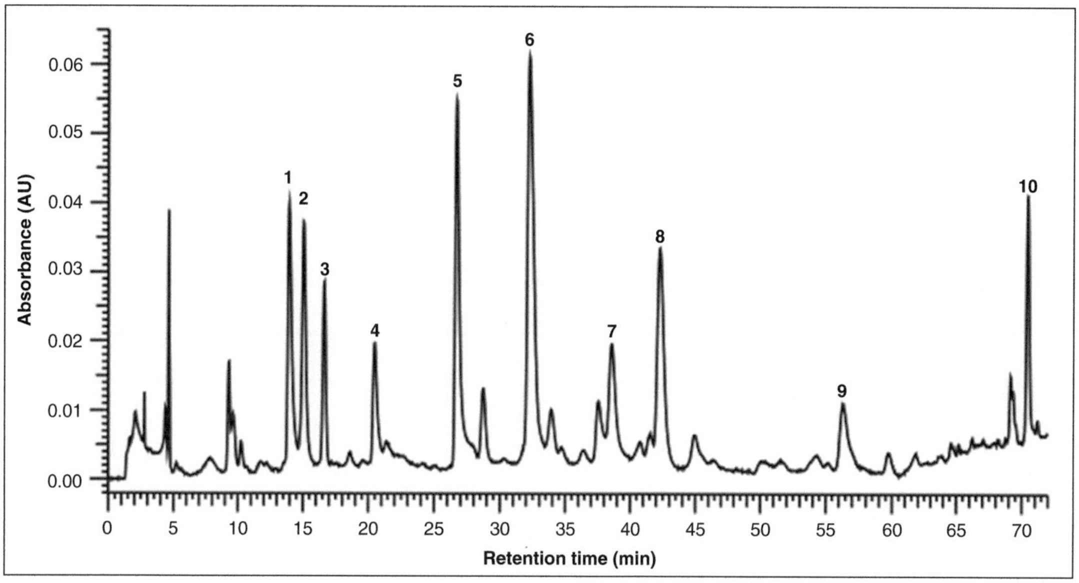

Fig. 4a HPLC-fingerprint analysis of the methanol extract of Herba Artemisiae annuae, sample 2

4. Description of Fig. 4a:

The HPLC-graph of the MeOH-extract of Artemisia annua is characterised by relative high concentrations of chlorogenic acid (peak **1**) and several other chlorogenic acid isomers (dicaffeoylquinic acids, peak **4–7**) according to their similar UV-spectra and the detection in TLC at Fig. 3 (samples 1–3). The peak 8 can be assigned to the cumarins isofraxidin or scopoletin. Artemisinin (peak **9**) could be identified with a reference compound. Its low concentration in the Herba extract can be explained with the low portion of leaves in the herbal drug sample. Artemisinin was isolated only from leaf material [30]. Peak **10** may be one compound of the essential oil.

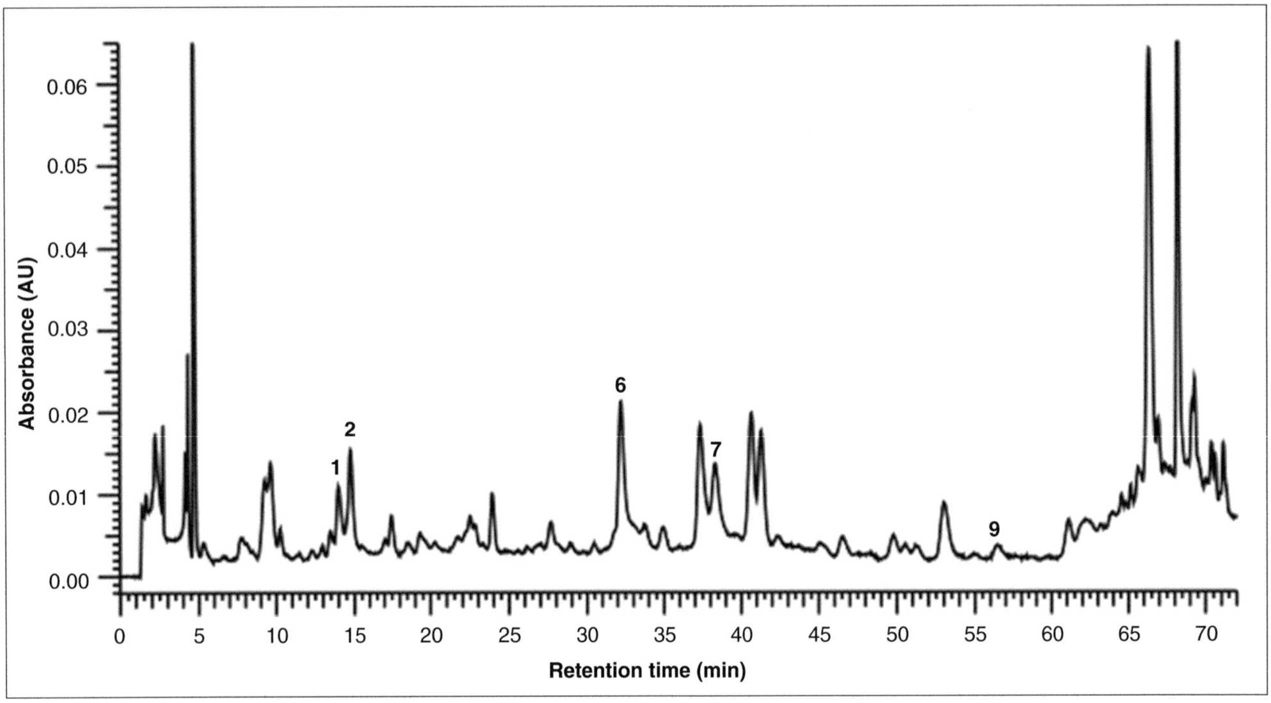

Fig. 4b HPLC-fingerprint analysis of the methanol extract of Folium Artemisiae argyi, sample 4

Description of Fig. 4b:

This figure shows only low concentrations of the compounds **1**,**2** and **6**,**7** inclusive Artemisinin (**9**) but a high concentration of the lipophilic terpenoids or acetylenic acids.

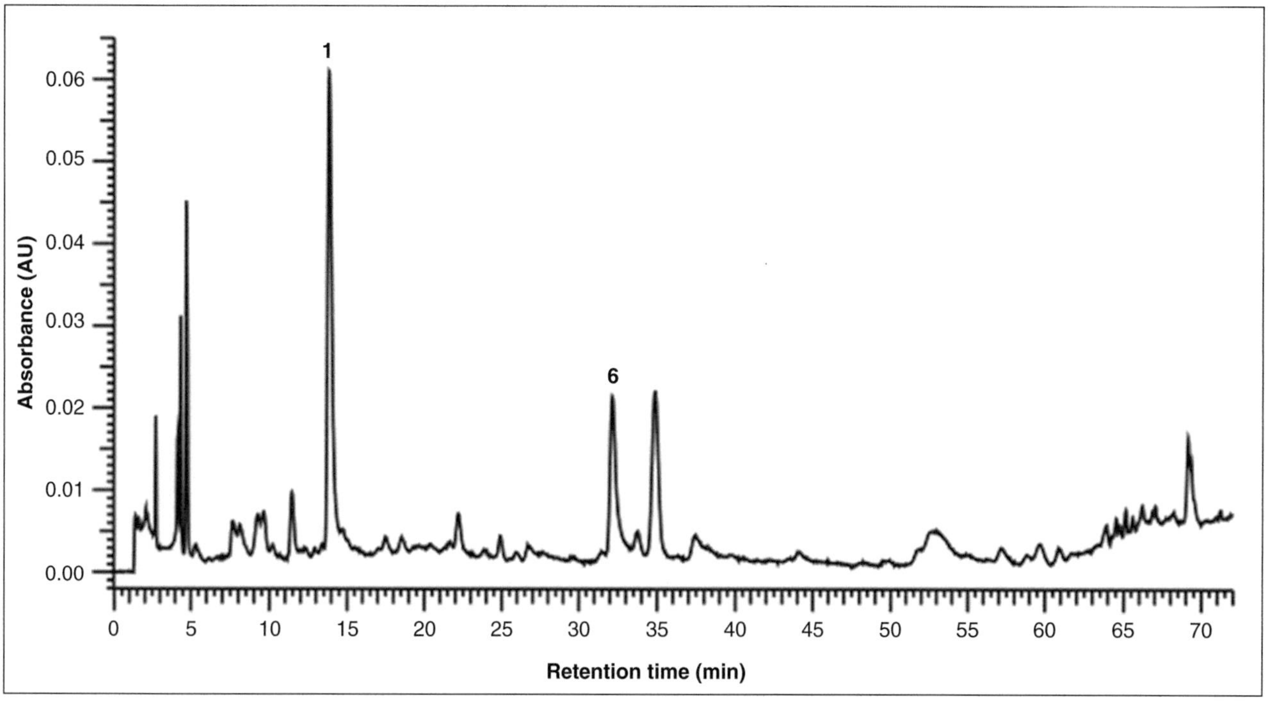

Fig. 4c HPLC-fingerprint analysis of the methanol extract of **Herba Artemisiae scopariae**, sample 7

Description of Fig. 4c:

Artemisia scoparia MeOH extract contains only chlorogenic acid (peak **1**) and one chlorogenic acid isomer (peak **6**).

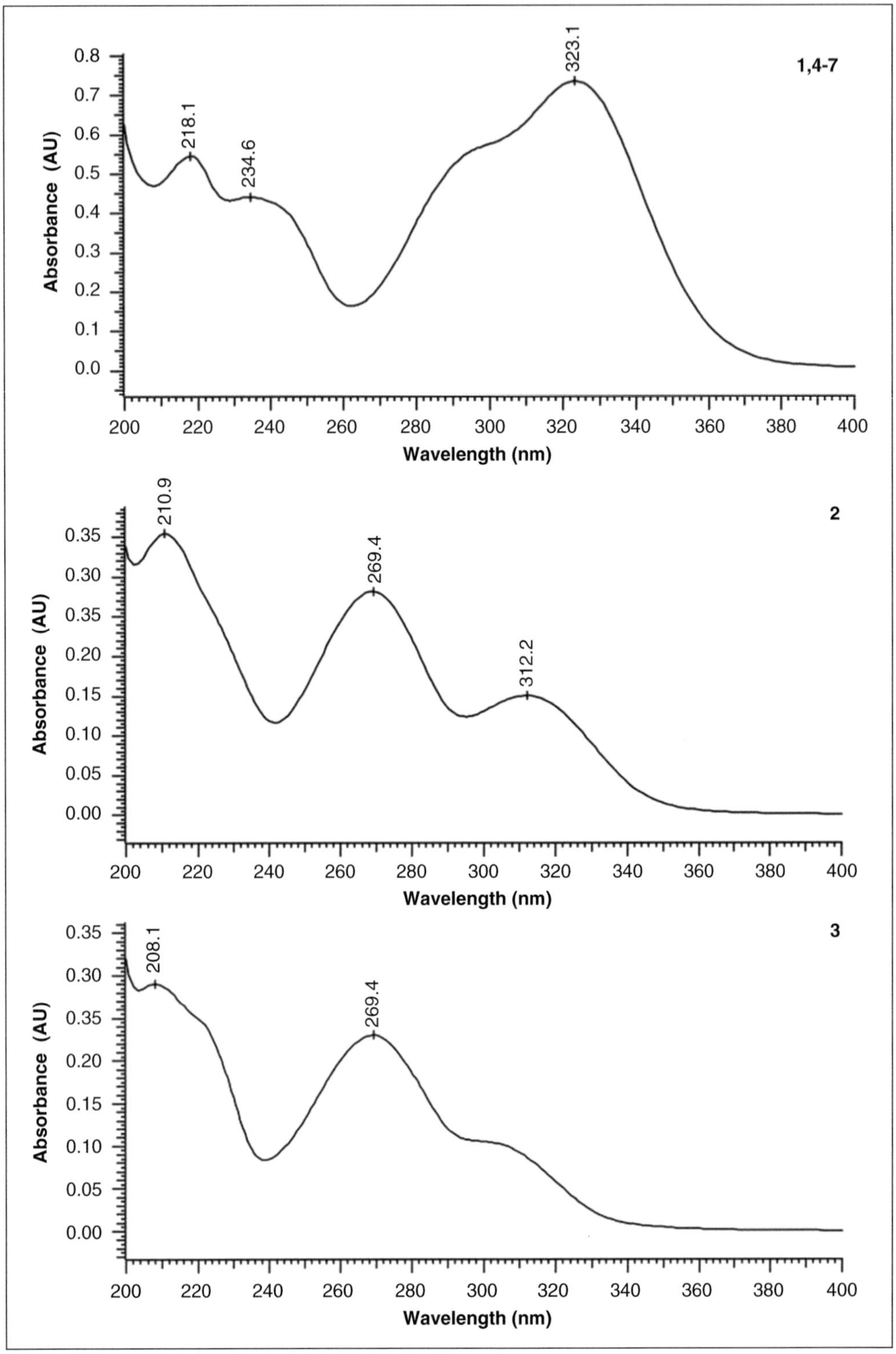

Fig. 5 On line UV-spectra of the main peaks of the different Artemisia extracts

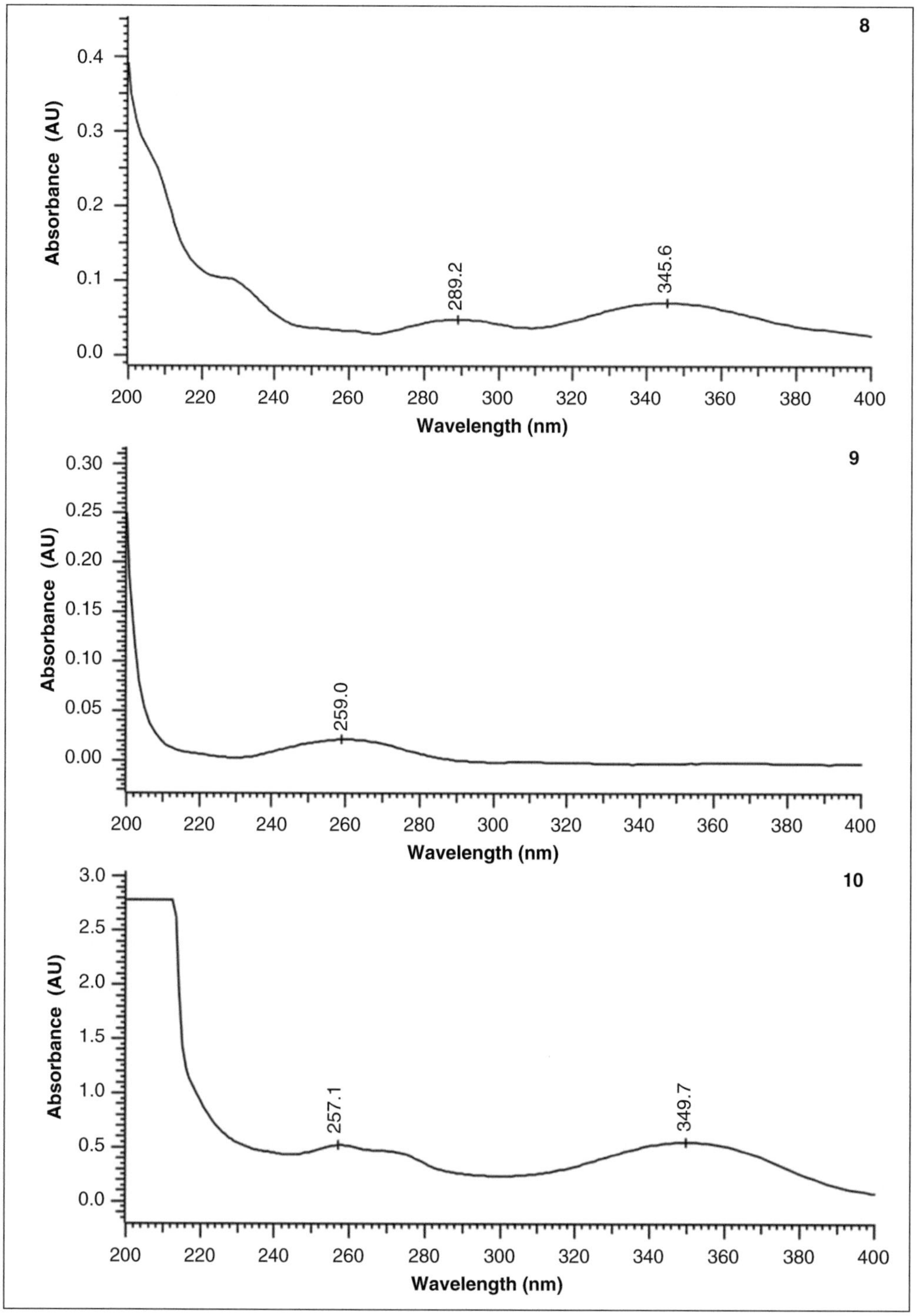

Fig. 5 (continued)

Note

The Chinese Pharmacopeia 2010 describes for Folium Artemisiae argyi a cineole content not less than 0.050 %, calculated with reference to the dried drug. [1]

Conclusion

The extract authentication of the three Artemisia species, *A. annua*, *A. argyi* and *A. scoparia* is very difficult because all three species do not contain any prominent marker compound in high concentrations which without phytochemical enrichment could be not unequivocally identified.

References

1. Pharmacopoeia of the people's republic of China, English edition, vol. I. People's Medical Publishing House, Beijing (2010)
2. Zhang, Z.Z.: An illustrated Chinese materia medica in Hong Kong. Hong Kong, School of Chinese Medicine/Hong Kong Baptist University (2004)
3. (a) Flora of China, *Artemisia annua*, FOC 20-21, 691 (www.eFloras.org). (b) Flora of China, *Artemisia argyi*, FOC 20-21, 697 (www.eFloras.org)
4. Hempen, C.-H., Fischer, T.: A materia medica for Chinese medicine, english edition. Elsevier GmbH, Urban & Fischer, Munich (2009)
5. Tang, W., Eisenbrand, G.: Handbook of Chinese medicinal plants – chemistry, pharmacology, toxicology, vol. 1. WILEY-VCH Verlag GmbH & Co KgaA, Weinheim (2011)
6. Song, Y., Desta, K.T., Kim, G.S., Lee, S.J., Lee, W.S., Kim, Y.H., Jin, J.S., Abd El-Aty, A.M., Shin, H.C., Shim, J.H., Shin, S.C.: Polyphenolic profile and antioxidant effects of various parts of *Artemisia annua* L. Biomed. Chromatogr. **30**(4), 588–595 (2015)
7. Kim, M.H., Seo, J.Y., Liu, K.H., Kim, J.S.: Protective effect of *artemisia annua* L. Extract against galactose-induced oxidative stress in mice. PLoS One **9**(7), e101486 (2014)
8. Abolaji, A.O., Eteng, M.U., Ebong, P.E., Brisibe, E.A., Dar, A., Kabir, N., Choudhary, M.I.: A safety assessment of the antimalarial herb *artemisia annua* during pregnancy in wistar rats. Phytother. Res. **27**(5), 647–654 (2013)
9. Appalasamy, S., Lo, K.Y., Ch'ng, S.J., Nornadia, K., Othman, A.S., Chan, L.K.: Antimicrobial activity of artemisinin and precursor derived from in vitro plantlets of *artemisia annua* L. Biomed. Res. Int. 2014 Article ID 215872 (2014)
10. Bhakuni, R.S., Jain, D.C., Sharma, R.P., Kumar, S.: Secondary metabolites of *Artemisia annua* and their biological activity. Curr. Sci. **80**(1), 35–48 (2001)
11. Oh, Y.C., Jeong, Y.H., Kim, T.K., Cho, W.K., Ma, J.Y.: Anti-inflammatory effect of Artemisiae annuae herba in lipopolysaccharide-stimulated RAW 264.7 cells. Pharmacogn. Mag. **10**(Suppl 3), S588–S595 (2014)
12. Ma, C., Wang, H., Lu, X., Xu, G., Liu, B.: Metabolic fingerprinting investigation of *Artemisia annua* L. in different stages of development by gas chromatography and gas chromatography-mass spectrometry. J. Chromatogr. A **1186**(1–2), 412–419 (2008)
13. Weathers, P.J., Towler, M.J.: Changes in key constituents of clonally propagated *Artemisia annua* L. during preparation of compressed leaf tablets for possible therapeutic use. Ind. Crops Prod. **62**, 173–178 (2014)
14. Zheng, G.Q.: Cytotoxic terpenoids and flavonoids from *artemisia annua*. Planta Med. **60**(1), 54–57 (1994)
15. Li, Y.J., Guo, Y., Yang, Q., Weng, X.G., Yang, L., Wang, Y.J., Chen, Y., Zhang, D., Li, Q., Liu, X.C., Kan, X.X., Chen, X., Zhu, X.X., Kmoníèková, E., Zídek, Z.: Flavonoids casticin and chrysosplenol D from *Artemisia annua* L. inhibit inflammation in vitro and in vivo. Toxicol. Appl. Pharmacol. **286**(3), 151–158 (2015)
16. Tu, Y.Y., Ni, M.Y., Zhong, Y.R., Li, L.N., Cui, S.L., Zhang, M.Q., Wang, X.Z., Ji, Z., Liang, X.T.: Studies on the constituents of *Artemisia annua* part II. Planta Med. **44**(3), 143–145 (1982)
17. Zhang, Y.H., Xue, M.Q., Bai, Y.C., Yuan, H.H., Zhao, H.L., Lan, M.B.: 3,5-Dicaffeoylquinic acid isolated from *Artemisia argyi* and its ester derivatives exert anti-Leucyl-tRNA synthetase of Giardia lamblia (GlLeuRS) and potential anti-giardial effects. Fitoterapia **83**, 1281–1285 (2012)
18. Adams, M., Efferth, T., Bauer, R.: Activity-guided Isolation of Scopoletin and Isoscopoletin, the Inhibitory active principles towards CCRF-CEM Leukemia cells and multi-drug resistant CEM/ADR5000 cells, from artemisiae argyi. Planta Med. **72**(9), 862–864 (2006)

19. Zhang, W.J., You, C.X., Yang, K., Chen, R., Wang, Y., Wu, Y., Geng, Z.F., Chen, H.P., Jiang, H.Y., Su, Y., Lei, N., Ma, P., Du, S.S., Deng, Z.W.: Bioactivity of essential oil of *Artemisia argyi* Lévl. et Van. and its main compounds against *Lasioderma seericorne.* J. Oleo Sci. **63**(8), 829–837 (2014)

20. Seo, J.M., Kang, H.M., Son, K.H., Kim, J.H., Lee, C.W., Kim, H.M., Chang, S.I., Kwon, B.M.: Antitumor activity of flavones isolated from *Artemisia argyi*. Planta Med. **69**(3), 218–222 (2003)

21. Yoshikawa, M., Shimada, H., Matsuda, H., Yamahara, J., Murakami, N.: Bioactive constituents of Chinese natural medicines I. New Sesquiterpene Ketones with Vasorelaxant Effect from Chinese Moxa, the processed leaves of *Artemisia argyi* LEVL. et VANT.: Moxartenone and Moxartenolide. Chem. Pharm. Bull. **44**(9), 1656–1662 (1996)

22. Guan, W., Li, S., Yan, R., Huang, Y.: Comparison of composition and antifungal activity of *Artemisia argyi* Lévl. et Vant inflorescence essential oil extracted by hydrodistillation and supercritical carbon dioxide. Nat. Prod. Res. **20**(11), 992–998 (2006)

23. Huang, H.C., Wang, H.F., Yih, K.H., Chang, L.Z., Chang, T.M.: Dual bioactivities of essential oil exracted from the leaves of *Artemisia argyi* as an Antimelanogenic versus Antioxidant agent and chemical composition analysis by GC/MS. Int. J. Mol. Sci. **13**(11), 14679–14697 (2012)

24. Khan, M., Yu, B., Rasul, A., Al Shawi, A., Yi, F., Yang, H., Ma, T.: Jaceosidin induces apoptosis in U87 glioblastoma cells through G2/M phase arrest. Evid. Based Complement Alternat. Med. 2012 Article ID 703034 (2012)

25. Li, N., Mao, Y., Deng, C., Zhang, X.: Separation and identification of volatile constituents in *Artemisia argyi* flowers by GC-MS with SPME and steam distillation. J. Chromatogr. Sci. **46**(5), 401–405 (2008)

26. Lim, J.C., Park, S.Y., Nam, Y., Nguyen, T.T., Sohn, U.D.: The protective effect of Eupatilin against hydrogen peroxide-induced injury involving 5-Lipoxygenase in feline Esophageal Epithelial cells. Korean J. Physiol. Pharmacol. **16**(5), 313–320 (2012)

27. Diawara, H.Z., Gbaguidi, F., Evrard, B., Leclercq, J.Q., Moudachirou, M., Debrus, B., Hubert, P., Rozet, E.: Validation, transfer and measurement uncertainty estimation of an HPLC-UV method for the quantification of artemisinin in hydro alcoholic extracts of *Artemisia annua* L. J. Pharm. Biomed. Anal. **56**(1), 7–15 (2011)

28. Wagner, H., Bauer, R., Melchart, D., Xiao, P.G., Staudinger, A.: Chromatographic fingerprint analysis of herbal medicines Vol. II (Herba Artemisiae scopariae). Springer, Wien/New York (2011)

29. Wagner, H., Bladt, S.: Plant drug analysis: a thin layer chromatography atlas, 2nd edn. Springer Verlag, Berlin (1996)

30. Klayman, D.L., Lin, A.J., Acton, N., Scovill, J.P., Hoch, J.M., Milhous, W.K., Theoharides, A.D., Dobek, A.S.: Isolation of Artemisinin (Qinghaosu) from *Artemisia annua* growing in the United States. J. Nat. Prod. **47**(4), 715–717 (1984)

Herba Ephedrae – *Mahuang*

Pharmacopoeia: [1] Pharmacopoeia of the People's Republic of China, English Edition Vol. I, 2010

Official drug: [1, 2] Ephedra is the dried herbaceous stem of *Ephedra sinica* Stapf, *Ephedra intermedia* Schrenk et C.A. Mey. or *Ephedra equisetina* Bge. (Fam. Ephedraceae).

The drug is collected in autumn and dried in the sun.

Origin: [2–4] Mainly in Chinese provinces such as Shanxi, Hebei, Gansu and Shaanxi, and in Xinjiang, Inner Mongolia and Ningxia.

Description of the drug: [1] *Herb of Ephedra sinica*

Slenderly cylindrical, infrequently branched, 1–2 mm in diameter. Some with a few brown woody stems. Externally pale green to yellowish-green, with fine longitudinal ridges, slightly rough. Nodes distinct, internodes 2–6 cm long. Scaly leaves on the nodes membranous, 3–4 mm long, with 2 lobes (rarely 3), acutely triangular, apex greyish-white, reversed, base tubular and reddish-brown in colour. Texture light, fragile, and easily broken, fracture slightly fibrous with greenish-yellow edge and subrounded reddish-brown pith. Odour, slightly aromatic; taste, astringent and slightly bitter.

Herb of Ephedra intermedia

Frequently branched, 1.5–3 mm in diameter, rough. Membranous scaly leaves 2–3 mm long, with 3 lobes (rarely 2), apex acute. Fracture showing a triangular rounded pith.

Herb of Ephedra equisetina

Frequently branched, 1–1.5 mm in diameter, without rough feeling. Internodes 1.5–3 cm long. Membranous scaly leaves 1–2 mm long, with 2 lobes (rarely 3), the upper part short-triangular, greyish-white, apex infrequently reversed, base brownish-red to brownish-black.

Processing: [1] *Herba Ephedrae (processed with honey)*

The sections of Herba Ephedrae are stir baked as described under the method for stir-baking with honey (Appendix II D, Chinese Pharmacopoeia (2010)) until it is not sticky to fingers, using 20 kg of refined honey per 100 kg of Herba Ephedrae.

Medicinal use: [4–12] Used for clinical treatment of bronchial asthma, nasal congestion, anaphylactic reaction, hypotension, common cold, hay fever and upper respiratory allergies.

It is also commonly used in some western countries as dietary supplement and as slimming aid.

H. Wagner et al. (eds.), *Chromatographic Fingerprint Analysis of Herbal Medicines*, Vol. 4,
DOI 10.1007/978-3-319-32328-2_11,

Effects and indications of Herba Ephedrae according to Traditional Chinese Medicine [1–3, 5, 13]	
Taste:	Pungent, mild bitter
Temperature:	Warm
Channels entered:	*Orbis pulmonalis, o. vesicalis*
Effects (functions):	To promote sweating and dissipate cold, diffuse the lung to relieve panting, and promote urination to alleviate edema
Symptoms and indications:	Common cold caused by wind-cold, oppression in the chest, panting and cough, edema caused by wind Herba Ephedrae (processed with honey) can moisten the lung to suppress cough, and can be used for panting and cough after relieving exterior pattern

Toxicity: [13] Strong sweating may occur if administered in high doses or over long periods of time. Blood pressure may be raised, restlessness, tremor, arrhythmias and mydriasis may occur. Symptoms of ephedrine poisoning are: sweating, raised temperature, epigastric pain, dry heaves, vomiting, seizures.

Main constituents: - **Alkaloids** [2–16]

(-)-Ephedrine, (+)-pseudoephedrine, pseudomethylephedrine, (-)-methylephedrine, (-)-norephedrine, (+)-methyl-pseudoephedrine, (+)-norpseudoephedrine, benzoyl-pseudoephedrine, 2,3,5,6-tetramethylpyrazine, ephedroxane, 3,4-dimethyl-5-phenyl-oxazolidin-2-one

- **Essential oil** [5, 8, 16]

α,α,4-trimethyl-3-cyclohexan-1-methanol; α/β-terpineol, 1-acetyl-1,3-dimethyl-3-cyclohexane; 6,10,14-trimethyl-2-pentadecanone; 3,7,11,15-tetramethyl-2-hexadecen-1-ol; docosane, limonene, (+)-4-carene, 1-methyl-4-(1-methylethenyl)-cyclohexanol, *n*-hexadecanoic acid

Minor constituents: - 4-(2-eicosyloxycarbonyl-vinyl)-benzoic acid, 4-(2-docosyloxycarbonyl-vinyl)-benzoic acid, benzo [a] pyrene (1,6-dinitropyrene, 3,9-dinitrofluoranthene), phenolics (eg catechin, epicatechin, epigallocatechin, o-coumaric acid glucoside, vitexin-2'-O-rhamnoside, quercetin-3-D-galactoside (= hyperoside), kaempferol- rhamnoside), lipids [6, 11, 14, 16]

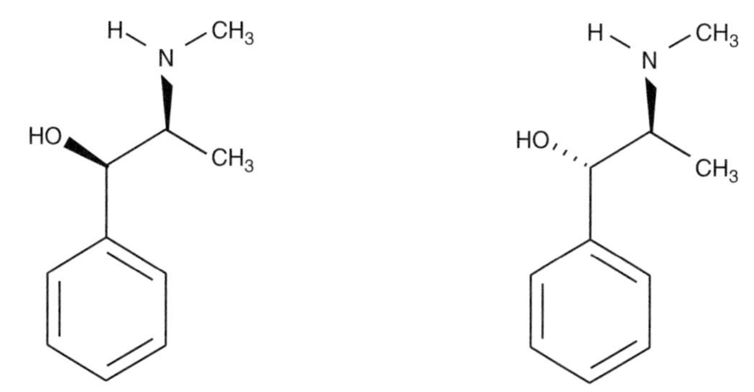

Fig. 1 Formulae of the two main alkaloids of Herba Ephedrae [5]

Reported pharmacology:

- anti-inflammatory [10, 13]
- sympathomimetic [13]
- diaphoretic [5, 9, 12, 14, 15]
- antihistaminic [5]
- diuretic [5, 14, 15]
- hypoglycemic [5]
- antiangiogenic activity [14]
- anticoagulative [5]
- immunosuppressive [5]
- antioxidative [5]
- antiviral [5, 8]
- antiobesity agent [5]
- cardiovascular effects [5]
- antiasthmatic [5, 6, 9, 12, 14, 15]
- antimutagenic/anticarcinogenic [14]
- antitumor activity [14]
- antipyretic [9]
- respiratory effects [9]
- antitussive [9]

TLC Fingerprint Analysis

	Drug samples	**origin**
1	Herba Ephedrae / *Ephedra sp.*	Sample of commercial drug obtained from firm China Medica (Charge: 12 0519)
2	Herba Ephedrae / *Ephedra sp.*	Province Shanxi (China)
3	Herba Ephedrae / *Ephedra sp.*	Province Hebei (China)
4	Herba Ephedrae / *Ephedra sp.*	Sample of commercial drug, obtained from TCM-Clinic Bad Kötzting (Charge: K 07.06.1999)
5	Herba Ephedrae / *Ephedra sp.*	Sample of commercial drug, obtained from TCM-Clinic Bad Kötzting (Charge: K 20.12.2000)

Reference compound of Fig. 2		**R*f***
T	Ephedrine	0.44

1. Extraction: 1 g powdered drug is extracted under reflux with 1 ml 10 % NH_3 and 10 ml methanol for 30 min. The extract is filtered, evaporated to dryness and the residue is dissolved in 1.5 ml methanol.
2. Reference compound: 1.0 mg is dissolved in 1.0 ml methanol
3. Separation parameters:

Plate: HPTLC Silica gel 60 F_{254}, Merck

Applied amounts: Herba Ephedrae extracts: each 5 µl

Reference compound: 10 µl

Solvent system: Dichloromethane + methanol + ammonia solution (25 %) (8 + 2 + 0.2)

Detection: 2 % ethanolic Ninhydrine solution

The plate is sprayed with 8 ml reagent and heated at 110 °C for 10 min. The plate is evaluated in VIS.

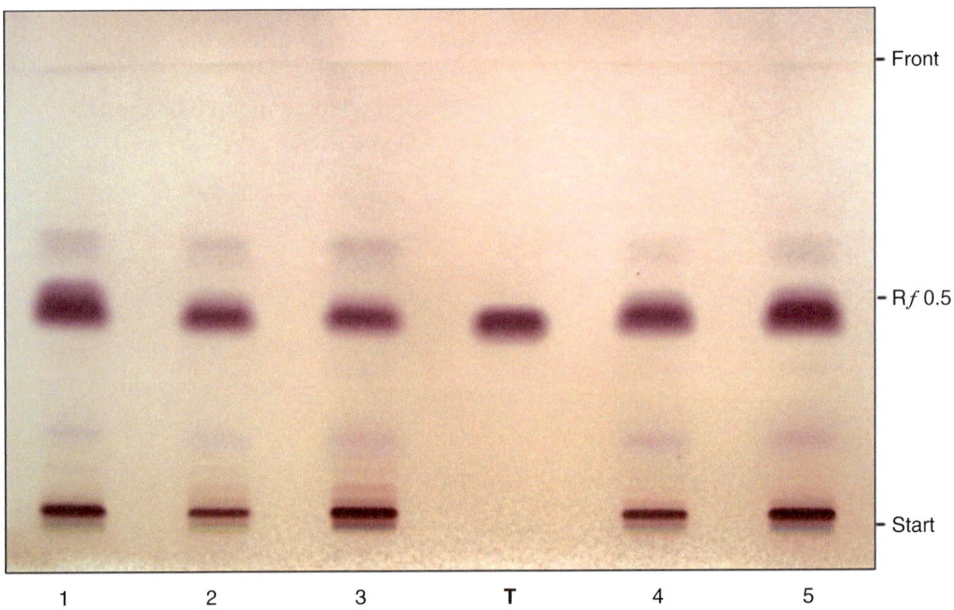

Fig. 2 Thin layer chromatogram of the NH_3- Methanol extracts of Herba Ephedrae, sprayed with Ninhydrin reagent (VIS)

4. Description:

   All five Ephedrae herb extracts show at R*f*=0.47 a broad brown zone of Ephedrine accompanied by a weak zone above Ephedrine at R*f*=0.58 and a strong on the solvent start. Their identities could be not destined.

HPLC-Fingerprint Analysis

1. Extraction:	1 g powdered drug is extracted under reflux with 1 ml 10% NH_3 and 10 ml methanol for 30 min. The extract is filtered, evaporated to dryness and the residue is dissolved in 1.5 ml methanol. filtered over Millipore®, Type 0.45 µm and injected into the HPLC-apparatus.
2. Injection volume:	Herba Ephedrae extracts: each 20 µl
	Ephedrin 1 mg/ml MeOH: 30 µl
3. HPLC parameter:	
Apparatus:	MERCK HITACHI D-6000 A Interface
	MERCK HITACHI L-4500 A Diode Array Detector
	MERCK HITACHI AS-2000 Autosampler
	MERCK HITACHI L-6200 A Intelligent Pump
Separation column:	LiChroCART® 250-4 LiChrospher® 60 RP select B (5 µm), Merck
Precolumn:	LiChroCART® 4-4 LiChrospher® 60 RP select B (5 µm), Merck
Solvent system:	A: 2.0 g hexanesulfonic acid/1 l water (Millipore Ultra Clear UV plus® filtered)+H_3PO_4 85% (pH=3.0)
	B: acetonitrile (VWR)
Gradient:	5% B for 5 min,
	5–65% B in 30 min
	total runtime: 30 min
Flow:	1.0 ml/min
Detection:	210 nm

Retention times of the main peaks

Peak	Rt (min)	Compound
1	1.7	Not identified
2	10.9	Not identified
3	12.9	Flavonoid
4	14.2	Flavonoid
5	14.9	Ephedrine
6	15.7	Not identified
7	21.1	Not identified

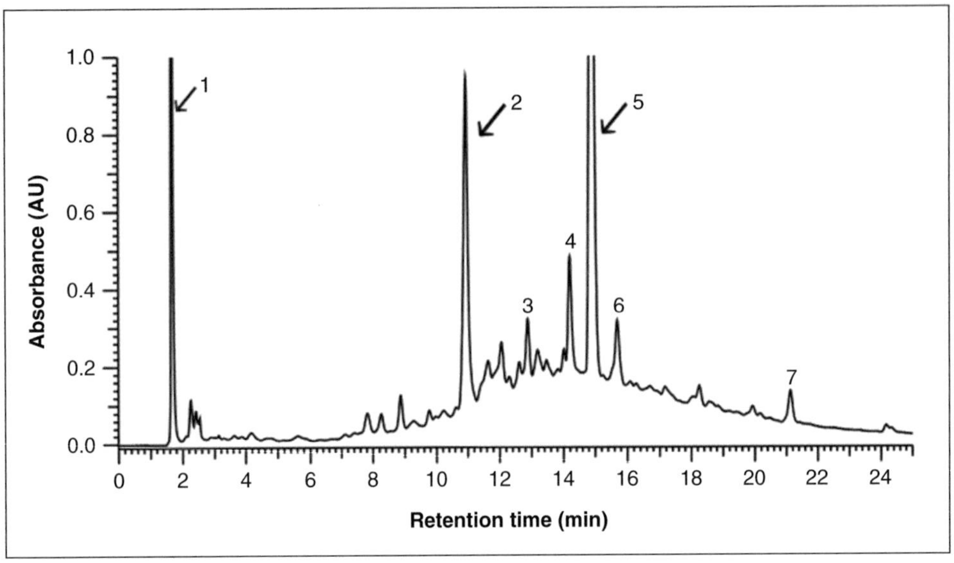

Fig. 3a HPLC-fingerprint analysis of the methanol extract of Herba Ephedrae, sample 1

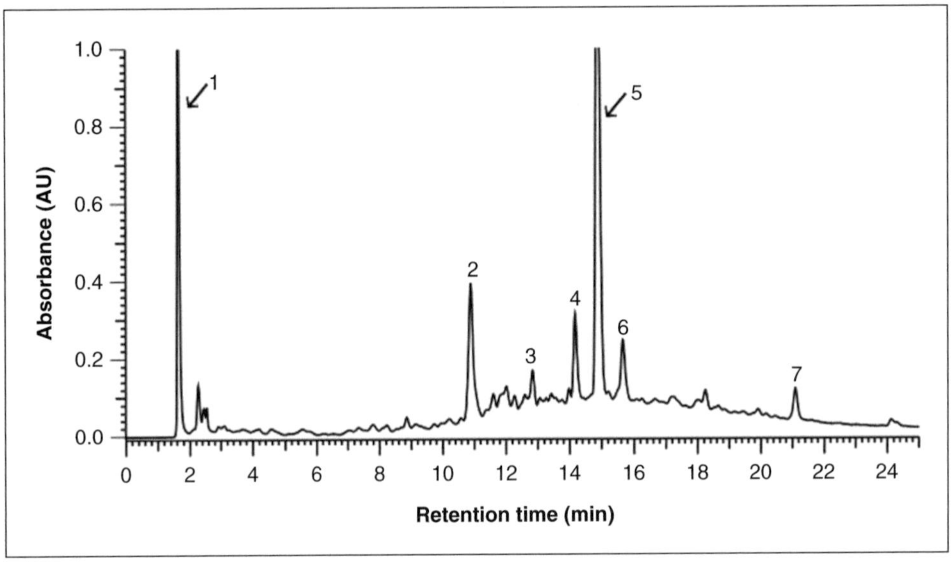

Fig. 3b HPLC-fingerprint analysis of the methanol extract of Herba Ephedrae, sample 3

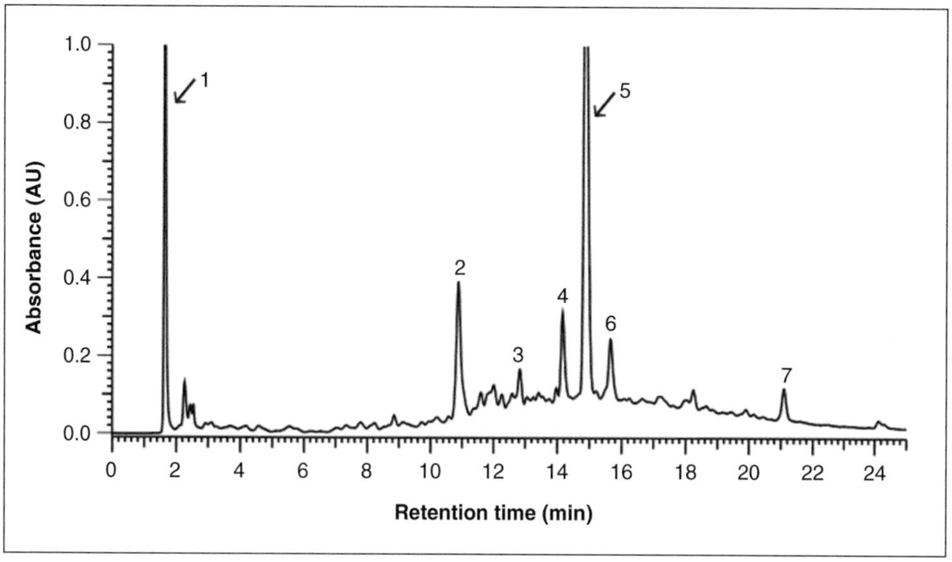

Fig. 3c HPLC-fingerprint analysis of the methanol extract of Herba Ephedrae, sample 4

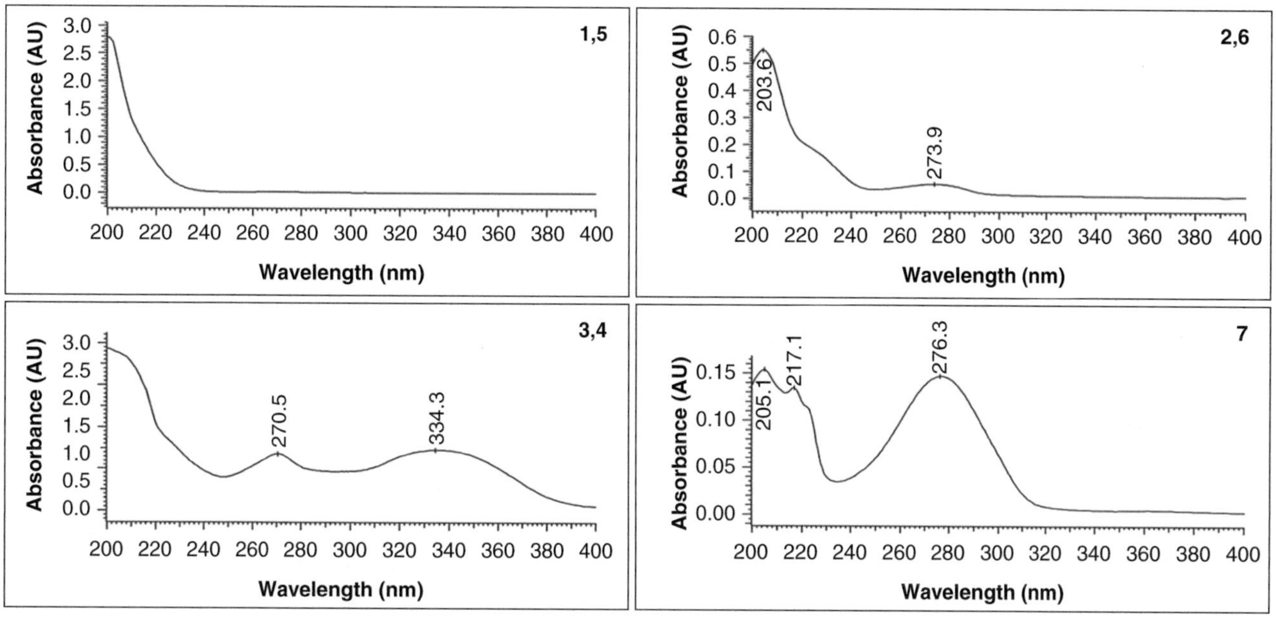

Fig. 4 On line UV-spectra of the main peaks of Herba Ephedrae extracts

4. Description of the HPLC- fingerprint:

   The fingerprint is characterized by an assembly of five distinct peaks in the Rt- range between Rt 10.3 and 16.2 (No. **2, 3, 4, 5** and **6**) with Ephedrine in the center at Rt = 14.8. The other peaks might be derivatives of Ephedrine as described on page 108, under "Main constituents" or flavonoids (No. **3** and **4**). The peak **1** at Rt = 1.7 could not be identified. Peak No. **7** might be derive from a caffeic acid derivative.

Note According to the Chinese Pharmacopeia 2010 Herba Ephedrae contains not less than 0.80 % of the total amount of ephedrine hydrochloride and pseudoephedrine hydrochloride, calculated with reference to the dried drug. [1]

Conclusion

Although apart from Ephedrae herb the accompanying constituents could be not difficultly determined, the authenticity of Ephedra extract can be easily confirmed by TLC and HPLC.

References

1. Pharmacopoeia of the people's republic of China, English edition Vol. I, People's Medical Publishing House, Beijing (2010)
2. Paulus, E., Ding, Y.-H.: Handbuch der traditionellen chinesischen Heilpflanzen, Karl F. Haug Verlag, Heidelberg (1987)
3. Zhang, Z.Z.: An illustrated Chinese materia medica in Hong Kong. School of Chinese Medicine/Hong Kong Baptist University, Hong Kong (2004)
4. Ganzera, M., Lanser, C., Stuppner, H.: Simultaneous determination of *Ephedra sinica* and *Citrus aurantium* var. *amara* alkaloids by ion-pair chromatography. Talanta **66**(4), 889–894 (2005)
5. Tang, W., Eisenbrand, G.: Handbook of Chinese medicinal plants – chemistry, pharmacology, toxicology, vol. 1. WILEY-VCH Verlag GmbH & CoKgaA, Weinheim (2011)
6. Schaneberg, B.T., Crockett, S., Bedir, E., Khan, I.A.: The role of chemical fingerprinting: application of *Ephedra*. Phytochemistry **62**(6), 911–918 (2003)
7. Ranieri, T.L., Ciolino, L.A.: Rapid selective screening and determination of ephedrine alkaloids using GC-MS footnote mark. Phytochem. Anal. **19**(2), 127–135 (2008)
8. Zhong, L.Y., Gong, Q.F., Zhu, J., Chen, F.: Comparative analysis of essential oil components in *Ephedra* and its processed products by GC-MS. J. Chin. Pharmaceut. Sci. **19**(1), 67–73 (2010)
9. Deng, D., Deng, H., Zhang, L., Su, Y.: Determination of ephedrine and pseudoephedrine by field-amplified sample injection capillary electrophoresis. J. Chromatogr. Sci. **52**(4), 357–362 (2014)
10. Xiao, Y.G., Wang, Q.H., Liang, J., Yang, B.Y., Li, G.Y., Kuang, H.X.: Development and application of a rapid and efficient CZE method coupled with correction factors for determination of monosaccharide composition of acidic hetero-polysaccharides from *Ephedra sinica*. Phytochem. Anal. **22**(2), 103–111 (2011)
11. Lv, M., Chen, J., Gao, Y., Sun, J., Zhang, Q., Zhang, M., Xu, F., Zhang, Z.: Metabolomics based on liquid chromatography with mass spectrometry reveals the chemical difference in the stems and roots derived from *Ephedra sinica*. J. Sep. Sci. **38**(19), 3331–3336 (2015)
12. Li, H.X., Ding, M.Y., Lv, K., Yu, J.Y.: Separation and determination of ephedrine alkaloids and tetramethylpyrazine in ephedra sinica stapf by Gas chromatography-mass spectrometry. J. Chromatogr. Sci. **39**(9), 370–374 (2001)
13. Hempen, C.-H., Fischer, T.: A materia medica for Chinese medicine, english edition. Elsevier GmbH, Urban & Fischer, Munich (2009)
14. Nam, N.H., Lee, C.W., Hong, D.H., Kim, H.M., Bae, K.H., Ahn, B.Z.: Antiinvasive, antiangiogenic and antitumour activity of *Ephedra sinica* extract. Phytother. Res. **17**(1), 70–76 (2003)
15. Lv, M., Sun, J., Wang, M., Huang, W., Fan, H., Xu, F., Zhang, Z.: GC-MS based metabolomics study of stems and roots of *Ephedra sinica*. J. Pharm. Biomed. Anal. **114**, 49–52 (2015)
16. Tellez, M.R., Khan, I.A., Schaneberg, B.T., Crockett, S.L., Rimando, A.M., Kobaisy, M.: Steam distillation-solid-phase microextraction for the detection of *Ephedra sinica* in herbal preparations. J. Chromatogr. A **1025**(1), 51–56 (2004)

Herba Violae – *Zihuadiding*

Pharmacopoeia: [1]	Pharmacopoeia of the People's Republic of China, English Edition Vol. I, 2010
Official drug: [1]	Tokyo Violet Herb is the dried herb of *Viola yedoensis* Makino (Fam. Violaceae). The drug is collected in spring and autumn, removed from foreign matter and dried in the sun.
Origin: [2]	southern China, Japan and Korea
Description of the drug: [1]	Frequently crumped into masses. The main roots long- conical, 1–3 mm in diameter, pale yellowish- brown, with fine longitudinal wrinkles. Leaves basal, greyish- green, when whole, lanceolate or oval lanceolate, 1.5–6 cm long, 1–2 cm wide; apex obtuse, base truncate or somewhat cordate, margin obtusely, serrate, both surfaces pubescent; petioles slender, 2–6 cm long, the upper part with distinct narrow wings. Pedicels slender tubular. Capsules elliptical or 3-splitted; seeds numerous, pale brown. Odour slight.
Pretreatment of the raw drug: [1]	Foreign matters are eliminated, washed clean, cut into pieces and dried.
Medicinal use: [3]	It is used to treat many skin diseases, i.e eczema, impetigo, acne, pruritus cradle cap, furthermore used to treat upper respiratory tract infections with fever.

Effects and indications of Herba Violae according to Traditional Chinese Medicine [1, 3–5]

Taste:	Bitter and pungent
Temperature:	Cold
Channels entered:	*Orbis cardialis, orbis hepaticus*
Effects (functions):	Cools heat, disinfects and detoxifies, Eczema, burns and unclerations, breaks up lumps, clears damp- heat, cools blood
Symptoms and indications:	Inflammation, red and swollen eyes, painful throat obstruction, ear pain and swelling, mumps

H. Wagner et al. (eds.), *Chromatographic Fingerprint Analysis of Herbal Medicines,* Vol. 4,
DOI 10.1007/978-3-319-32328-2_12, © Springer International Publishing Switzerland 2016

Main constituents:

- **Flavonoid glycosides:** [6–8]
Flavone di-C-glycosides: Apigenin 6-C-β-D- glucosyl-8-C-α-L-rhamnoside (Violanthin) apigenin 6,8-di-C-α-L-arabinopyranoside, apigenin 6-C-α-L-arabinopyranosyl-8-C-β-D-glucopyranoside (isoschaftoside), apigenin 6-C-β-D-glucopyranosyl-8-C-α-L-arabinopyranoside (schaftoside), apigenin 6-C-β-D-glucopyranosyl-8-C-β-L-arabinopyranoside (neoschaftoside), apigenin 6,8-di-C-β-D-glucopyranoside (vicenin-2), apigenin 6-C-α-L-arabinopyranosyl-8-C-β-D-xylopyranoside, apigenin 6-C-β-D-xylopyranosyl-8-C-α-L-arabinopyranoside, luteolin-8-C-glucoside (orientin), luteolin 6-C-β-D-glucopyranoside (isoorientin) and luteolin 6-C-α-L-arabinopyranosyl-8-C-β-D-glucopyranoside (isocarlinoside), isovetexin (saponaretin), kaempferol-3-O-β-D-glucosyl(1→2)-β-D-glucosyl-7-O-α-L-rhamnoside

- **Coumarins:** [9–11] Esculetin, scopoletin, isoscopoletin, 6-hydroxymethyl-3-pyridinol,5,5-bi (6,7- dihydroxycoumarin), 6,6,7,7-tetrahydroxy-5,8-bicoumarin, dimeresculetin, euphorbetin

- **Sesquiterpens:** [5, 10] Yedoensin A, yedoensin B, versicolatone B, loliolide, dehydrololiolide, madolin W, aristoyonnolin E, and madolin Y

- **Cyclotides:** [12, 13] Peptide E, cycloviolacin Y5, and cyclovolacin VY1

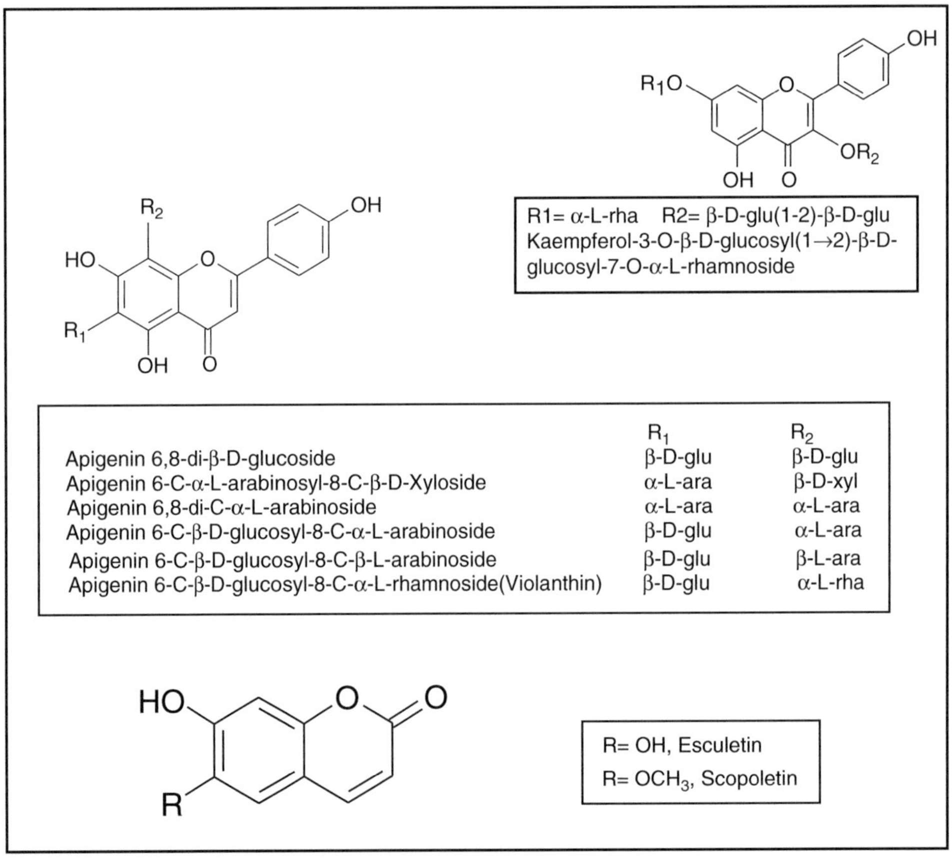

Fig. 1 Formulae of the main constituents of Herba Violae [4, 6, 7]

Reported pharmacological activities:

-α- glucosidase inhibition activity [7]

-anti- HIV effect [6, 13, 14]

-anti- inflammatory activity [7, 15]

-antimicrobial activity [7, 16]

-antioxidant activity [7]

-sedative effect [7]

-antihypertensive effect [7]

-anti- influenza A H1N1 activity (cycloviolacin Y5) [12, 15]

-anti-coagulant activity [5]

TLC Fingerprint Analysis

Drug samples		Origin
1	Herba Violae / *Viola yedoensis*	Sample of commercial drug obtained from firm HerbaSinica
2	Herba Violae / *Viola yedoensis*	Sample of commercial drug, obtained from TCM-Clinic Bad Kötzting (Charge: 30.08.2000)
3	Herba Violae / *Viola yedoensis*	Sample of commercial drug, obtained from TCM-Clinic Bad Kötzting (Charge:19.06.1995)
4	Herba Violae / *Viola yedoensis*	Province Hebei (China)
5	Herba Violae cum Floribus conc. / *Viola tricolor* as reference drug	Firm Caeser & Lorenz GmbH (Germany)

Reference compounds of Fig. 2		R*f*
T1	Scopoletin	0.95
T2	Orientin	0.65
T3	Isoorientin	0.54
T4	Luteolin-7-O-glucosid	0.68
T5	Violanthin	0.34
T6	Esculetin	0.94
T7	Vitexin	0.73
T8	Schaftoside	0.29
T9	Esculin	0.50
T10	Rutin	0.43
T11	Saponaretin	0.61
T12	Vicenin-1	0.30

1. Extraction: 1.0 g of the powdered drug with 10 ml methanol are ultrasonicated for 30 min with 10 ml methanol. The extract is filtered and the filtrate evaporated to dryness. The residue is dissolved in 2 ml methanol and filtered over Chromafil® filtration unit, type 0-20 µm/25 mm
2. Reference compounds: Each 1 mg is dissolved in 1 ml methanol
3. Separation parameters:
   - Plate: HPTLC Silica gel 60 F_{254}, Merck
   - Applied amounts: Herba Violae extracts: 10 µl each
Reference compounds: 5 µl each
   - Solvent system: ethyl acetate + glacial acetic acid + formic acid + water
(20 + 2.2 + 2.2 + 2.6)
   - Detection: Natural products – Polyethylene glycol Reagent (NP/PEG)
I: 1 % diphenylboric acid-β-ethylamino ester
(= diphenylboryloxyethylamine, NP) in methanol
II: 5 % polyethylene glycol-4000 (PEG) in ethanol
The plate is sprayed first with solution I and then with solution II. The evaluation is carried out in UV 366 nm.

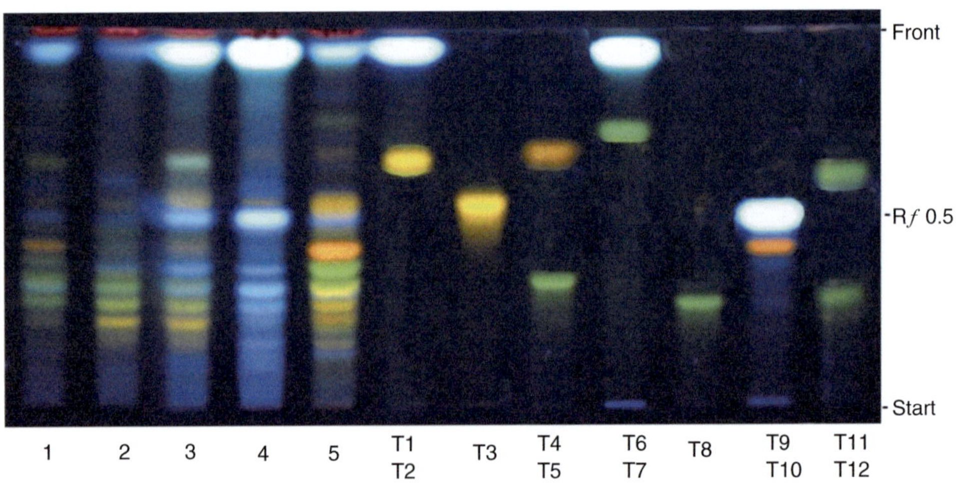

Fig. 2 Thin layer chromatogram of the methanol extracts of Herba Violae sprayed with NP-PEG reagent (UV 366 nm)

Description:

The Herba extract samples 1–4 show due to a quite different leaf-root-and flower composition a very heterogeneous zone pattern. In the most representative extract sample 3 appears on the solvent front a mixture of scopoletin/esculetin **T1**, **T6** followed in the middle R*f*-range the green fluorescent zone of saponaretin at R*f*=0.70 **(T11)** and a further orange zone of isoorientin (**T3**), esculin (**T9**) and rutin (**T10**). The following downwards zones are the flavonoid- C- glycosides violanthin (**T5**) and vicenin-1 (**T12**). Extract sample 1 and 2 possess a similar zone profile only in the deep R*f*- range. The extract sample 4 contains only a mixture of the cumarins scopoletin and esculetin (**T1**+**T6**) inclusive a green fluorescents one, probably violanthin (**T5**).

The extract sample 5 derives from the species *Viola tricolor* which is not official in the Chinese Pharmacopoeia and only known in western countries. This species is a low growing species with abundant amount of flowers with a high concentration of flavonoids.

HPLC-Fingerprint Analysis

1. Sample preparation: 1.0 g of the powdered drug with 10 ml methanol are ultrasonicated for 30 min. The extract is filtered and the filtrate evaporated to dryness. The residue is dissolved in 2 ml methanol and filtered over Chromafil® filtration unit, type 0-20.
2. Injection volume: Herba Violae extracts: 20 µl each
   Reference compounds: 10 µl each
3. HPLC parameter:

Apparatus:	MERCK HITACHI D-6000 A Interface MERCK HITACHI L-4500 A Diode Array Detector MERCK HITACHI AS-2000 Autosampler MERCK HITACHI L-6200 A Intelligent Pump
Separation column:	LiChroCART® 250 -4 LiChrospher® 100 RP-18 (5 µm), Merck
Precolumn:	LiChroCART® 4-4 LiChrospher® 100 RP-18 (5 µm), Merck
Solvent System:	A: 0.0001 % aq. H_3PO_4 (Millipore Ultra Clear UV plus® filtered) B: Acetonitrile (VWR)
Gradient:	0–5 % B in 5 min, flow: 1.0 ml/min 5–30 % B in 35 min, flow: 0.8 ml/min 30 % B for 5 min, flow: 0.8 ml/min total run time: 45 min
Detection:	340 nm

Retention times of the main peaks

Peak	Rt (min)	Compound
1	21.1	Esculin
2	26.0	Esculetin
3	27.5	Not identified flavone
4	28.2	Vicenin-1
5	29.3	Not identified flavone
6	30.1	Schaftoside
7	30.9	Isoorientin
8	31.2	Orientin
9	32.5	Violanthin
10	33.2	Vitexin
11	34.0	Rutin
12	35.3	Luteolin-7-glucoside
13,14	36.7, 37.6	Not identified flavones
B (Sample 4)	33.8	Scopoletin
A,C (sample 4)	31.0, 40.4	Not identified coumarins

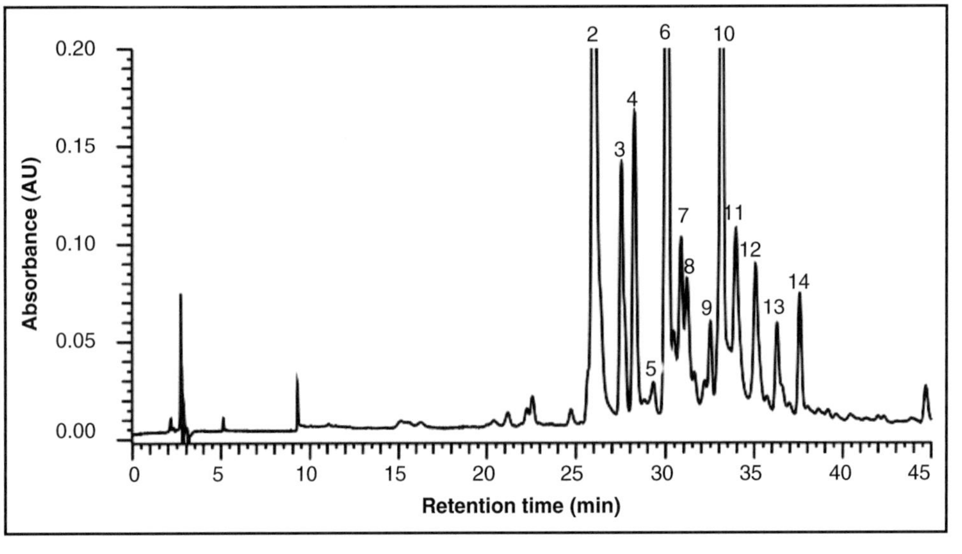

Fig. 3a HPLC-fingerprint analysis of the methanol extract of Herba Violae, sample 1

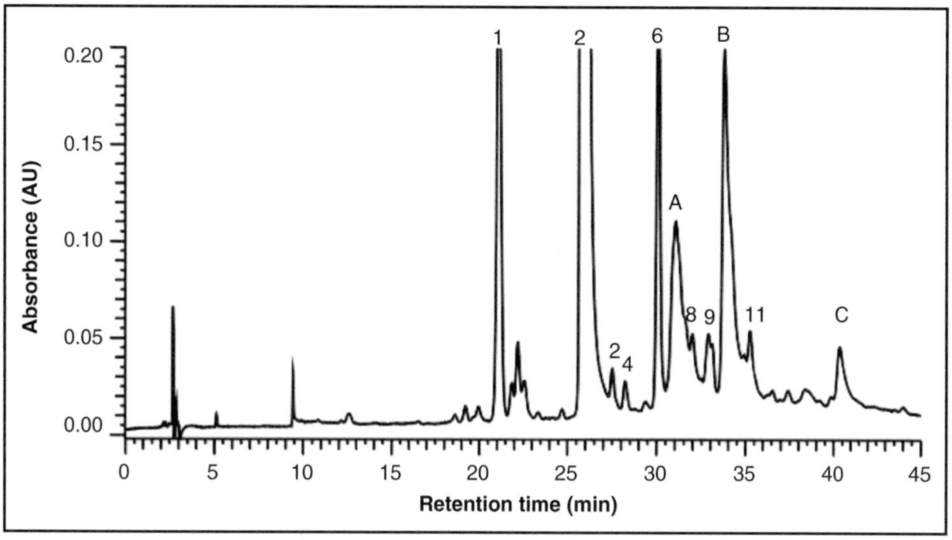

Fig. 3b HPLC-fingerprint analysis of the methanol extract of Herba Violae, sample 4

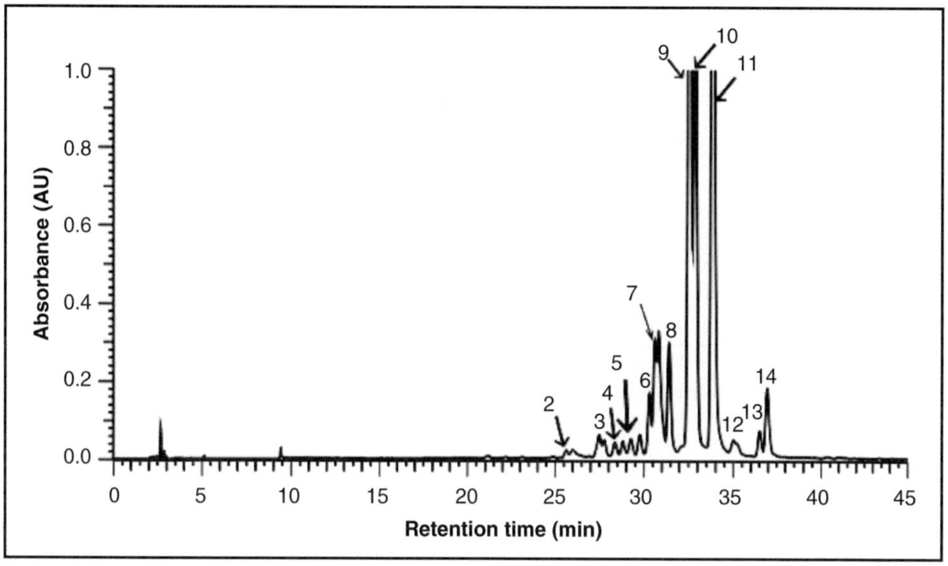

Fig. 3c HPLC-fingerprint analysis of the methanol extract of Herba Violae cum Floribis conc., sample 5

4. Description:

   Fig 3a: The Herba Violae extract sample 3 is characterized by the main peaks **1** (esculetin), **6** (schaftoside) and **10** (vitexin). This peak profile corresponds with the TLC Fig. 2a (extract sample 3**).**

   Fig. 3b: The Herba Violae extract sample 4 is characterized by the main peaks **1** (esculin), peak **2** (esculetin), peak 6 (schaftoside) and **B** (scopoletin). This main peak profile corresponds with that of TLC Fig. 2a (extract sample 4)

   Fig. 3c: The *Violae tricolor* extract sample 5 is characterized by the peaks **9** (violanthin), **10** (vitexin) and **11** (rutin). This peak profile corresponds again with the TLC Fig. 2a of extract sample 5.

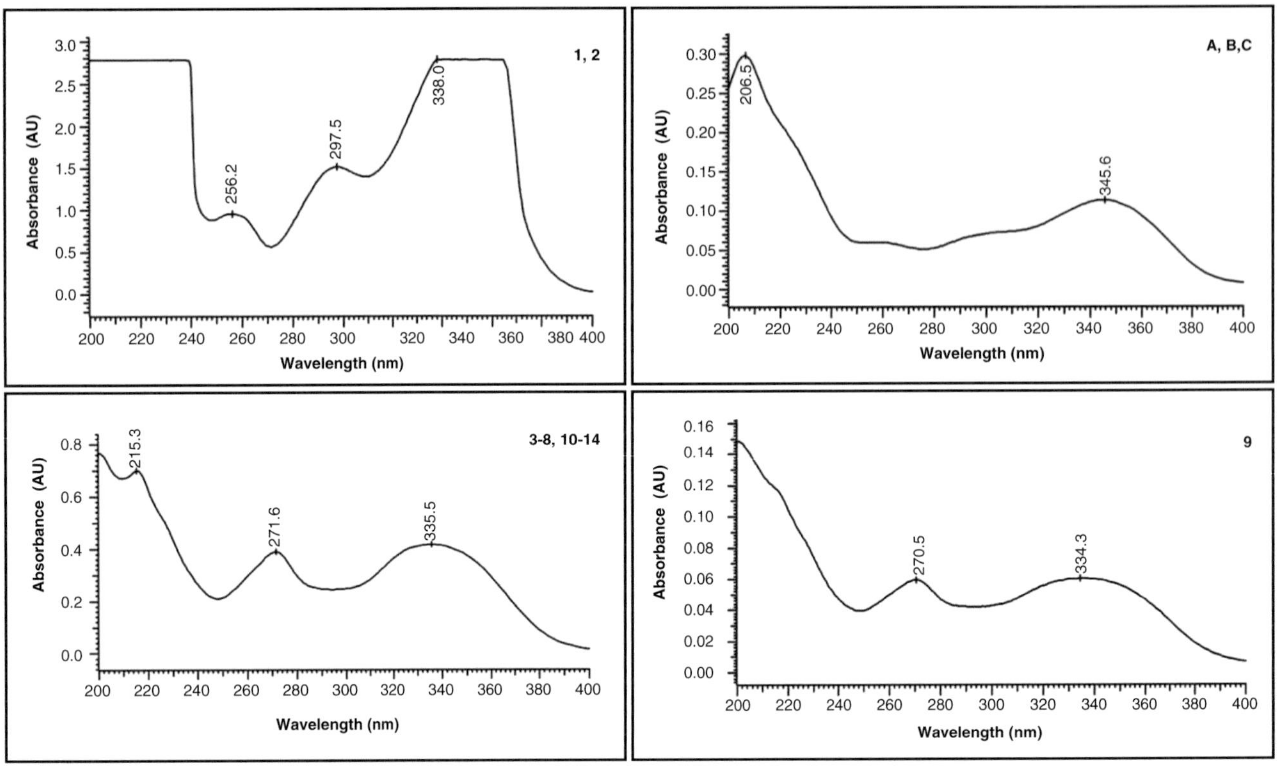

Fig. 4 On line UV-spectra of the main compounds (peaks) of Herba Violae

Conclusion

Although the authentication of Chinese Herba drugs is often difficult because of the changing amounts of leaves, stems, flowers and roots in the various available herbal drug samples, the identification of the characteristic coumarins and flavone- C- glycosides is sufficient indication for the botanical authentication of Herba Violae.

References

1. Pharmacopoeia of the people's republic of China, English edition Vol. I, People's Medical Publishing House, Beijing (2010)
2. Porkert, M.: Klinische Chinesische Pharmakologie, Verlag für Medizin Dr. Ewald Fischer GmbH, Heidelberg (1972)
3. Hempen, C.-H., Fischer, T.: Leitfaden Chinesische Phytotherapie 2. Auflage Urban & Fischer, , Munich (2007)
4. Hong Kong Chinese Materia Medica Standards, vol. 2. Chinese Medicine Division – Department of Health – Government of the Hong Kong Special Administrative Region – the People's Republic of China, Hong Kong (2008)
5. Du, D., Cheng, Z., Chen, D.: Anti-complement sesquiterpenes from *Viola yedoensis*. Fitoterapia **101**, 73–79 (2015)
6. Xie, C., Veitch, N.C., Houghton, P.J., Simmonds, M.S.: Flavone C-glycosides from *Viola yedoensis* MAKINO. Chem. Pharm. Bull. (Tokyo) **51**(10), 1204–1207 (2003)
7. Cao, J., Yin, C., Qin, Y., Cheng, Z., Chen, D.: Approach to the study of flavone di-C-glycosides by high performance liquid chromatography-tandem ion trap mass spectrometry and its application to characterization of flavonoid composition in *Viola yedoensis*. J. Mass Spectrom. **49**(10), 1010–1024 (2014)
8. Wagner, H., Rosprim, L., Düll, P.: Die Flavon-C- Glycoside von *Viola Tricolor* L. Zeitschrift für Naturforschung Letters, Band 27b, Heft 8 (1972)
9. Wichtl, M.: Teedrogen und Phytopharmaka, 4. Auflage Wissenschaftliche Verlagsgesellschaft mbH , Stuttgart (2002)

10. Huang, J., Yang, J., Xue, Q., Yu, L., Zhang, D.: Studies on chemical constituents from herbs of Viola yedoensis. Zhongguo Zhong Yao Za Zhi **34**(9), 1114–1116 (2009)
11. Zhou, H.Y., Hong, J.L., Shu, P., Ni, Y.J., Qin, M.J.: A new dicoumarin and anticoagulant activity from *Viola yedoensis* Makino. Fitoterapia **5**(80), 283–285 (2009)
12. Liu, M.Z., Yang, Y., Zhang, S.X., Tang, L., Wang, H.M., Chen, C.J., Shen, Z.F., Cheng, K.D., Kong, J.Q., Wang, W.: A cyclotide against influenza A H1N1 virus from *Viola yedoensis*. Yao XueXue Bao. **49**(6), 905–912 (2014)
13. Wang, C.K., Colgrave, M.L., Gustafson, K.R., Ireland, D.C., Goransson, U., Craik, D.J.: Anti-HIV cyclotides from the Chinese medicinal herb *Viola yedoensis*. J. Nat. Prod **71**(1), 47–52 (2007)
14. Nagan, F., Chang, R.S., Tabba, H.D., Smith, K.M.: Isolation, purification and partial characterization of an active anti-HIV compound from the Chinese medicinal herb *viola yedoensis*. Antiviral Res. **10**(1–3), 107–116 (1988)
15. Li, W., Xie, J.Y., Li, H., Zhang, Y.Y., Cao, J., Cheng, Z.H., Chen, D.F.: *Viola yedoensis* liposoluble fraction ameliorates lipopolysaccharide-induced acute lung injury in mice. M. J. Chin. Med. **40**(5), 1007–1018 (2012)
16. Sun, Y., Du, L., Zhou, L., Zhang, W., Miao, F., Yang, X., Geng, H.: Study of antibacterial active components from *Viola Yedoensis*. Zhongguo Zhong Yao Za Zhi **36**(19), 2666–2671 (2011)

Lignum Sappan – *Sumu*

Pharmacopoeia: [1]	Pharmacopoeia of the People's Republic of China, English Edition Vol. I, 2010
Official drug: [1]	Sappan Wood is the dried heart wood of *Caesalpinia sappan* L. (Fam. Caesalpinaceae). The drug is collected mostly in autumn, removed from the white sap wood and dried.
Origin: [2–11]	Cultivated mainly in the Chinese provinces such as Hainan, Taiwan, Fujian, Sichuan, Yunnan, Guangdong, Guangxi and Guizhou. Available and exported also from Cambodia, India, Indonesia, Laos, Malaysia, Myanmar, Taiwan, Thailand, Vietnam and America.
Description of the drug: [1]	Long cylindrical or semicylindrical owing to cutting lengthwise, 10–100 cm long, 3–12 cm in diameter. Externally yellowish-red to brownish-red, with traces of knife cutting, usually showing longitudinal cracks. Texture compact and hard. Transversely cut surface somewhat lustrous, annual rings distinct, sometimes dark brown, loose and bright-dotted pith visible. Odour, slight; taste, slightly astringent.
Medicinal use:	The medicinal use is multivalent. Mainly used for the prevention and treatment of cardiovasculary and inflammatory diseases, as antidiabetic, against menorrhagia, as immunomodulatory drug, antithrombotic and antioxidants and several other internal and external complaints.

Effects and indications of Lignum Sappan according to Traditional Chinese Medicine [1, 3, 6–13]

Taste:	Sweet, salty, pungent
Temperature:	Neutral
Channels entered:	*Orbis cardialis, o. hepaticus, o. lienalis*
Effects (functions):	To activate blood, to eliminate stasis, disperse swelling and relieve pain
Symptoms and indications:	Injuries from falls and tights, fracture and sinew injury, stasis and stagnation, swelling and pain, amenorrhea and dysmenorrhea, postpartum stasis and obstruction, stabbing pain in the chest and the abdomen, swelling and painful abscesses, caruncles, ulcers, tetanus and cellulitis. As expectorant and emmenagouge. Treatment of diarrhea, diabetes, epilepsy and traumatic diseases, leprosy and dysentery

H. Wagner et al. (eds.), *Chromatographic Fingerprint Analysis of Herbal Medicines*, Vol. 4,
DOI 10.1007/978-3-319-32328-2_13, © Springer International Publishing Switzerland 2016

Main constituents:

Homoisoflavonoids/Phenolic compounds [5–11, 14–29]	Brazilin, brazilein, brazilane, tetraacetylbrazilin, 3′-O-methylbrazilin, 4′-O-methylbrazilin, hematoxylin, sappanchalcone, 3-deoxysappanchalcone, sappanone A + B, 3-hydroxysappanone B, 3-deoxysappanone B, 3′-deoxysappanone B, sappanol, episappanol; 3′-deoxy-4-O-methylsappanol; 3′-deoxy-4-O-methylepisappanol; 3′-O-methylsappanol; 3′-O-methylisosappanol; 3′, 4-O-methylsappanol, 4-di-O-methylepisappanol, caesalpin J + P, 7,3′,4′-trihydroxy-3-benzyl-2H-chromene, caesalpiniaphenol G + H, quercetin-3,7-di-O-methyl ether
Protosappanins/dibenzoxin derivatives [10, 11, 14, 16, 20, 23, 26, 27, 29]	Protosappanin A-D, E-1 and E-2, isoprotosappanin B, protosappanin A dimethyl acetal, protosappanin C dimethyl acetal, caesappin A + B, 10-O-methyprotosappanin B, 10-O-methyliso-protosappanin B, neoprotosappanin

Minor constituents:

- essential oil (ocimene), sappanin (=4-(3,4-dihydroxyphenyl)benzene-1,3-diol) [2]
- triterpenoids, steroids, oxygen heterocycles [7, 10, 13]
- 3′-deoxy-4-O-methylepisappanol, palmitic acid, (+)-lyoniresinol, (+)-(8*S*,8′*S*)-bisdihydrosiringenin [16]
- 3,8-dihydroxy-4,10-dimethoxy-7-oxo-[2] benzopyrano[4,3-b] benzopyran; rhamnetin; 3, 7-dihydroxy-chroman-4-one; dimethyladipate [17]
- 2-methoxy-3-hydroxyxanthone [19]
- campesterol, stigmasterol, β-sitosterol, daucosterin (= daucosterol), hematein [17, 20, 26]
- (3R,4S)-3-(4′-hydroxybenzyl)-3,4-dihydro-2″-3″-dimethyl-3H-[1,3]dioxolo[4,5-c]chromen-7-ol; (3R,4S)-3-(3′-methoxy-4′-hydroxybenzyl)-3,4-dihydro-2″,3″-dimethyl-3H-[1,3]dioxolo[4,5-c]chromen-7-ol; (αS)-α,2′,4,4′-tetrahydroxydihydrochalcone [21]
- isoliquiritigenin 2′-methyl ether, pluchioc acid [22]
- 4-(7-hydroxy-2,2-dimethyl-9βH-1,3,5-troxa-cyclopenta[α]naphthalene-3-lymethyl)-benzene-1,2-diol [24]

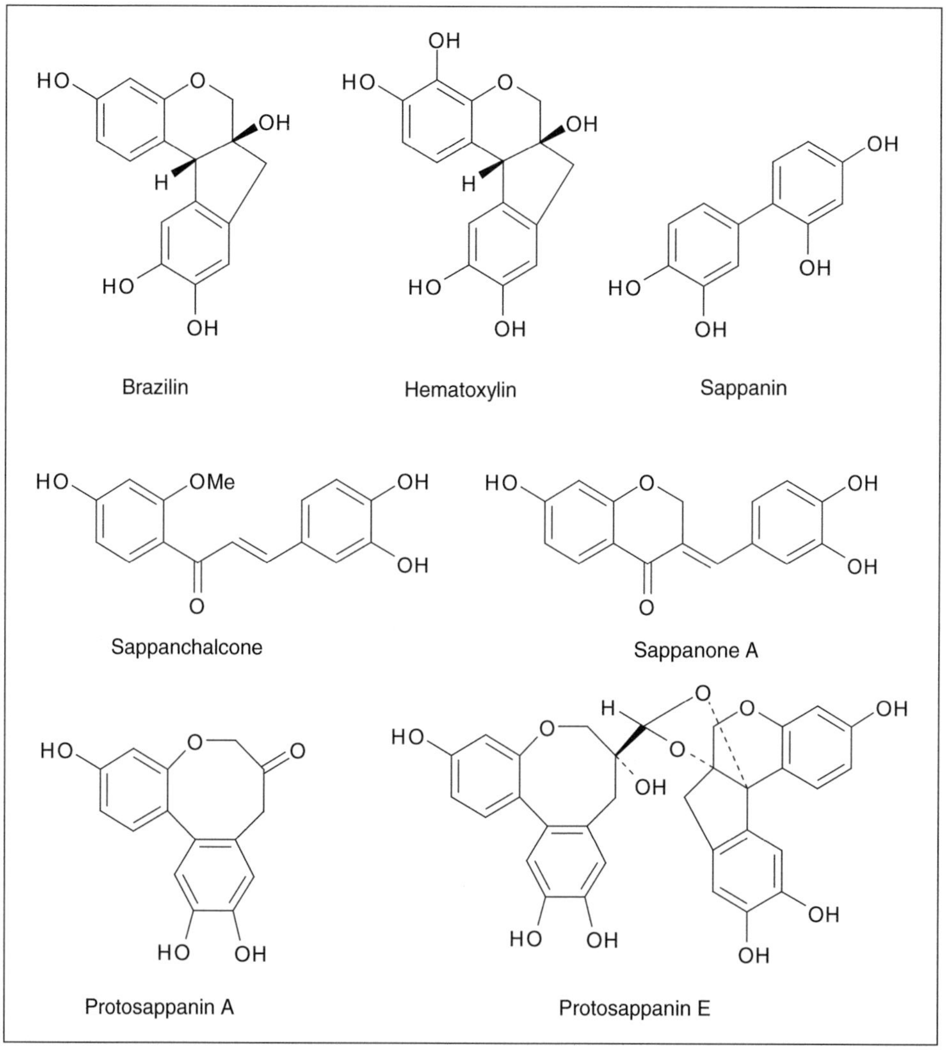

Fig. 1 Formulae of the main compounds of Lignum Sappan [15]

Reported pharmacology:

- analgesic [5, 8, 14, 15, 20, 22, 26, 30, 31]
- anti-inflammatory [5, 7–15, 20–22, 25–27, 30–40]
- positive inotropic [12]
- anti-hypercholesteremic [14, 20, 25, 41]
- hypoglycemic activity [5, 7, 9, 10, 13, 33, 38, 39]
- anti-platelet aggregation [5, 33]
- anti-oxidative [6, 9–11, 22, 31, 32, 34, 35, 38, 39]
- anti-convulsant [7, 10, 13, 34, 38, 41]
- anti-tumoral [5, 9, 38, 42]
- anti-allergic [21, 23, 27, 40]
- anti-edemic [15]
- hemodynamic [15]
- antibiotic [12]
- sedative [12, 14, 20]
- hypnotic [12]
- antineoplastic [12]
- anti-hepatotoxicity/hepatoprotective [5, 7, 10, 13, 33, 36, 37]
- antibacterial [5, 8, 15, 21, 35]
- anti-microbial [6, 9, 22, 25, 26, 33–35]
- anti-viral [9, 34]
- vasorelaxation [6, 33, 34, 36, 37, 39]
- modulation of immune function/immunomodulative/ immune-suppressive [7, 10, 11, 13, 16, 25, 34, 36, 38, 39, 42]
- anti-complementary [6–8, 10, 13, 20, 22, 25, 30, 38]
- cytotoxic [7, 13, 31]
- anti-atherosclerosis [9]
- spasmolytic [9, 39]
- depressing/supporting effects on central nervous system inhibitory activity in CNS [20, 25, 41]
- anti-influenza [21, 23, 24, 27, 40]
- neuroprotective [21, 23, 27, 40, 42]
- anti-proliferative [32, 34]
- aldose reductase inhibition/lens-aldose reductase inhibitory effect [33, 36, 37]

TLC-Fingerprint Analysis

	Drug samples	Origin
1	Lignum Sappan / *Caesalpinia sappan*	Sample of commercial drug (China Medica, origin: Yunnan)
2	Lignum Sappan / *Caesalpinia sappan*	Province Vietnam, China
3	Lignum Sappan / *Caesalpinia sappan*	Province Guangxi, China
4	Lignum Sappan / *Caesalpinia sappan*	Province Guangdong, China
5	Lignum Sappan / *Caesalpinia sappan*	Sample of commercial drug (TCM-Clinic, Bad Kötzting, 28717012010)
6	Lignum Sappan / *Caesalpinia sappan*	Sample of commercial drug (TCM-Clinic, Bad Kötzting, 30.08.2000)
7	Lignum Sappan / *Caesalpinia sappan*	Sample of commercial drug (TCM-Clinic, Bad Kötzting, 26.04.2004)
8	Lignum Sappan / *Caesalpinia sappan*	Sample of commercial drug (TCM-Clinic, Bad Kötzting, 01.06.2013)

Reference compounds of Fig. 2, 3a/b		R*f*
T1	Brazilin	0.59
T2	Hematoxylin	0.36

1. Extraction: 0.5 g powdered drug are extracted with 5 ml methanol under reflux for 15 min. The extract is filtered, evaporated to dryness and the residue is dissolved in 0.5 ml methanol.

2. Reference compounds: 1.0 mg is dissolved in 1.0 ml methanol

3. Separation parameters:

Plate:	HPTLC Silica gel 60 F_{254}, Merck
Applied amounts:	Lignum Sappan extracts: each 10 µl; Brazilin: 5 µl; Hematoxylin: 10 µl
Solvent system:	Chloroform + acetone + formic acid (12 + 6 + 1.5)
Direct evaluation:	The plate is dried and evaluated the next day under UV 254 nm (Fig. 2).
Detection:	10 % ethanolic Sulphuric acid The plate is sprayed with 8 ml reagent and heated at 110 °C for 10 min. The plate is evaluated in VIS (Fig. 3a) and under UV 366 nm (Fig. 3b).

4. Description of Fig. 2:

This TLC gives an excellent and homogeneous overview on the chemical composition of the homoisoflavonoids and protosappanins in all 8 Lignum Sappan extract samples.

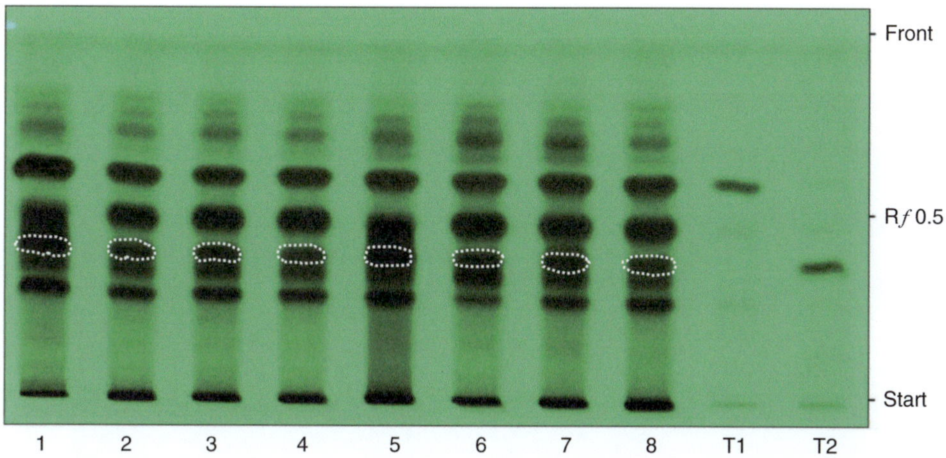

Fig. 2 TLC of the methanol extracts of Lignum Sappan, without chemical treatment (UV 254 nm)

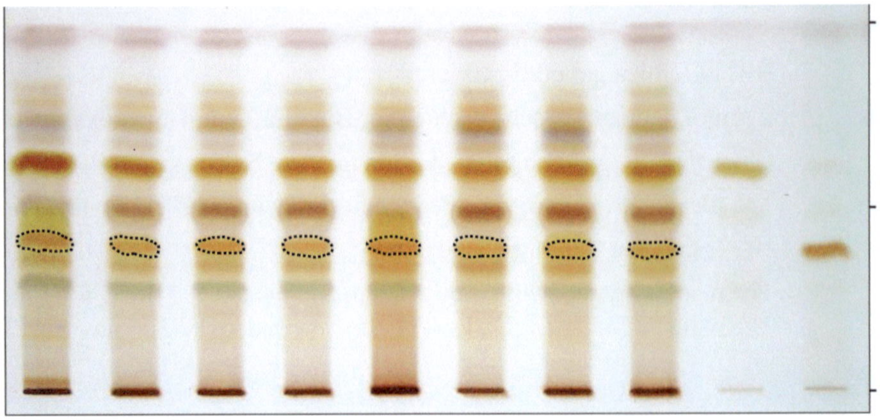

Fig. 3a TLC of the methanol extracts of Lignum Sappan, sprayed with 10 % ethanolic sulphuric acid (VIS)

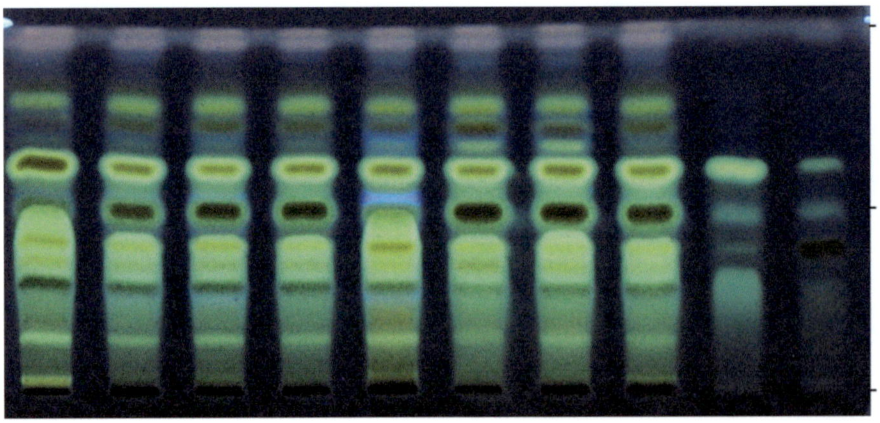

Fig. 3b TLC of the methanol extracts of Lignum Sappan, sprayed with 10 % ethanolic sulphuric acid (UV 366 nm)

Description of Fig. 3a and 3b:

Fig. 3a shows in VIS the sprayed TLC of the 8 methanol extract samples of Lignum Sappan. In the R*f*-range between R*f* 0.35 and 0.75 are detected the red/orange homoisoflavonoids with Brazilin (**T1**, R*f*=0.59) and Hematoxylin (**T2**, R*f*=0.36). The orange zones at R*f*=0.70 and 0.75 can be assigned to the Sappanchalcone and Sappanone A. Protosappanin A and E may be visible under UV 366 nm in deeper R*f*-range in Fig. 3b as weak green/yellow fluorescent zones. In this Figure the homoisoflavonoids appear also with green/yellow fluorescent between R*f*=0.36 and 0.60.

HPLC-Fingerprint Analysis

1. Extraction: 0.5 g powdered drug are extracted with 5 ml methanol under reflux for 15 min. The extract is filtered, evaporated to dryness and the residue is dissolved in 0.5 ml methanol. The extract is filtered over Millipore® filtration unit, Type 0.45 µm.

2. Injection volume: Lignum Sappan extracts: each 5 µl

3. HPLC parameter:

Apparatus: MERCK HITACHI D-6000 A Interface
MERCK HITACHI L-4500 A Diode Array Detector
MERCK HITACHI AS-2000 Autosampler
MERCK HITACHI L-6200 A Intelligent Pump

Separation column: LiChroCART® 250-4 LiChrospher® 100 RP-18 (5 µm), Merck

Precolumn: LiChroCART® 4-4 LiChrospher® 100 RP-18 (5 µm), Merck

Solvent system: A: 0.0001 % phosphoric acid/water (Millipore Ultra Clear UV plus® filtered)
B: acetonitrile (VWR)

Gradient: 5–45 % B in 40 min,
Total runtime: 40 min

Flow: 1.0 ml/min

Detection: 285 nm

Retention times of the main peaks

Peak	**Rt (min)**	**Compound**
1	13.5	Hematoxylin
2	16.3	Brazilin

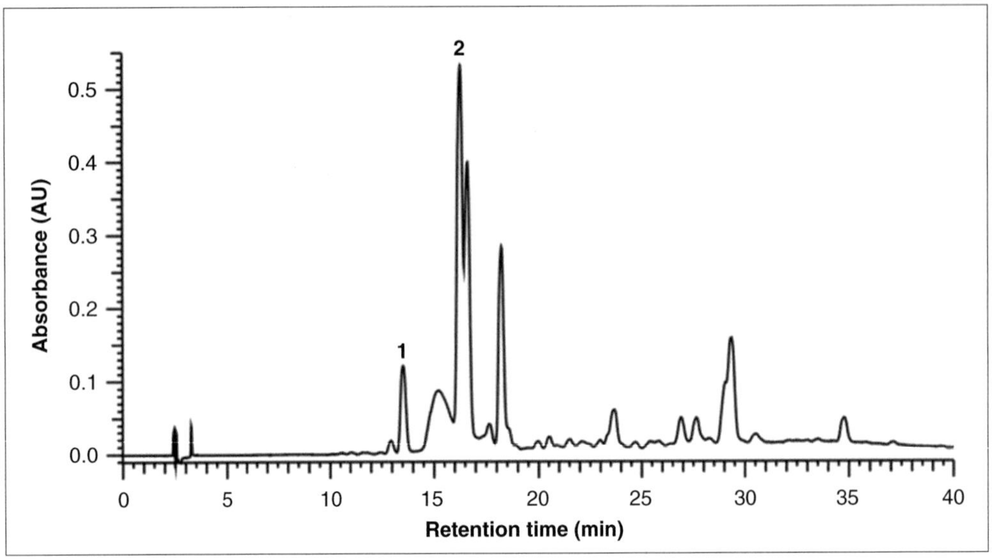

Fig. 4a HPLC-fingerprint analysis of the methanol extract of Lignum Sappan (sample 2)

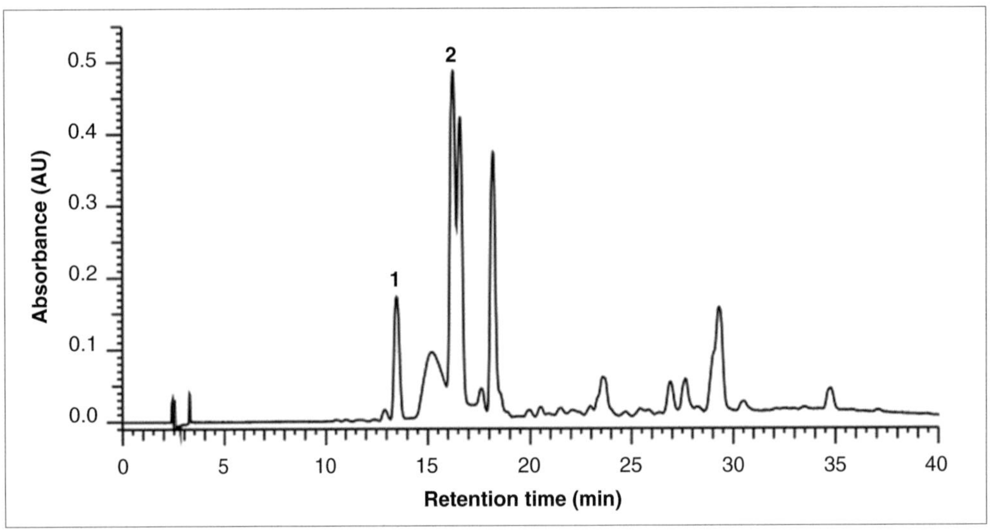

Fig. 4b HPLC-fingerprint analysis of the methanol extract of Lignum Sappan (sample 4)

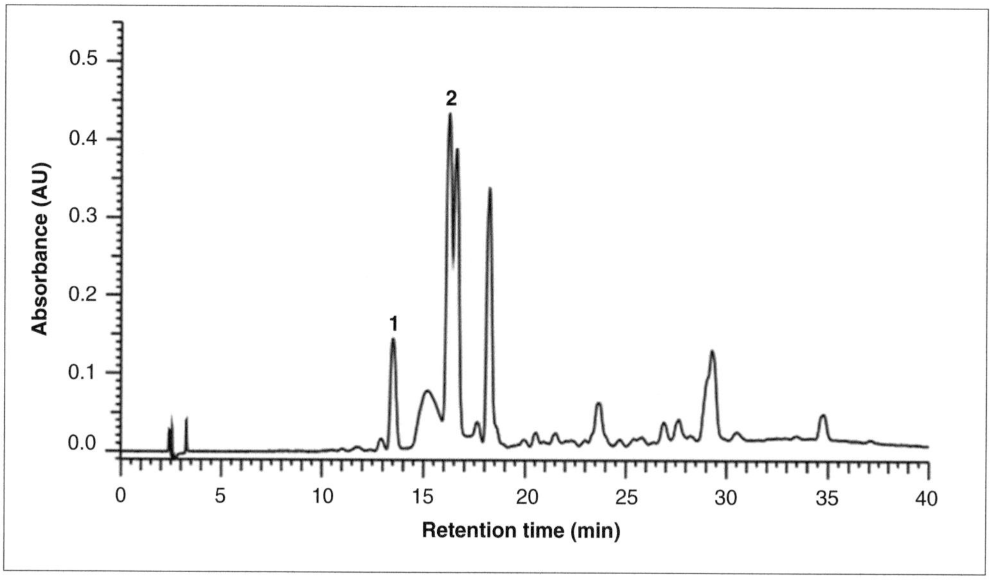

Fig. 4c HPLC-fingerprint analysis of the methanol extract of Lignum Sappan (sample 8)

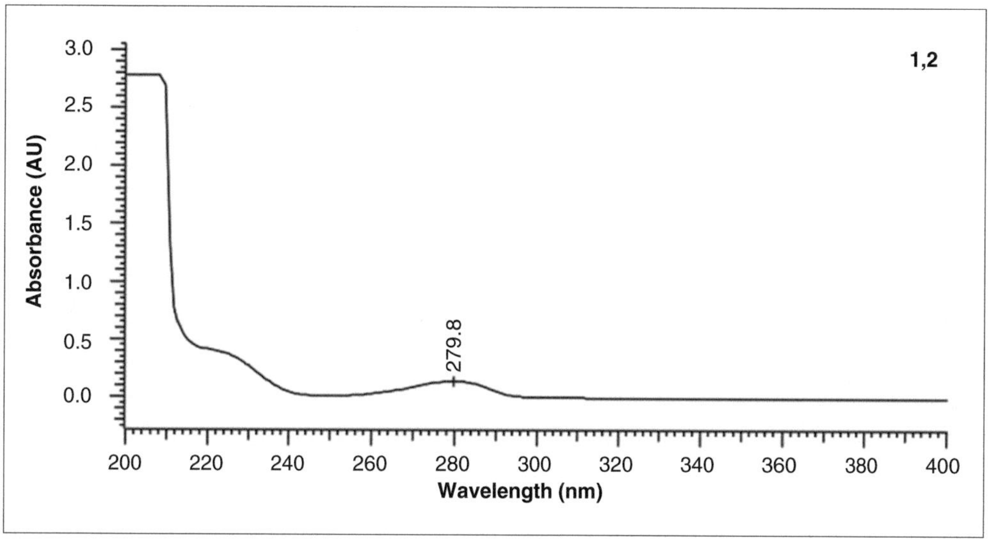

Fig. 5 On line UV-spectra of the main peaks of Lignum Sappan extracts

4. Description of the HPLC-Figures:

   The methanol extract sample 2, 4 and 8 detected at 285 nm possess in the Rt-range between 13.0 and 20.0 their main peaks of Hematoxylin (**1**, Rt = 13.5) and Brazilin (**2**, Rt = 16.3). Hematoxylin and Brazilin are characterized by UV_{max} at 279 nm (Fig. 5). The other peaks of the not assignable homoisoflavonoids in the Rt-range 23.0 and 35.0 possess UV_{max} at 254 and 283 nm. The sappanchalcones can be identified by their longwave maxima at 359 nm.

Note: The Chinese Pharmacopeia 2010 describes for Lignum Sappan a brazilin content not less than 0.5 % and for protosapponin B not less than 0.5 %, calculated with reference to the dried drug. [1]

Conclusion

The Lignum Sappan extracts can be easily authenticated by their characteristic Brazilin and Hematoxylin constituents.

References

1. Pharmacopoeia of the people's republic of China, English edition, vol. I. People's Medical Publishing House, Beijing (2010)
2. Zhang, Z.Z.: An illustrated Chinese materia medica in Hong Kong. School of Chinese Medicine/Hong Kong Baptist University, Hong Kong (2004)
3. Porkert, M.: Klinische Chinesische Pharmakologie, Verlag für Medizin Dr. Ewald Fischer, Heidelberg (1978)
4. Flora of China, *Caesalpinia sappan*, FOC 10, 45 (www.eFloras.org)
5. Jia, Y.Y., Li, Y., Song, Y., Zhao, J., Dou, F., Sun, Y., Wen, A.D.: A simple high-performance liquid chromatography method for the determination of brazilin and its application to a pharmacokinetic study in rats. J. Ethnopharmacol. **151**(1), 108–113 (2014)
6. Yodsaoue, O., Cheenpracha, S., Karalai, C., Ponglimanont, C., Tewtrakul, S.: Anti-allergic activity of principles from the roots and heartwood of *Caesalpinia sappan* on antigen-induced β-hexosaminidase release. Phytother. Res. **23**(7), 1028–1031 (2009)
7. Badami, S., Moorkoth, S., Rai, S.R., Kannan, E., Bhojraj, S.: Antioxidant activity of *Caesalpinia sappan* heartwood. Biol. Pharm. Bull. **26**(11), 1534–1537 (2003)
8. Ye, M., Xie, W.D., Lei, F., Meng, Z., Zhao, Y.N., Su, H., Du, L.J.: Brazilein, an important immunosuppresive component from *Caesalpinia sappan* L. Int. Immunopharmacol. **6**(3), 426–432 (2006)
9. Liang, C.H., Chan, L.P., Chou, T.H., Chiang, F.Y., Yen, C.M., Chen, P.J., Ding, H.Y., Lin, R.J.: Brazilein from *Caesalpinia sappan* L. antioxidant inhibits adipocyte differentiation and induces apoptosis through caspase-3 activity and anthelmintic activites against *Hymenolepis nana* and *Anisakis simplex*. Evid. Based Complent. Alternat. Med. Article ID 864892 (2013)
10. Wang, Z., Sun, J.B., Qu, W., Guan, F.G., Li, L.Z., Liang, J.Y.: Caesappin A and B, two novel protosappanins from *Caesalpinia sappan* L. Fitoterapia **92**, 280–284 (2014)
11. Chu, M.J., Wang, Y.Z., Itagaki, K., Ma, H.X., Xin, P., Zhou, X.G., Chen, G.Y., Li, S., Sun, S.Q.: Identification of active compounds from *Caesalpinia sappan* L. extracts suppressing IL-6 production in RAW 264.7 cells by PLS. J. Ethnopharmacol. **148**(1), 37–44 (2013)
12. Hempen, C.-H., Fischer, T.: Leitfaden Chinesische Phytotherapie 2. Auflage Urban & Fischer, Munich (2007)
13. Badami, S., Rai, S.R., Moorkoth, S., Rajan, S., Suresh, B.: Pharmacognostical evaluation of *Caesalpinia sappan* heartwood. Anc. Sci. Life. **32**(2), 100–107 (2003)
14. Tang, W., Eisenbrand, G.: Chinese drugs of plant origin. Springer, Berlin (1992)
15. Tang, W., Eisenbrand, G.: Handbook of Chinese medicinal plants – chemistry, pharmacology, toxicology, vol. 1. WILEY-VCH Verlag GmbH & CoKgaA, Weinheim (2011)
16. Fu, L.C., Huang, X.A., Lai, Z.Y., Hu, Y.J., Liu, H.J., Cai, X.L.: A New 3-Benzylchroman derivative from Sappan Lignum (*Caesalpinia sappan*). Molecules **13**(8), 1923–1930 (2008)
17. Shu, S.H., Han, J.L., Du, G.H., Qin, H.L.: A new flavonoid from heartwood of *Caesalpinia sappan*. Zhongguo Zhong Yao Za Zhi **33**(8), 903–905 (2008)
18. Zhao, H., Bai, H., Wang, Y., Li, W., Koike, K.: A new homoisoflavan from *Caesalpinia sappan*. J. Nat. Med. **62**(3), 325–327 (2008)
19. Zhao, H., Wang, X., Li, W., Koike, K., Bai, H.: A new minor homoisoflavonoid from *Caesalpinia sappan*. Nat. Prod. Res. **28**(2), 102–105 (2014)
20. Oh, S.R., Kim, D.S., Lee, I.S., Jung, K.Y., Lee, J.J., Lee, H.K.: Anticomplementary activity of constituents from the heartwood of *Caesalpinia sappan*. Planta Med. **64**(5), 456–458 (1998)

21. Min, B.S., Cuong, T.D., Hung, T.M., Min, B.K., Shin, B.S., Woo, M.H.: Compounds from the heartwood of *Caesalpinia sappan* and their anti-inflammatory activity. Bioorg. Med. Chem. Lett. **22**(24), 7436–7439 (2012)

22. Jeong, G.S., Lee, D.S., Kwon, T.O., Lee, H.S., An, R.B., Kim, Y.C.: Cytoprotective constituents of the heartwood of *Caesalpinia sappan* on glutamate-induced oxidative damage in HT22 cells. Biol. Pharm. Bull. **32**(5), 945–949 (2009)

23. Hung, T.M., Hai, N.X., Nhan, N.T., Quang, T.T., Quan, T.L., Cuong, T.D., Dang, N.H., Dat, N.T.: Cytotoxic activity of new phenolic compounds from vietnamese *Caesalpinia sappan*. Biosci. Biotechnol. Biochem. **77**(12), 2378–2382 (2013)

24. Jeong, H.J., Kim, Y.M., Kim, J.H., Kim, J.Y., Park, J.Y., Park, S.J., Ryu, Y.B., Lee, W.S.: Homoisoflavonoids from *Caesalpinia sappan* displaying viral neurominidases inhibition. Biol. Pharm. Bull. **35**(5), 786–790 (2012)

25. Niranjan Reddy, V.L., Ravikanth, V., Jansi Lakshmi, V.V.N.S., Suryanarayan Murty, U., Venkateswarlu, Y.: Inhibitory activity of homoisoflavonoids from *Caesalpinia sappan* against *Beauveria bassiana*. Fitoterapia **74**(6), 600–602 (2003)

26. Kim, K.J., Yu, H.H., Jeong, S.I., Cha, J.D., Kim, S.M., You, Y.O.: Inhibitory effects of *Caesalpinia sappan* on growth and invasion of methicillin-resistant *Staphylococcus aureus*. J. Ethnopharmacol. **91**(1), 81–87 (2004)

27. Cuong, T.D., Hung, T.M., Kim, J.C., Kim, E.H., Woo, M.H., Choi, J.S., Lee, J.H., Min, B.S.: Phenolic compounds from *Caesalpinia sappan* heartwood and their anti-inflammatory activity. J. Nat. Prod. **75**(12), 2069–2075 (2012)

28. He, W., Fang, T., Zhang, K., Tu, P.: Vasorelaxation effects of homoisoflavonoids from *Caesalpinia sappan* in rat thoracic aortic rings. Zhongguo Zhong Yao Za Zhi **34**(6), 731–734 (2009)

29. Nguyen, M.T.T., Awale, S., Tezuka, Y., Tran, Q.L., Kadota, S.: Xanthine oxidase inhibitors from the heartwood of vietnames *Caesalpinia sappan*. Chem. Pharm. Bull. **53**(8), 984–988 (2005)

30. Zhao, Y.N., Pan, Y., Tao, J.L., Xing, D.M., Du, L.J.: Study on cardioactive effects of brazilein. Pharmacology **76**(2), 76–83 (2006)

31. Shu, S.H., Deng, A.J., Li, Z.H., Qin, H.L.: Two novel biphenyl dimers from the heartwood of *Caesalpinia sappan*. Fitoterapia **82**(5), 762–766 (2011)

32. Lee, D.Y., Lee, M.K., Kim, G.S., Noh, H.J., Lee, M.H.: Brazilin inhibits growth and induces apoptosis in human glioblastoma cells. Molecules **18**(2), 2449–2457 (2013)

33. Zhang, J., Abdel-Mageed, W.M., Liu, M., Huang, P., He, W., Li, L., Song, F., Dai, H., Liu, X., Liang, J., Zhang, L., Caesanines, A.-D.: New cassane diterpenes with unprecedented N bridge from *Caesalpinia sappan*. Org. Lett. **15**(18), 4726–4729 (2013)

34. Srilakshmi, V.S., Vijayan, P., Raj, P.V., Dhanaraj, S.A., Chandrashekhar, H.R.: Hepatoprotective properties of *Caesalpinia sappan* Linn. heartwood on carbon tetrachloride induced toxicity. Indian J. Exp. Biol. **48**(9), 905–910 (2010)

35. Tong, X.Z., Zhu, H., Shi, Y., Xu, H.T., Wang, B., Zhao, J.H.: An LC/MS/MS method for simultaneous quantitation of two homoisoflavones: Protosappanin B and brazilin with hypoglycemic activity in rat plasma and its application to a comparative pharmacokinetic study in normal and streptozotocin-treated rats. J. Ethnopharmacol. **148**(2), 682–690 (2013)

36. Washiyama, M., Sasaki, Y., Hosokawa, T., Nagumo, S.: Anti-inflammatory constituents of Sappan Lignum. Biol. Pharm. Bull. **32**(5), 941–944 (2009)

37. Xu, P., Guan, S., Feng, R., Tang, R., Guo, D.: Separation of four homoisoflavonoids from *Caesalpinia sappan* by high-speed countercurrent chromatography. Phytochem. Anal. **23**(3), 228–231 (2012)

38. Wu, J., Hou, J.B., Zhang, M.M., Zou, Y.P., Yu, B.: Protosappanin A, an Immunosuppresive constituent from a Chinese herb, prolongs graft survival and attenuates acute rejection in rat heart allografts. Transplant. Proc. **40**(10), 3719–3722 (2008)

39. Mitani, K., Takano, F., Kawabata, T., Allam, A.E., Ota, M., Takahashi, T., Yahagi, N., Sakurada, C., Fushiya, S., Ohta, T.: Suppression of melanin synthesis by the phenolic constituents of Sappanwood (*Caesalpinia sappan*). Planta Med. **79**(1), 37–44 (2013)

40. Sasaki, Y., Suzuki, M., Matsumoto, T., Hosokawa, T., Kobayashi, T., Kamata, K., Nagumo, S.: Vasorelaxant activity of Sappan Lignum constituents and extracts on rat aorta and mesenteric artery. Biol. Pharm. Bull. **33**(9), 1555–1560 (2010)

41. Baek, N.I., Jeon, S.G., Ahn, E.M., Hahn, J.T., Bahn, J.H., Jang, J.S., Cho, S.W., Park, J.K., Choi, S.Y.: Anticonvulsant compounds from the wood of *Caesalpinia sappan* L. Arch. Pharm. Res. **23**(4), 344–348 (2000)

42. Tao, L.Y., Li, J.Y., Zhang, J.Y.: Brazilein, a compound isolated from *Caesalpinia sappan* Linn., induced growth inhibition in breast cancer cells via involvement of GSK-3β/β-Catenin/cyclin D1 pathway. Chem. Biol. Interact. **206**(1), 1–5 (2013)

Radix Gentianae macrophyllae – *Qinjiao*

Pharmacopoeia: [1] Pharmacopoeia of the People's Republic of China, English Edition Vol. I, 2010

Official drug: [1, 2] Largeleaf Gentian Root is the dried root of *Gentiana macrophylla* Pall., *Gentiana straminea* Maxim., *Gentiana crassicaulis* Duthie ex Burk. or *Gentiana dahurica* Fisch. (Fam. Gentianaceae).

According to the description the former three species are known as "Qinjiao" and "Mahuajiao", respectively. The latter is known as "Xiaoqinjiao".

The drug is collected in spring and autumn, removed from soil; "Qinjiao" and "Mahuajiao" are softened in the sun, either piled up until the drug becomes reddish-yellow or greyish-yellow on the surface, spread out and dried in the sun; or dried in the sun directly after collecting; as for "Xiaoqinjiao", the black bark is rubbed off while fresh and dried in the sun.

Origin: [3–7] North of China (Gansu, Henan, Shanxi, Hebei, Shaanxi, Heilongjiang, Liaoning, Nei Mongol, Ningxia, Shandong). It is also grown in the continental Yunnan, Szuchuan. Siberia, Mongolia and Kazakhstan.

Description of the drug: [1]

Qinjiao

Subcylindrical, the upper part thick and the lower part thin, twisted, 10–30 cm long, 1–3 cm in diameter. Externally yellowish-brown or greyish-yellow, with longitudinal or twisted wrinkles. Remains of stem bases and fibrous pericladium occurring at the apex. Texture hard and fragile, easily broken, fracture slight oily, bark yellow or brownish-yellow, wood yellow. Odour, characteristic; taste, bitter and slightly astringent.

Mahuajiao

Subconical, frequently expanded by some small roots gathered, up to 7 cm in diameter. Externally dark brown, rough, with fissures, showing reticulated pits. Texture lax and fragile, easily broken, fracture frequently rotten-wood-shaped.

Xiaoqinjiao

Subconical or subcylindrical, 8–15 cm long, 0.2–1 cm in diameter. Externally brownish-yellow. Main root frequently 1, having fibrous pericladia on the remains of stem bases, often branched at the lower part. Fracture yellowish-white.

Medicinal use: [7–14] Treatment of diverse diseases, such as diabetes, hypertension, apoplexy, paralysis, osteoarthritis and rheumatism.

H. Wagner et al. (eds.), *Chromatographic Fingerprint Analysis of Herbal Medicines*, Vol. 4,
DOI 10.1007/978-3-319-32328-2_14,

Effects and indications of Radix Gentianae macrophyllae according to Traditional Chinese Medicine [1–5, 8–11, 13, 15, 16]

Taste:	Pungent and bitter
Temperature:	Neutral, cold
Channels entered:	*Orbis stomachi, oo. hepaticus et felleus*
Effects (functions):	To dispel wind-dampness, clear dampness-heat, relieve impediment pain, relieve deficiency heat
Symptoms and indications:	Painful impediment caused by wind-dampness, hemiplegia caused by wind-stroke, hypertonicity of the sinews and vessels, sore pain in joint and bone, dampness-heat jaundice, bone-steam ing and tidal fever, infantile malnutrition with accumulation and fever. Treatment of hepatitis, constipation, stomachic and choleretic ailments

Main constituents: - **Secoiridoid glycosides** [1, 8–11, 13, 14, 16–21]

Gentiopicroside (= gentiopicrin), 6'-O-β-D-glucosyl-gentiopicroside, 6'-O-β-D-glucopyranosylgentiopicroside, loganic acid, sweroside, swertiamarin, swertiapunimarin (=6'-O-β-D-glucopyranosylsweroside), trifloroside, rindoside

- **Triterpenoids and sterols** [1, 8, 13, 21], Roburic acid, α-amyrin, oleanolic acid, β-sitosterol, β-sitosterol-3-O-gentiobioside, daucosterol, stigmasterol

Minor constituents: - alkaloids (gentianine, gentianidin), essential oil [4, 12, 15]

- chromenes (2-methoxyanofinic acid, macrophylloside C,D) [8]

- disaccharide (gentiobiose) [8]

- benzoic acid derivative (methyl 2-hydroxy-3-(1-β-D-glucopyranosyl)oxybenzoate) [8]

- flavonoids/flavone derivatives (kurarinone, kushenol I, vitexin, isovitexin, macrophylloside A+B, isoorientin, isoorientin-4'-O-glucoside) [8, 10, 13]

- 5-carboxyl-3,4-dihydrogen-1H-2-benzopyran-1-one, erythrocentauric acid [21]

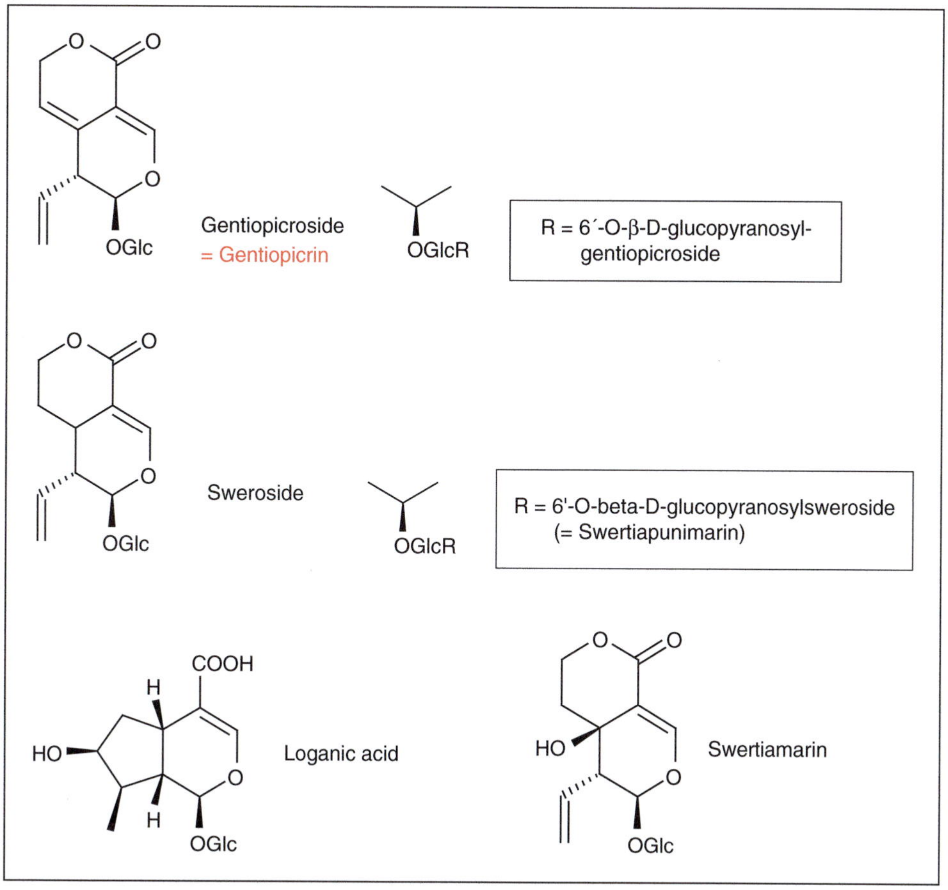

Fig. 1 Formulae of the main compounds of Radix Gentianae macrophyllae [18]

Pharmacology:

- antidiabetic [7, 9, 10]
- anti-inflammatory [9, 11, 12, 15, 16]
- antihepatotoxic/hepatoprotective [9, 11, 14, 16]
- antihistamine activity [9, 16]
- antioxidant [11]
- antiulcerogenic activity [12]
- anti-analgesic [12]
- anti-hypertensive [15]
- sedative [15]
- antifungal [9, 11, 16]

TLC Fingerprint Analysis

	Drug samples	**Origin**
1	Radix Gentianae macrophyllae / *Gentiana sp.*	Province Neimeng, China
2	Radix Gentianae macrophyllae / *Gentiana sp.*	Province Yunnan, China
3	Radix Gentianae macrophyllae / *Gentiana sp.*	Province Qinghai, China
4	Radix Gentianae macrophyllae / *Gentiana sp.*	Sample of commercial drug obtained from China Medica (origin: Abazhou, Sichuan)
5	Radix Gentianae macrophyllae / *Gentiana sp.*	Sample of commercial drug obtained from HerbaSinica (origin: Gui zhou)
6	Radix Gentianae macrophyllae / *Gentiana sp.*	Sample of commercial drug obtained from TCM-Clinic Bad Kötzting (imported 2000)
7	Radix Gentianae macrophyllae / *Gentiana sp.*	Sample of commercial drug obtained from TCM-Clinic Bad Kötzting (imported 2004)

	Reference compounds of Fig. 2a/b	**R*f***
T1	Oleanolic acid	0.98
T2	Gentiopicroside	0.55
T3	Loganic acid	0.24
T4	Sweroside	0.51

1. Extraction: 1.0 g powdered drug is extracted with 10 ml ethanol under reflux for 30 min. The extract is filtered, evaporated to dryness and the residue is dissolved in 1 ml ethanol.

2. Reference compounds: 1.0 mg is dissolved in 1.0 ml methanol

3. Separation parameters:

Plate: HPTLC Silica gel 60 F_{254}, Merck

Applied amounts: Radix Gentianae macrophyllae extracts: each 10 µl; Reference compounds: each 10 µl

Solvent system: Ethyl acetate + methanol + water (10 + 2 + 1)

Detection: 10% ethanolic sulphuric acid

The plate is sprayed with 8 ml reagent and heated at 110 °C for 10 min. The plate is evaluated in VIS (Fig. 2a) and under UV 366 nm (Fig. 2b).

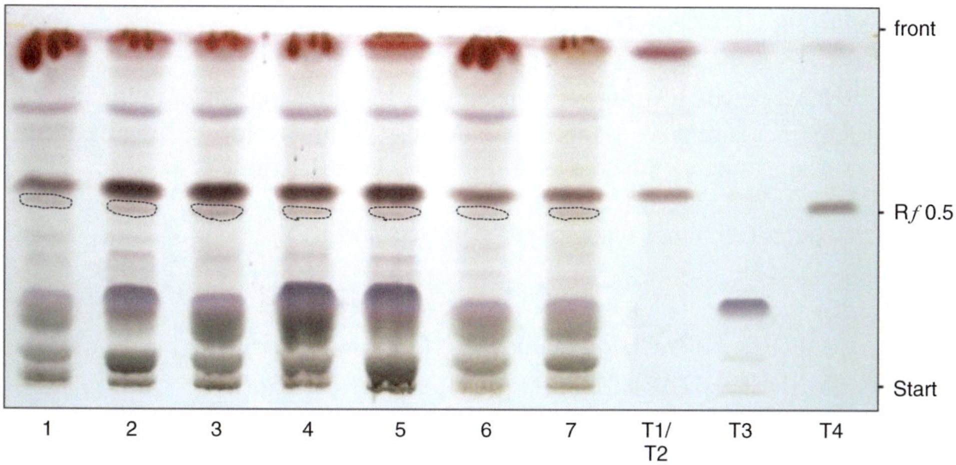

Fig. 2a Thin layer chromatogram of the ethanol extracts of Radix Gentianae macrophyllae, sprayed with 10% ethanolic sulphuric acid (VIS)

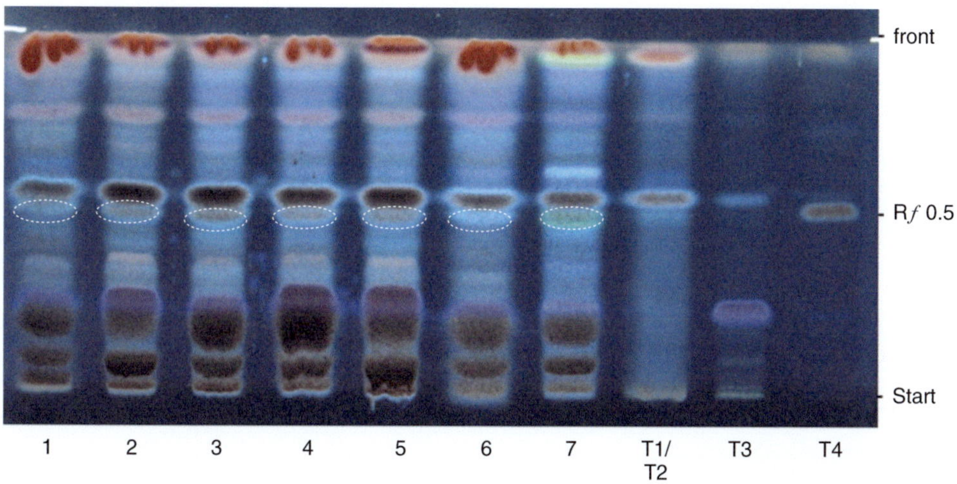

Fig. 2b Thin layer chromatogram of the ethanol extracts of Radix Gentianae macrophyllae, sprayed with 10% ethanolic sulphuric acid (UV 366 nm)

Description of Fig. 2a:

- Red-brown zones on the solvent front at R*f*=0.98; oleanolic acid (**T1**) with sterols.
- A weak purple zone at R*f*=0.78; not identified, probably, triterpene- or sterol-monoglycosides.
- A strong brown zone at R*f*=0.55 (gentiopicroside=**T2**) and directly below at R*f*=0.51 sweroside (**T4**).
- In the deep R*f*-range are detectable at R*f*=0.24 loganic acid (**T3**) and between R*f*=0.25 and the solvent start several strong brown zones which can be assigned to the corresponding secoiridoid diglyosides.

Description of Fig. 2b:

This TLC recorded at UV 366 nm provides a very similar TLC of compounds as those of Fig. 2a but with a dark blue fluorescent background.

HPLC-Fingerprint Analysis

1. Extraction: 1.0 g powdered drug is extracted with 10 ml ethanol under reflux for 30 min. The extract is filtered, evaporated to dryness and the residue is dissolved in 1 ml ethanol. The extract is filtered over Chromafil®, Type 0.20 µm.

2. Injection volume: Radix Gentianae macrophyllae extracts: each 15 µl

3. HPLC parameter:

Apparatus: MERCK HITACHI D-6000 A Interface
MERCK HITACHI L-4500 A Diode Array Detector
MERCK HITACHI AS-2000 Autosampler
MERCK HITACHI L-6200 A Intelligent Pump

Separation column: LiChroCART® 250-4 LiChrospher® 100 RP-18 (5 µm), Merck

Precolumn: LiChroCART® 4-4 LiChrospher® 100 RP-18 (5 µm), Merck

Solvent system: A: 0.1 % acetic acid/water (Millipore Ultra Clear UV plus® filtered)
B: acetonitrile (VWR)

Gradient: 5–50 % B in 20 min,
50–100 % B in 5 min,
100 % B for 10 min
total runtime: 35 min

Flow: 1.0 ml/min

Detection: 254 nm

Retention times of the main peaks

Peak	Rt (min)	Compound
1	9.3	Loganic acid
2/3	10.4/10.6	Not identified
4a	11.7	Gentiopicroside
4b	11.7	Sweroside
5	15.8	Not identified
6	24.2	Not identified

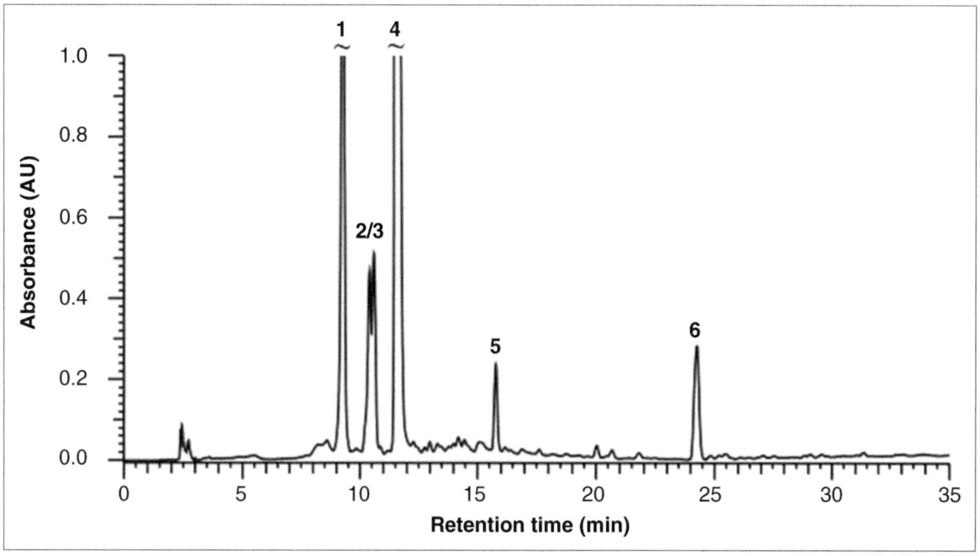

Fig. 3a HPLC-fingerprint analysis of the ethanol extract of Radix Gentianae macrophyllae, sample 2

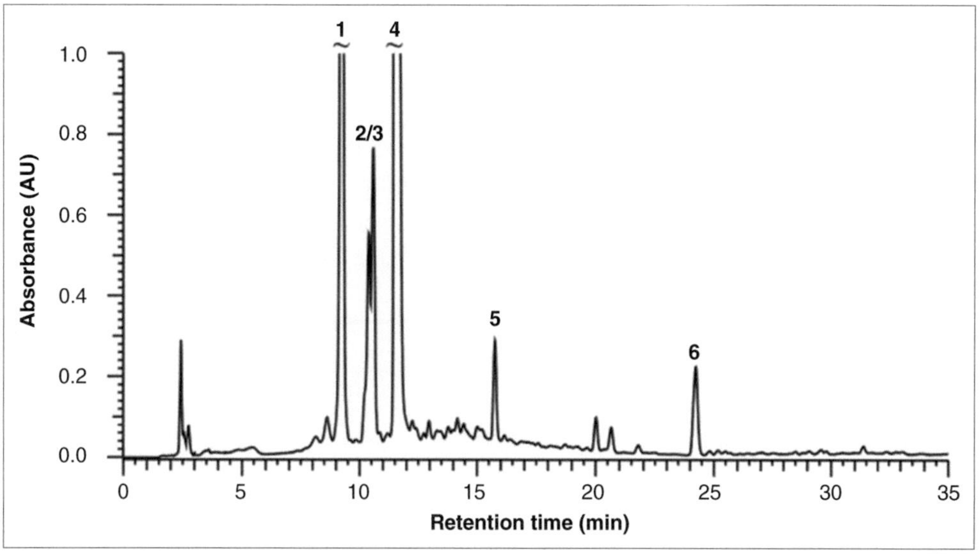

Fig. 3b HPLC-fingerprint analysis of the ethanol extract of Radix Gentianae macrophyllae, sample 5

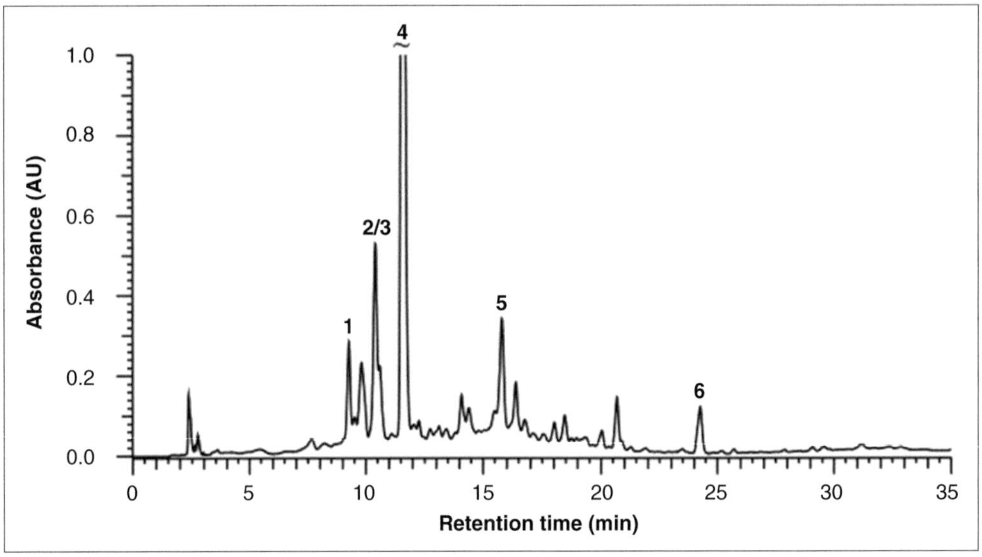

Fig. 3c HPLC-fingerprint analysis of the ethanol extract of Radix Gentianae macrophyllae, sample 6

4. Description of the HPLC-Figures:

The HPLC-graphs of Radix Gentianae macrophyllae extracts of sample 2 and 5 show the same peak profile with loganic acid (**1**) at Rt=9.3, unidentified peaks **2/3** at Rt=10.4 and 10.6 which may be assigned to the secoiridoid 6'-O-glucosides of gentiopicroside and sweroside or swertiamarin. Peak **4** at Rt=11.7 consists of a mixture of gentiopicroside (**a**) and sweroside (**b**). The peaks **5** and **6** with different UV-spectra could be not assigned to any of the other constituents reported in the literature.

In extract sample 6 the peak magnitude of loganic acid (**1**) and that of the compounds of **2/3** correspond with the lower concentrations of the diglycosides as shown in the TLC in Fig. 2a.

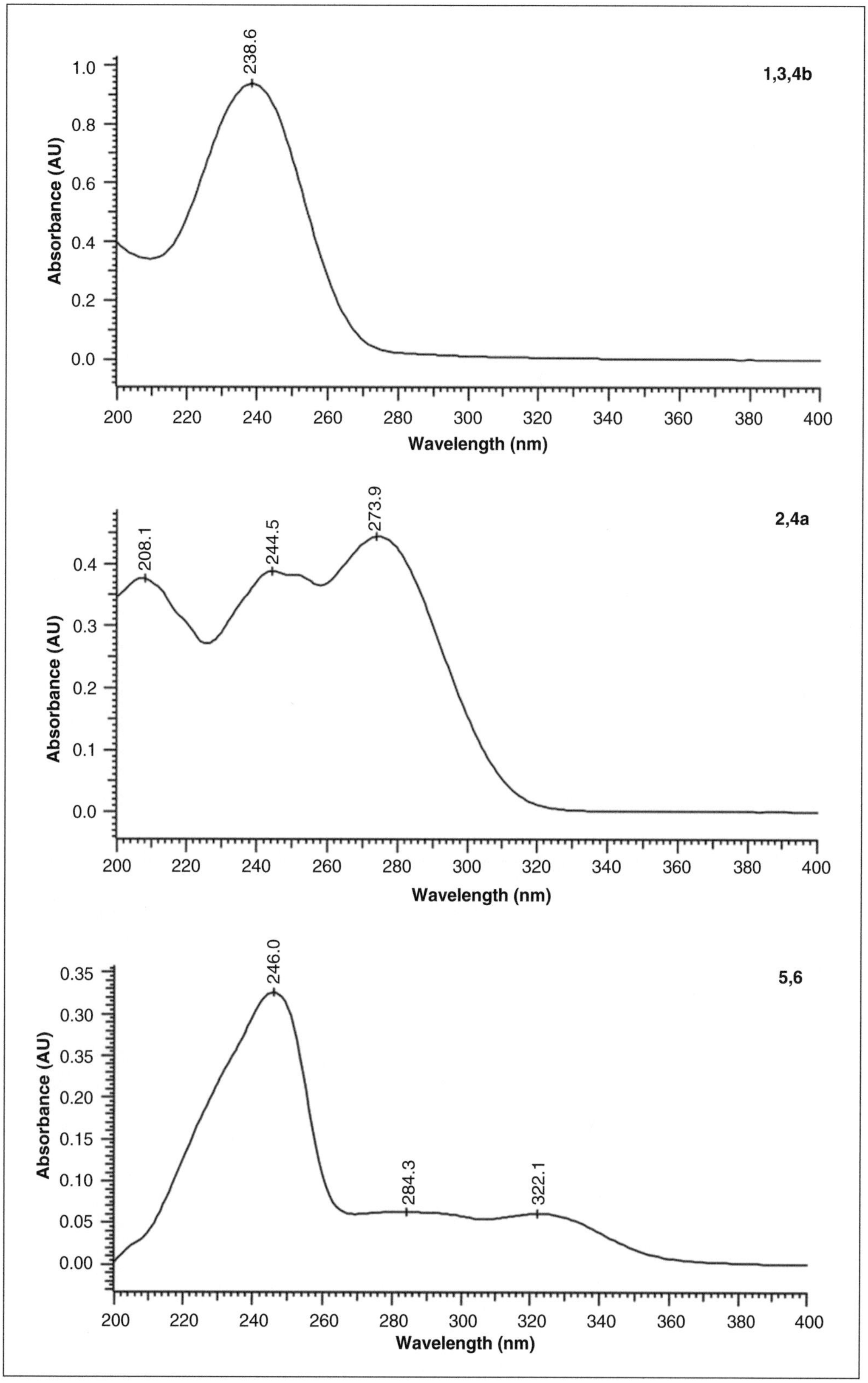

Fig. 4 On line UV-spectra of the main peaks of Radix Gentianae macrophyllae extracts

Note: According to the Chinese Pharmacopeia 2010 Radix Gentianae macrophyllae contains not less than 2.5% of the total amount of gentiopicrin and loganin, calculated with reference to the dried drug. [1]

Conclusion

The bitter tasting herbal drug Radix Gentianae macrophyllae can be definitely authenticated by the described TLC and HPLC.

References

1. Pharmacopoeia of the people's republic of China, English edition, vol. I. People's Medical Publishing House, Beijing (2010)
2. Geng, J., Huang, W., Ren, T., Ma, X.: Materia Medica der Chinesischen Arzneimitteltherapie (Bd. 2: Praxis der Chinesischen Arzneimitteltherapie), Verlag für Ganzheitliche Medizin Dr. Erich Wühr GmbH, Kötzting/Bayer. Wald (1993)
3. Porkert, M.: Klinische Chinesische Pharmakologie, Verlag für Medizin Dr. Ewald Fischer, Heidelberg (1978)
4. Paulus, E., Ding, Y.-H.: Handbuch der traditionellen chinesischen Heilpflanzen, Karl F. Haug Verlag, Heidelberg (1987)
5. Zhang, Z.Z.: An illustrated Chinese materia medica in Hong Kong. School of Chinese Medicine/Hong Kong Baptist University, Hong Kong (2004)
6. Flora of China, *Gentiana macrophylla*, FOC 16, 37 (www.eFloras.org)
7. Yu, F., Yu, F., Li, R., Wang, R.: Inhibitory effects of the *Gentiana macrophylla* (Gentianaceae) extract on rheumatoid arthritis of rats. J. Ethnopharmacol. **95**(1), 77–81 (2004)
8. Tan, R.X., Wolfender, J.L., Zhang, L.X., Ma, W.G., Fuzzati, N., Marston, A., Hostettmann, K.: Acyl Secoiridoids and antifungal constituents from *Gentiana macrophylla*. Phytochemistry **42**(5), 1305–1313 (1996)
9. Hua, W., Zhang, P., He, Y., Cui, L., Kong, W., Wang, Z.: An insight into the genes involved in secoiridoid biosynthesis in *Gentiana macrophylla* by RNA-seq. Mol. Biol. Rep. **41**(7), 4817–4825 (2014)
10. Jia, N., Li, Y., Wu, Y., Xi, M., Hur, G., Zhang, X., Cui, J., Sun, W., Wen, A.: Comparison of the anti-inflammatory and analgesic effects of *Gentiana macrophylla* Pall. and *Gentiana straminea* Maxim., and identification of their active constituents. J. Ethnopharmacol. **144**(3), 638–645 (2012)
11. Zhou, D., Hou, Q., Si, Q., Liu, J., Yang, H.: Concentrations of the active constituents of the Tibetan folk medicine qinjiao (Gentiana sect. Cruciata) within and between taxonomic species across the Qinghai-Tibetan plateau. Chem. Biodivers. **7**(8), 2088–2094 (2010)
12. Kwak, W.J., Kim, J.H., Ryu, K.H., Cho, Y.B., Jeon, S.D., Moon, C.K.: Effects of gentianine on the production of pro-inflammatory cytokines in male sprague-dawley rats treated with lipopolysaccharide (LPS). Biol. Pharm. Bull. **28**(4), 750–753 (2005)
13. Su, Q., Shang, P., Zhang, Y., Jia, N., He, J., Zhao, W., Sun, W.: HPLC fingerprint and LC-TOF-MS analysis on extract from roots of *Gentiana macrophylla*. Chinese Herbal Med. **4**(3), 245–251 (2012)
14. Wang, C.H., Cheng, X.M., He, Y.Q., White, K.N., Bligh, S.W., Branford-White, C.J., Wang, Z.T.: Pharmacokinetic behavior of gentiopicroside from decoction of radix Gentianae, *Gentiana macrophylla* after oral administration in rats: a Pharmacokinetic comparison with Gentiopicroside after oral and intravenous administration alone. Arch. Pharm. Res. **30**(9), 1149–1154 (2007)
15. Hempen, C.-H., Fischer, T.: A materia medica for Chinese medicine, english edition. Elsevier GmbH, Urban & Fischer, Munich (2009)
16. Yin, H., Zhao, Q., Sun, F.M., An, T.: Gentiopicrin-producing endophytic fungus isolated from *Gentiana macrophylla*. Phytomedicine **16**(8), 793–797 (2009)
17. Tang, W., Eisenbrand, G.: Handbook of Chinese medicinal plants – chemistry, pharmacology, toxicology, vol. 1. WILEY-VCH Verlag GmbH & Co. KgaA, Weinheim (2011)
18. Cao, X.Y., Wang, Z.Z.: Simultaneous determination of four iridoid and secoiridoid glycosides and comparative analysis of radix gentianae macrophyllae and their related substitutes by HPLC. Phytochem. Anal. **21**(4), 348–354 (2010)
19. Liu, F.F., Wang, Y.M., Zhu, H.T., Wang, D., Yang, C.R., Xu, M., Zhang, Y.J.: Comparative study on "Long-Dan", "Qin-Jiao" and their adulterants by HPLC analysis. Nat. Prod. Bioprospect. **4**(5), 297–308 (2014)
20. Wu, J.R., Wu, L.H., Zhao, Z.L., Wang, Z.T.: Simultaneous determination of five iridoids in gentianae macrophyllae radix and their local variety by HPLC. Zhongguo Zhong Yao Za Zhi **39**(4), 715–720 (2014)
21. Chen, Q.L., Shi, Z.Y., Tu, G.Z., Sun, W.J.: Studies on the chemical constituents in root of *Gentiana macrophylla* from Shaanxi. Zhongguo Zhong Yao Za Zhi **30**(19), 1519–11522 (2005)

Radix Trichosanthis – *Tianhuafen*

Pharmacopoeia: [1]	Pharmacopoeia of the People's Republic of China, English Edition Vol. I, 2010
Official drug: [1]	Snakegourd root is the dried root of *Trichosanthes kirilowii* Maxim. or *Trichosanthes rosthornii* Harms (Fam. Cucurbitaceae) The drug is collected in autumn and winter, washed clean, peeled, cut into segment or sliced longitudinally, and dried.
Origin: [2]	Chinese Provinces: Jiangsu, Zhejiang, Shandong and Anhui, Henan
Description of the drug: [1]	Irregular cylindrical, spindle-shaped, or in pieces, 8–16 cm long, 1.5–5.5 cm diameter. Externally yellowish-white or pale brownish-yellow, with longitudinal wrinkles, rootlet scars and slightly concave transverse lenticel-like streak. Some remained with yellowish-brown outer bark. Texture compact, fracture white or yellowish, starchy, wood yellow somewhat, radially arranged in section and striated in longitudinally cut section. Odour, slight; taste, slightly bitter.
Pretreatment of the raw drug: [1]	The drug is collected in autumn and winter, washed clean, peeled, cut into segment or sliced longitudinally, and dried.
Medicinal use: [3]	Pulmonary and cardiovascular diseases, Angina pectoris, treatment of diabetes mellitus.

Effects and indications of Radix Trichosanthis according to Traditional Chinese Medicine [1–4]

Taste:	Bitter and sweet
Temperature:	Cold
Channels entered:	*Orbis pulmonalis, o. stomachi, o. intestini crassi*
Effects (functions):	To clear heat and purge fire, engender fluid to quench thirst, disperse swelling and expel pus
Symptoms and indications:	Vexation and thirst caused by heat disease, lung heat and dryness cough, interior heat wasting-thirst, sore and ulcer, swelling and toxin

H. Wagner et al. (eds.), *Chromatographic Fingerprint Analysis of Herbal Medicines*, Vol. 4,
DOI 10.1007/978-3-319-32328-2_15, © Springer International Publishing Switzerland 2016

Main constituents: [4, 5, 13, 14, 16]

- Triterpenoids	Cucurbitacin B, isocucurbitacin B, cucurbitacin D, isocucurbitacin D, 3-epi-isocucurbitacin B, dihydrocucurbitacin B, dihydroisocucurbitacin E, byronolic acid
- Proteins	Trichosanthin, karasurin, karasurin-A, karasurin-B and C
- Amino acids	L-citrullin, L-arginin, tryptophan
- Polysaccharides	Various Polysaccharides composed of glucose, galactose, fructose, mannose, xylose or arabinose and galactose

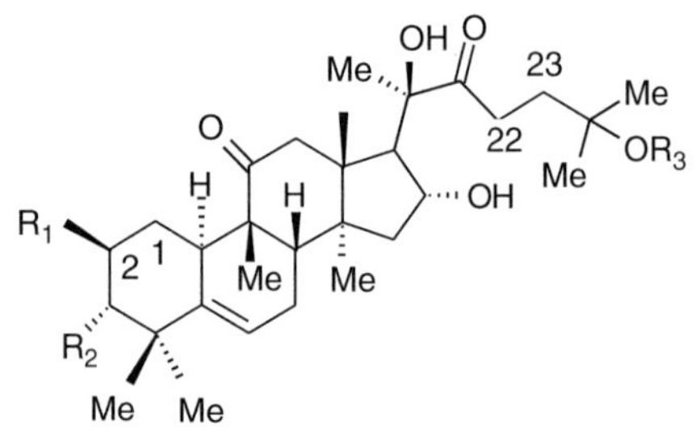

	R_1	R_2	R_3	$\Delta^{1/2}$	$\Delta^{22/23}$
Cucurbitacin B	OH	=O	$COCH_3$	-	+
Isocucurbitacin B	=O	OH	$COCH_3$	-	+
Dihydrocucurbitacin B	OH	=O	$COCH_3$	-	-
Dihydroisocucurbitacin B	=O	OH	$COCH_3$	-	-
Cucurbitacin D	H	=O	H	-	+
Isocucurbitacin D	=O	OH	H	-	+
Dihydrocucurbitacin E	OH	=O	$COCH_3$	+	-

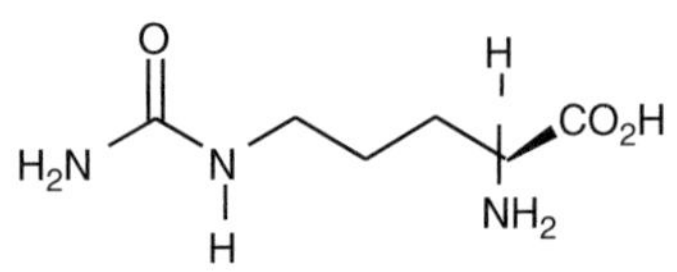

Fig. 1 Formulae of the main constituents of Radix Trichosanthis [5, 7]

Reported pharmacological activities:

- antioxidant (*in vitro* and *in vivo*) [6, 9]
- anti-inflammatory [11]
- immunomodulatory [8]
- effective against trophoblastic neoplasmen [10]
- diabetes prevention [10, 12]
- effective in abortion induction [3, 10]
- anti-hepatitis B viral effect [8]

TLC Fingerprint Analysis

	Drug samples	Origin
1	Radix Trichosanthis / *Trichosanthes kirilowii*	Province Hebei (China)
2	Radix Trichosanthis / *Trichosanthes kirilowii*	Province Henan (China)
3	Radix Trichosanthis / *Trichosanthes kirilowii*	Province Shanaxi (China)
4	Radix Trichosanthis / *Trichosanthes rosthornii*	Province (unknown)
5	Radix Trichosanthis / *Trichosanthes kirilowii*	Province (unknown)
6	Radix Trichosanthis / *Trichosanthes kirilowii* Maxim	Sample of commercial drug, obtained from TCM-Clinic Bad Kötzting (Charge: K20.12.2000)
7	Radix Trichosanthis / *Trichosanthes kirilowii* Maxim	Sample of commercial drug, obtained from TCM-Clinic Bad Kötzting (Charge: K08.10.2004)
8	Radix Trichosanthis / *Trichosanthes kirilowii* Maxim	Sample of commercial drug, obtained from China Medica
9	Radix Trichosanthis / *Trichosanthes kirilowii* Maxim	Sample of commercial drug, obtained from Herba Sinica (origin: Hebei)

1. **TLC-fingerprint analysis of Amino acids:**

Reference compounds of Fig. 2		R*f*
T 1	L-Citrullin	0.36
T 2	L-Argenin	0.22

1. Extraction: 2 g powdered drug with 10 ml 80 % ethanol are ultrasonicated for 30 min. and filtered. The filtrate is used as the test solution.
2. Reference compounds: Each 1 mg is dissolved in 0.5 ml ethanol
3. Separation parameters:

   Plate: HPTLC Silica gel 60 F_{254}, Merck

   Applied amounts: Radix Trichosanthis extracts: 10 µl each

   Reference compounds: 10 µl each

   Solvent system: n-Butanol + absolute ethanol + glacial acetic acid + water (12 + 4 + 4 + 2)

   Detection: Ninhydrin-Solution

   Dissolve 2 g Ninhydrin in 100 ml ethanol

   The plate is sprayed with the solution and heated at 100 °C for 5–10 min. The evaluation is carried out under VIS.
4. Description of TLC-fingerprint analysis Fig. 2:

   All Radix Trichosanthis samples are characterized by a deep red-violet zone of L-citrullin (**T1**) at R*f*=0.39. Another weak orange zone at R*f*=0.24 can be assigned to L-arginin (**T2**). Further not identified amino acids in lower concentration appear in the upper R*f* – range between R*f*=0.50 and 0.72.

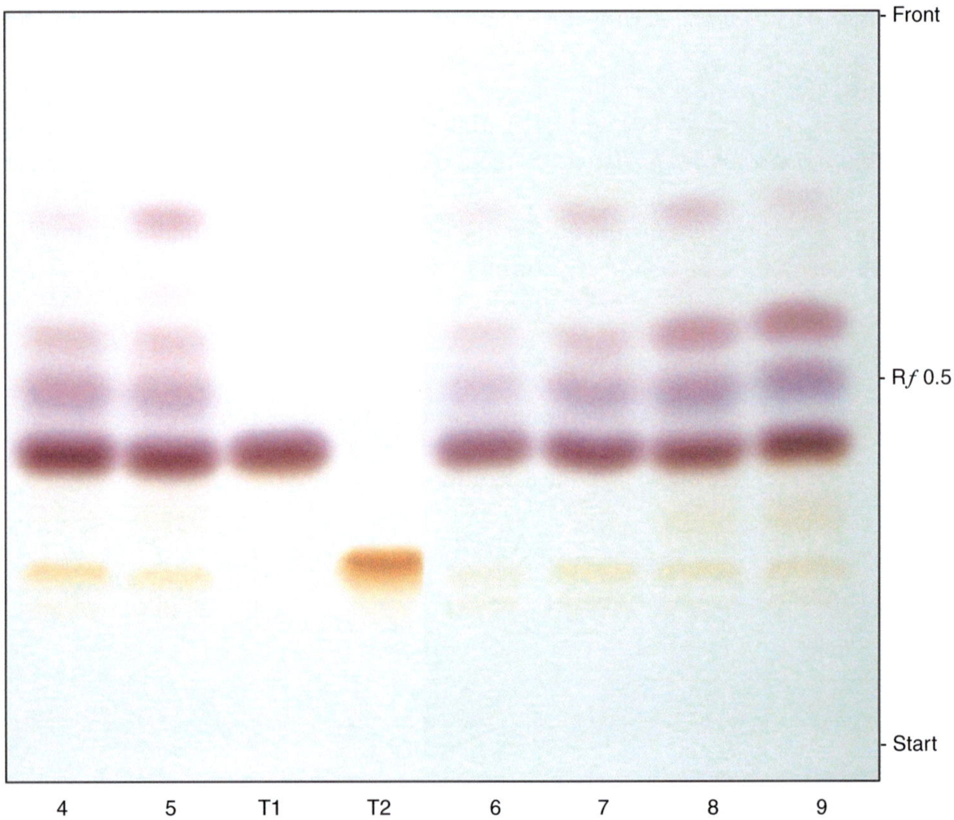

Fig. 2 Thin layer chromatogram of the 80 % ethanol extracts of Radix Trichosanthis sprayed with ninhydrin reagent (VIS)

2. **TLC-fingerprint analysis of Triterpenes:** [7]

Reference compounds of Fig. 3		**Rf of Fig. 3a**	**Rf of Fig. 3b, c**
T3	Cucurbitacin B	0.84	0.68
T4	Cucurbitacin D	0.72	0.32

1. Extraction: 2 g powdered drug are extracted with 15 ml ethanol under reflux for 15 min. The extract is cooled, filtered and evaporated to dryness. The residue is dissolved in 1 ml ethanol and filtered over Chromafil® filtration unit, type 0–20 µm/25 mm.
2. Reference compounds: Each 0.5 mg is dissolved in 0.5 ml ethanol
3. Separation parameters:

   Plate: HPTLC Silica gel 60 F_{254}, Merck

   Applied amounts: Radix Trichosanthis extracts: 30 µl each

   Reference compounds: 10 µl each

   Solvent system: a) chloroform + ethyl acetate + methanol (8 + 6 + 2)

   b) ethyl acetate + toluene (7.5 + 2.5)

   Detection: Vanillin – Phosphoric acid reagent:

   1 g vanillin is dissolved in a small amount of ethanol and filled up to 100 ml with 50 % aqueous phosphoric acid.

   The plate is sprayed with this solution, heated for 10 min at 105 °C and evaluated in VIS and at 366 nm.
4. Description of the Fig. 3a, 3b and 3c:

The TLC in Fig. 3a with solvent (a) shows in VIS in the R*f*–range between R*f*=0.65 and R*f*=0.85, 5–6 violet/light blue zones. In the same R*f*– range are detectable also the cucurbitacins **T3** and **T4** with distinct pink colour (very low concentration). The colour differences of the other zones may be due to overlapping of other triterpenoids.

In Fig. 3b the cucurbitacins B and D of the extracts 3 and 8 apear analoge to the reference compounds **T1** and **T2** in violet pink colours.

In the TLC of Fig. 3c with solvent (b) detected at UV 366 nm appear the cucurbitacins of the extracts and the reference with distinct characteristic carmin red fluorescent colour as reported in the literature [16].

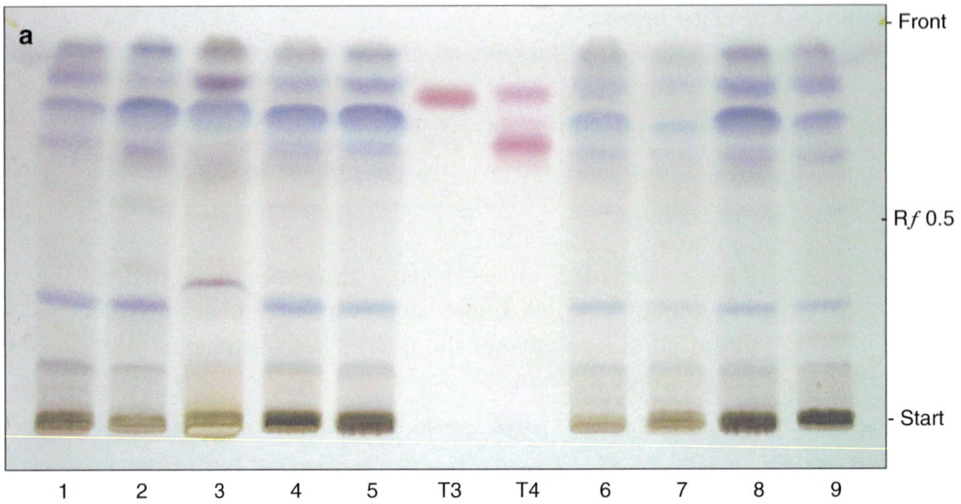

Fig. 3a The TLC of the ethanol extracts of Radix Trichosanthis sprayed with Vanillin-Phosphoric acid reagent (VIS)

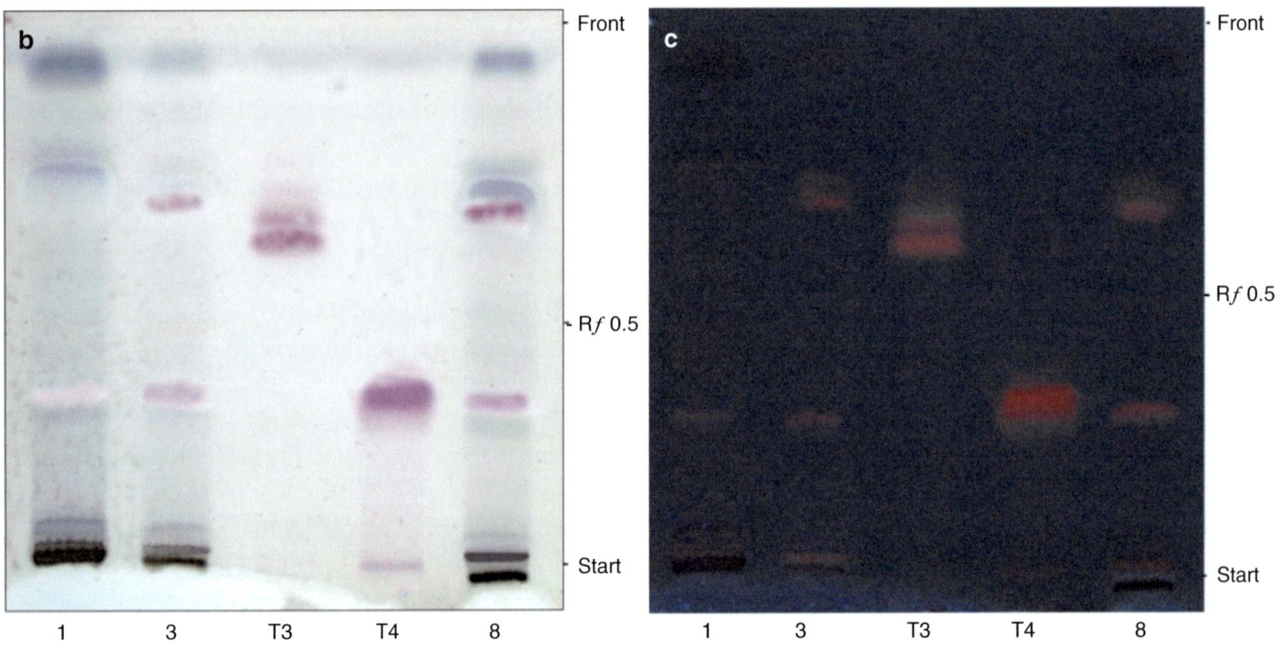

Fig. 3b, c Thin layer chromatogram of the ethanol extracts of Radix Trichosanthis sprayed with Vanillin-Phosphoric acid reagent and detected (**b**) under VIS (**c**) at UV 366 nm VIS

HPLC-Fingerprint Analysis

1. Sample preparation: 2 g powdered drug are extracted with 15 ml ethanol under reflux for 15 min. The extract is cooled, filtered and evaporated to dryness. The residue is dissolved in 1 ml ethanol and filtered over Chromafil® filtration unit, type 0–20 μm/25 mm
2. Injection volume: Trichosanthis Radix extracts: 50 μl each

   Reference compounds: 20 μl each
3. HPLC parameter:

   Apparatus: MERCK HITACHI D-6000 A Interface

   MERCK HITACHI L-4500 A Diode Array Detector

   MERCK HITACHI AS-2000 Autosampler

   MERCK HITACHI L-6200 A Intelligent Pump

   Separation column: LiChroCART® 250 -4 LiChrospher® 100 RP-18 (5 μm), Merck

   Precolumn: LiChroCART® 4-4 LiChrospher® 100 RP-18 (5 μm), Merck

   Solvent System: A: 0.001 % aq. H_3PO_4: (Millipore Ultra Clear UV plus® filtriert)

   B: Acetonitrile (VWR)

   Gradient: 0–40 % B in 40 min,

   40–100 % B in 20 min,

   100 % B for 12 min

   Total run time: 72 min

   Flow: 1 ml/min

   Detection: 230 nm

Retention times of the main peaks recorded at 230 nm[a]

Peak	Rt (min)	Compound
1	3.19	Amino acid
2	6.45	Amino acid
3	11.95	Tryptophan
4	14.34	Aromatic amino acid
5	20.39	Aromatic amino acid?
6	37.14	Cucurbitacin D
7	41.88	Cucurbitacin B
8	45–52	Cucurbitacins mixture
9	52.5, 53.11	Isocucurbitacin B + mixture
10	58.99	Dihydrocucurbitacin (E or B)

[a]L-citrullin is detectable only with other chromatographic condition [15]

4. Description of the HPLC-Fig. 4a, b and c:

   The peaks profile of Radix Trichosanthis can be divided into two Rt –ranges between Rt 2.0 and Rt 22.0 and between Rt=35.0 and 60.0. In the first appear the various aromatic amino acids with peaks No. **1, 2, 3, 4** and **5** inclusive tryptophan at Rt=11.95. The second contains cucurbitacins D and B at Rt=37.1, 41,9 respectively, the Peaks **8** and **9** can be assigned to isocucurbitacin B + mixture according to their UV-spectra with max at **232** and **235** nm. Peak **10** might be assignable to dihyrocucurbitacins (B or E) [17].

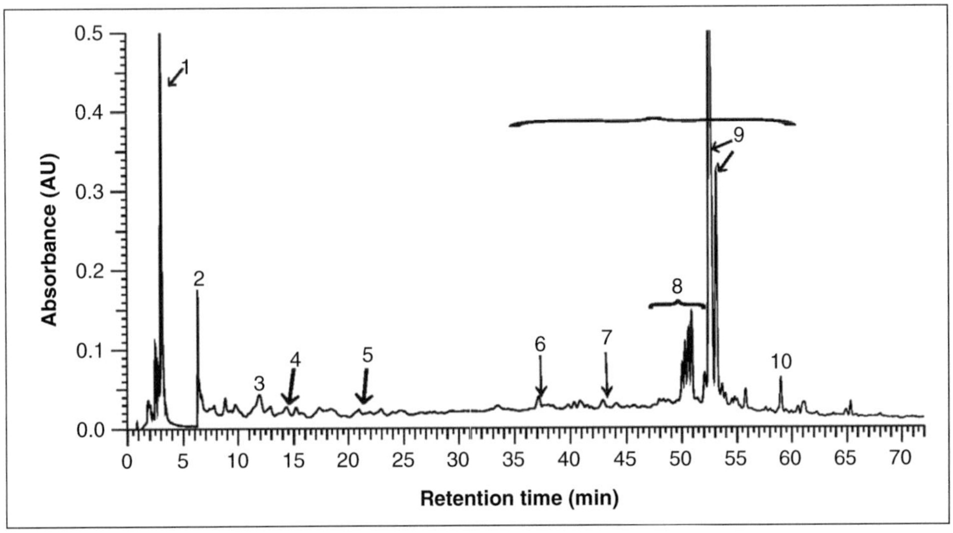

Fig. 4a HPLC-fingerprint analysis of the ethanol extract of Radix Trichosantis (sample 4)

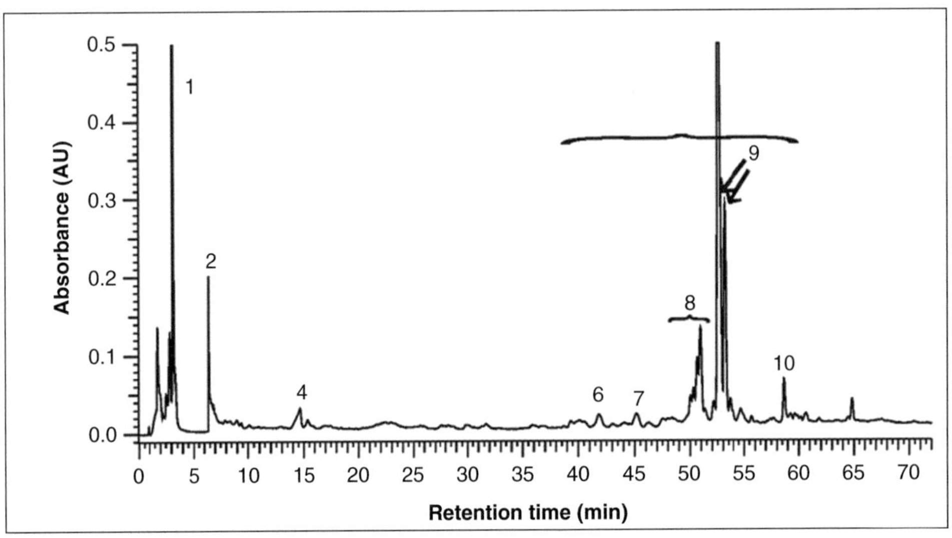

Fig. 4b HPLC-fingerprint analysis of the ethanol extract of Radix Trichosantis (sample 8)

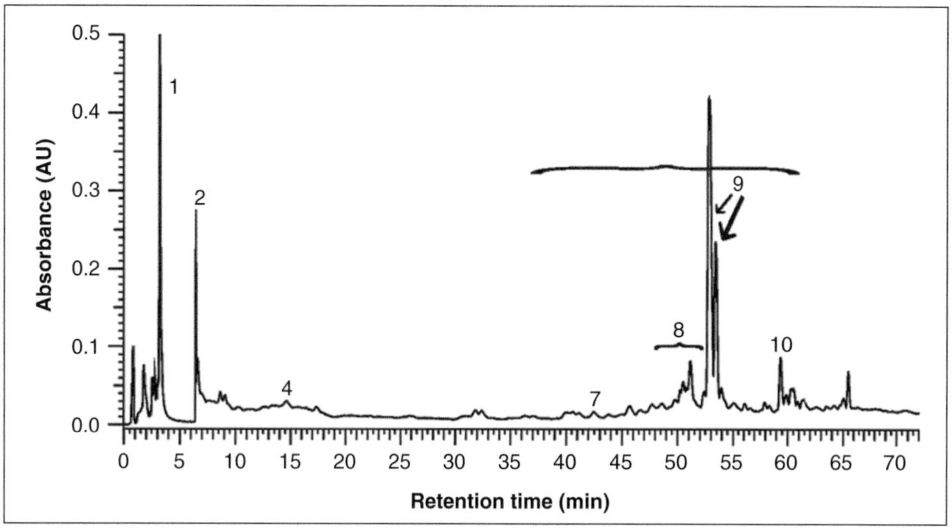

Fig. 4c HPLC-fingerprint analysis of the ethanol extract of Radix Trichosantis (sample 9)

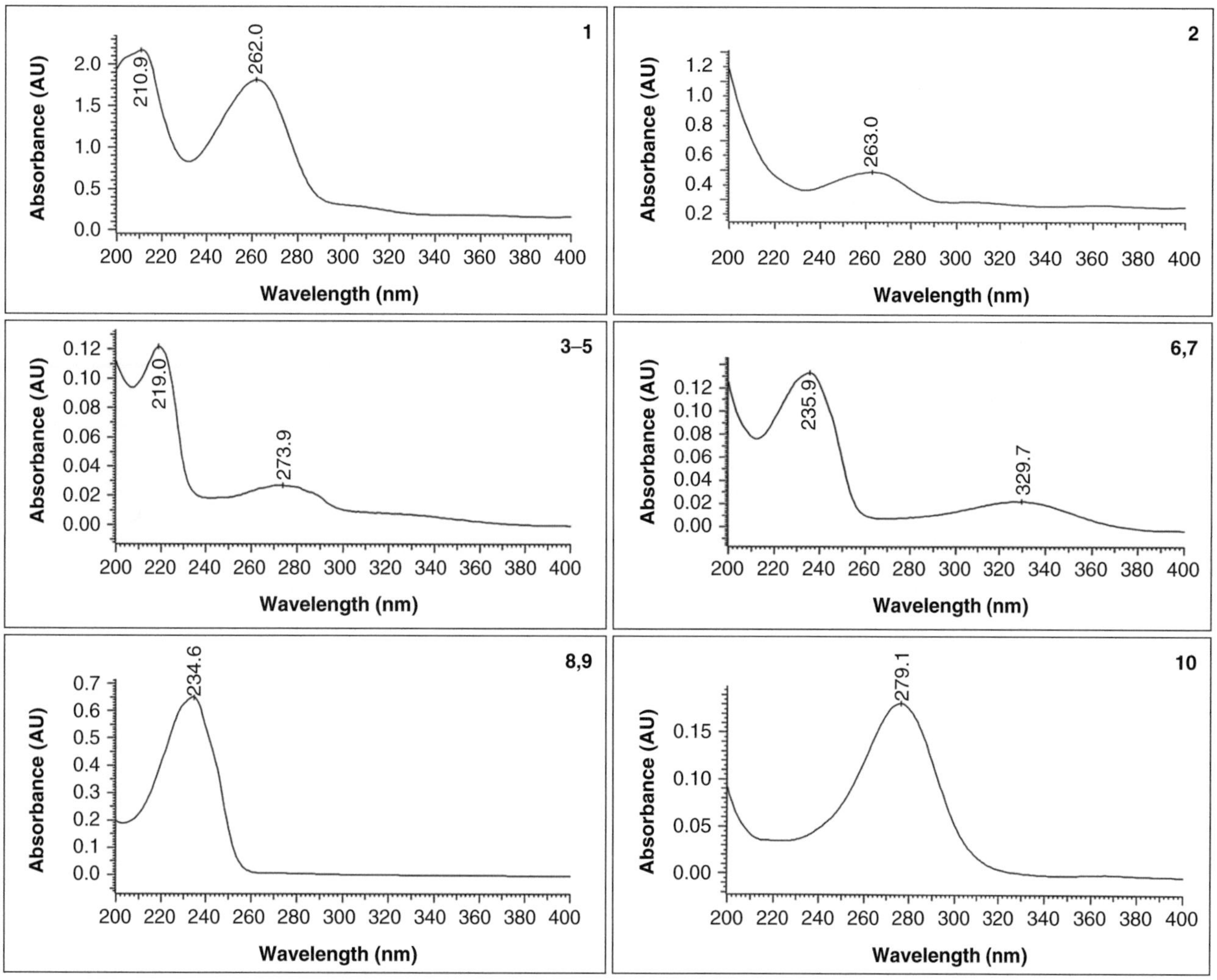

Fig. 5 On line UV-spectra of the main compounds (peaks) of Trichosanthis Radix

Conclusion

The authentication of the Radix Trichosanthes kirilowii or Tr. rosthornii extracts can be confirmed by the TLC and HPLC-detection of the characteristic cucurbitacins.

References

1. Pharmacopoeia of the people's republic of China, English edition, vol. I. People's Medical Publishing House, Beijing (2010)
2. Paulus, E., Ding, Y.-H.: Handbuch der traditionellen chinesischen Heilpflanzen, Karl F. Haug Verlag, Heidelberg (1987)
3. Tang, W., Eisenbrand, G.: Handbuch of Chinese medicinal plants. Wiley-VCH Verlag, Wheinheim (2011)
4. Tang, W., Eisenbrand, G.: Chinese drugs of plant origin: chemistry, pharmacology, and use tradionaland modern medicine. Springer-Verlag, Berlin, Heidelberg (1992)
5. Hempen, C.-H., Fischer, T.: Leitfaden Chinesische Phytotherapie 2. Auflage Urban & Fischer , Munich (2007)
6. Porkert, M.: Klinische Chinesische Pharmakologie, Verlag für Medizin Dr. Ewald Fischer GmbH, Heidelberg (1972)
7. Wagner, H., Bladt, S.: Plant drug analysis, a thin layer chromatography atlas, 2nd edn. Springer-Verlag, Wien (2001)
8. Wen D, Wang J, Yan H, Chen J, Xia K, Liu K, Zhang A: Effect of Radix Trichosanthis and Trichosanthin on hepatitis B virus in HepG2.2.15 Cells. J Nanosci Nanotechnol. 15(3), 2094–8, (2015)
9. Chen, Y., Miao, Y., Huang, L., Li, J., Zhao, Y., Yang, J., Zhou, W.: Antioxidant activities of saponins extracted from Radix Trichosanthis: an in vivo and in vitro evaluation. BMC Complement. 14(86), (2014)
10. No authors listed.: Studies on the mechanisms of abortion induction by Trichosanthin. Sci. Sin. **19**(6), 811–830 (1976)
11. Xie, W., Du, L.: Diabetes is an inflammatory disease: evidence from traditional Chinese medicine. Diabetes Obes. Metab. **13**(4), 289–301 (2011)
12. Li, W.L., Zheng, H.C., Bukuru, J., De Kimp, N.: Natural medicines used in traditional Chinese medical system for therapy of diabetes mellitus. J. Ethnopharmacol. **92**(1), 1–21 (2004)
13. Toyokawa, S., Takedo, T., Kato, Y., Wakabayashi, K., Ogihara, Y.: The complete amino acid sequence of an abortifacient protein, karasurin. Chem. Pharm. Bull. **39**(5), 12244–12249 (1991)
14. Toyokawa, S., Takedo, T., Ogihara, Y.: Isolation and characterization of a new abotifacient protein, Karasurin, from root tubers of *Trichosanthis kirilowii* MAX. *var japonicum* KITAM. Chem. Pharm. Bull. **39**(3), 716–719 (1991)
15. Jayaprakasha, Chidamba. K.N., Patil, B.S.: Rapid HPLC-UV method for quantification of L-citrulline in watermelon and its potential role on smooth muscle relaxation markers. Food Chem. **127**(1), 240–248 (2011)
16. Kondo, T., Inoue, M., Mizukami, H., Ogihara, Y.: Cytotoxic activity of bryonolic acid isolated from transformed hairy roots of *Trichosanthes kirilowii* var. *japonica*. Biol. Pharm. Bull. **18**(9), 726–729 (1995)
17. Bauer, R., Wagner, H.: Cucurbitacinhaltige Drogen; analyse and Standardisierung von Arzneidrogen und Phytopräparaten dursch Hochleistungsflüssigchromatographie (HPLC) und ander chromatographische Verfahren(II). Deutsche Apotheker Zeitung **123**(27), 1313–1321 (1983)

Ramulus Mori – *Sangzhi*

Pharmacopoeia: [1]	Pharmacopoeia of the People's Republic of China, English Edition Vol. I, 2010
Official drug: [1]	Mulberry Twig is the dried young branch of *Morus alba* L. (Fam. Moraceae) The drug is collected from late spring to early summer, removed from leaf, and dried in the sun, or cut into slices while fresh, and dried in the sun.
Origin: [2]	China, Indochina, Japan and Philippines
Description of the drug: [1]	Long cylindrical, branched occasionally, varying in length, 0.5–1.5 cm in diameter. Externally greyish-yellow or yellowish-brown, with numerous yellowish-brown dotted lenticels and fine longitudinal striations, with greyish-white and slightly semicircular leaf scars, and yellowish-brown axillary buds. Texture hard and tenacious, uneasily broken, fracture fibrous. Slices 0.2–0.5 cm thick, bark relatively narrow, wood yellowish-white, medullary rays radiate, pith white or yellowish-white. Odour, slight; taste, bitter.
Pre-treatment of the raw drug: [1]	The unsliced twigs are washed clean, soften thoroughly, cut into thick slices, and dried in the sun.
Medicinal use: [3–5]	The twigs of *Morus alba* have been applied for the treatment of joint swellings, rheumatic pains, arthritis, arthrosis and against spasms.

Effects and indications of Ramulus Mori according to Traditional Chinese Medicine [1, 3, 6]

Taste:	Neutral
Temperature:	Bitter
Channels entered:	*Orbis hepaticus*
Effects (functions):	To dispel wind-dampness, and disinhibit the joint
Symptoms and indications:	Wind-dampness impediment disease, numbness and sore pain in the joint, shoulder and arm

H. Wagner et al. (eds.), *Chromatographic Fingerprint Analysis of Herbal Medicines*, Vol. 4,
DOI 10.1007/978-3-319-32328-2_16, © Springer International Publishing Switzerland 2016

Main constituents:

- **Stilbenes** [5, 7–12]
  Resveratrol, oxyresveratrol, dihydrooxyresveratrol, (cis)-mulberroside A
- **Flavonoids/chalcones** [8, 11–14]
  Rutin, hyperosid (= hyperin), quercetin, isoquercitrin, kaempferol, dihydrokaempferol, morin (= 3,5,7,2′,4′-pentahydroxyflavone), dihydromorin, mulberrin (= kuwanon C), cyclomulberrin, butein, prenylflavonoids (morusin)
- **Coumarins** (7-hydroxycumarin, scopoletin) [8, 11]

Minor constituents: Tannin, betulinic acid, α-amyrin, moracin A-H, maclurin, saccharides (arabinose, xylose, stachyose, raffinose), 4-hydroxycinnamic acid [11, 12]

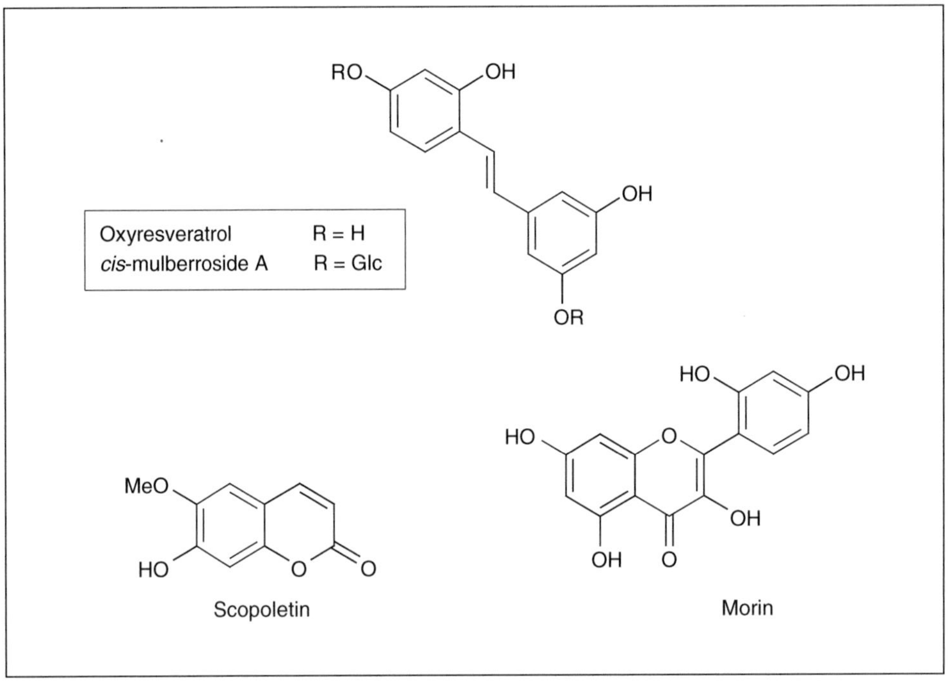

Fig. 1 Formulae of the main compounds of Ramulus Mori

Reported pharmacology:
- immune-stimulating [4, 5, 12]
- antineoplastic [4]
- antibiotic/antibacterial [4, 12]
- antihypertensive [4, 12]
- spasmolytic [4]
- anti-obesity [8]
- anti-aging [8]
- hepatoprotective [8]
- strong radical scavenging [8, 9]
- anti-oxidant [8–11, 15]
- anti-inflammation [5, 8–12]
- analgesic [12]
- anti-hyperlipidemia [5, 11, 15]
- anti-hyperglycemia [11, 15]
- neuroprotective [9, 10]
- anti-HIV [10]
- anti-viral [9]
- anti-herpetic [10]
- anti-atherogenic [15]

TLC Fingerprint Analysis

	Drug samples	Origin
1	Ramulus Mori / *Morus alba*	Beijing, China
2	Ramulus Mori / *Morus alba*	Province Hunan, China
3	Ramulus Mori / *Morus alba*	Province An Guo, China
4	Ramulus Mori / *Morus alba*	Sample of commercial drug, obtained from China Medica (origin: Langzhong)
5	Ramulus Mori / *Morus alba*	Sample of commercial drug, obtained from HerbaSinica (origin: Guangxi)
6	Ramulus Mori / *Morus alba*	Sample of commercial drug, obtained from TCM-clinic Bad Kötzting (Charge: K20.12.2000)
7	**Cortex Mori** / *Morus alba*[a]	Beijing, China
8	**Folium Mori** / *Morus alba*[a]	Beijing, China

[a]For comparison

1. TLC-fingerprint analysis of the flavonoids: [16]

Reference compounds of Fig. 2		R*f*
T1	Mulberrosid A	0.30
T2	Rutin	0.46
T3	Hyperosid	0.67
T4	Morin	0.96

1. Extraction: 1.0 g powdered drug is extracted with 7 ml ethanol (95 %) for 2 h, filtered and evaporated to dryness. The residue is dissolved in 1 ml ethanol
2. Reference compounds: 1.0 mg is dissolved in 1.0 ml methanol.
3. Separation parameters:
   Plate: HPTLC Silica gel 60 F_{254}, Merck
   Applied amounts: Ramulus Mori extracts: each 10 µl, Reference compounds: each 10 µl
   Solvent system: Ethyl acetate + formic acid + glacial acetic acid + water (20 + 2.2 + 2.2 + 5.2)
   Spray reagents: Natural products – Polyethylene glycol reagent (NP/PEG)
   I: 1 % diphenylboric acid-β-ethylamino ester
   (= diphenylboryloxyethylamine, NP) in methanol
   II: 5 % Polyethylene glycol-4000 (PEG) in ethanol (80 %)
   The plate is sprayed first with solution **I** and then with solution **II**.
   After 30 min the evaluation is carried out under UV 366 nm.
4. Description of Fig. 2:

The extract samples 1–6 are characterized by a fairly homogeneous pattern of 6–8 blue fluorescent zones in the R*f*-range 0.15–0.80. In this R*f*-range Mulberrosid A (**T1**) can be identified at R*f*=0.30. The flavonoid Morin (**T4**) could be detected as blue-green fluorescent zone at R*f*=0.96. In the extract samples 2 and 6 the main constituents are hardly visible. The flavonol-glycosides Rutin (**T2**) and Hyperosid (**T3**) could be only detected in extract sample 8 (Folium Mori) which was chromatographed for comparison. The extract sample 7 (Cortex Mori), however, shows a blue fluorescent zone profile which is similar to that of Ramulus Mori.

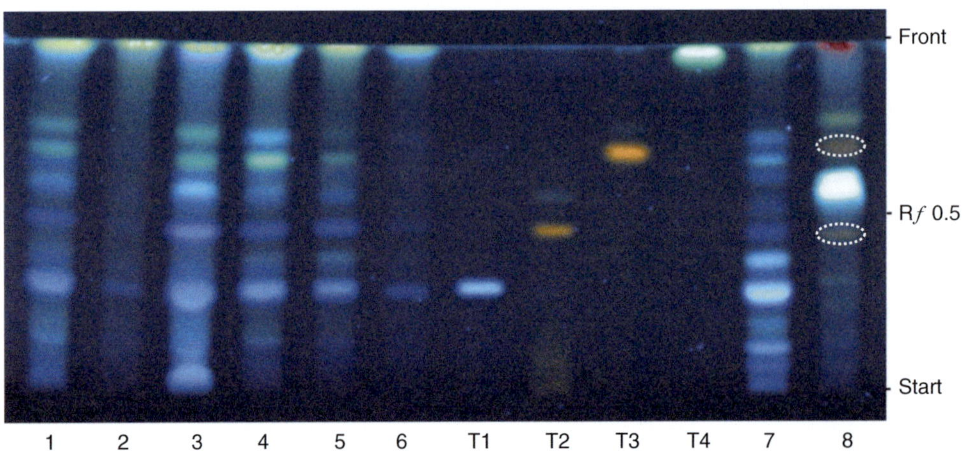

Fig. 2 Thin layer chromatogram of the methanol extracts of Ramulus Mori sprayed with NP/PEG reagent (UV 366 nm)

2. TLC-fingerprint analysis of stilbenes: [17a]

Reference compounds of Fig. 3a/b		**R*f***
T1	Mulberrosid A	0.13
T4	Morin	0.77
T5	Resveratrol	0.87
T6	Oxyresveratrol	0.86
T7	Morusin	0.91

1. Extraction: 1.0 g powdered drug is extracted with 7 ml ethanol (95%) for 2 h, filtered and evaporated to dryness. The residue is dissolved in 1 ml ethanol
2. Reference compound: 1.0 mg is dissolved in 1.0 ml methanol.
3. Separation parameters:
   Plate: HPTLC Silica gel 60 F_{254}, Merck
   Applied amounts: Ramulus Mori extracts: each 10 µl, Reference compounds: each 10 µl
   Solvent system: Ethyl acetate + methanol + water (20 + 2.6 + 2)
   Detection: 10% ethanolic sulphuric acid
   The plate is sprayed with 8 ml reagent and heated at 105 °C for 10 min. The evaluation is carried out in VIS (Fig. 3a) and under UV 366 nm (Fig. 3b).

4. Description of Fig. 3a/b:

In Fig. 3a appear the constituents in VIS with brown colour, in Fig. 3b with blue/green or yellow-brown colour. In the TLC the stilbene derivatives are detectable in VIS and under UV 366 nm at R*f*=0.87 (resveratrol, **T5**), at R*f*=0.86 (oxyresveratrol, **T6**) and at R*f*=0.13 (mulberoside A, **T1**). The flavonoid morin (**T4**) appears at R*f*=0.77 and morusin (**T7**) at R*f*=0.91.

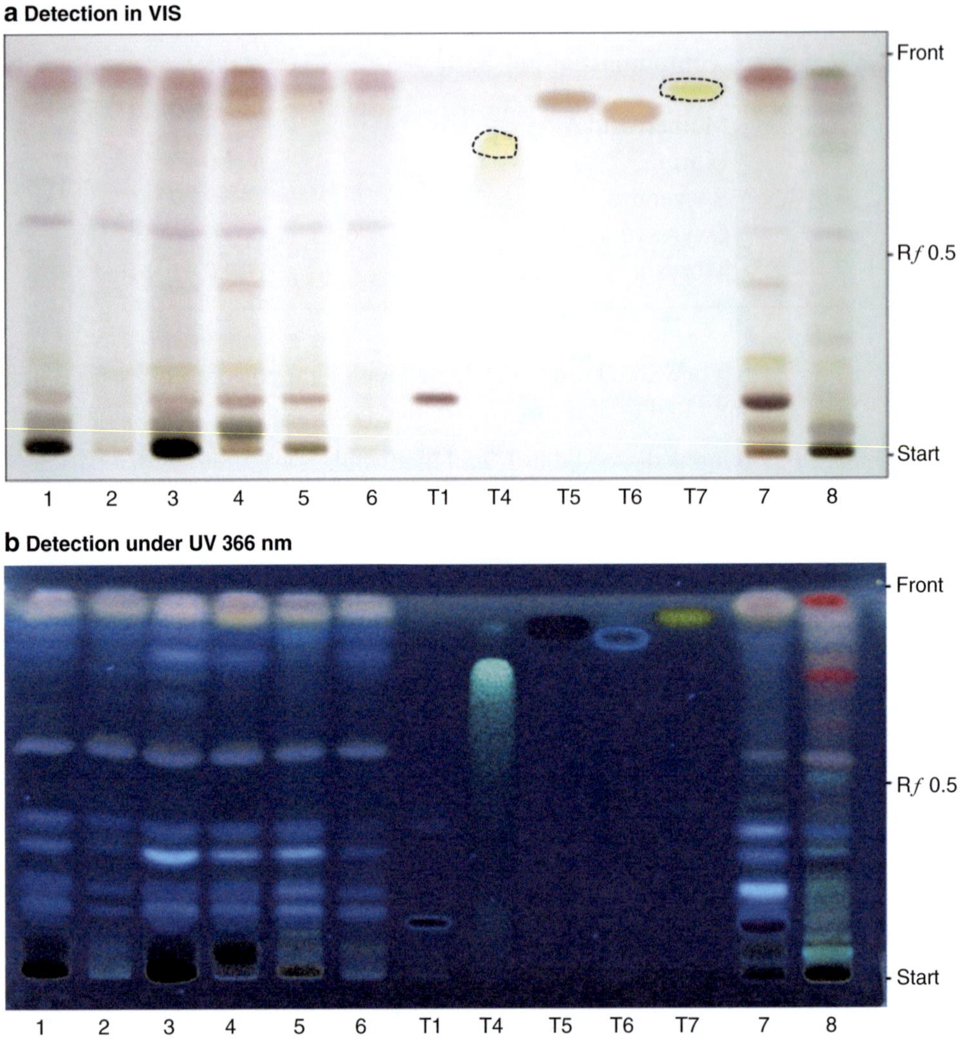

Fig. 3a/b Thin layer chromatogram of the methanol extracts of Ramulus Mori, sprayed with 10 % eth. H_2SO_4 (**a**=VIS, **b**=UV 366 nm)

HPLC-Fingerprint Analysis: [17b]

1. Extraction: 1.0 g powdered drug is extracted with 7 ml ethanol (95 %) for 2 h, filtered evaporated to dryness. The residue is dissolved in 1 ml ethanol and filtered over Chromafil® filtration unit, type 0-20 µm/25 mm.

2. Injection volume: Ramulus Mori extracts: each 20 µl
Cortex Mori extract: 20 µl
Folium Mori extract: 20 µl

3. HPLC – Parameter:

Apparatus:	MERCK HITACHI D-6000 A Interface
	MERCK HITACHI L-4500 A Diode Array Detector
	MERCK HITACHI AS-2000 Autosampler
	MERCK HITACHI L-6200 A Intelligent Pump
Separation column:	LiChroCART® 250-4 LiChrospher® 60 RP-select B (5 µm), Merck
Precolumn:	LiChroCART® 4-4 LiChrospher® 60 RP-select B (5 µm), Merck
Solvent System:	A: 0.1 % Phosphoric acid/Water (Millipore Ultra Clear UV plus® filtered)
	B: Acetonitrile (VWR)
Gradient:	0–40 % B in 33 min,
	40–60 % B in 12 min,
	60–100 % B in 5 min,
	100 % B for 20 min,
	Total runtime: 70 min
Flow:	0.6 ml/min
Detection:	260 nm

Retention times of the main peaks

Peak	**Rt (min)**	**Compound**
1	4.6	Unknown
2	5.7	Unknown
3	21.9	Unknown acid?
4	24.3	Unknown
5	24.5	Mulberrrosid A
6	27.7	Unknown
7	30.0	Unknown
8	33.2	4-hydroxycinnamic acid
9	53.6	Unknown
10	59.0	Unknown
11	65.5	Morusin
A	26.0 – 34.0	Flavonoids

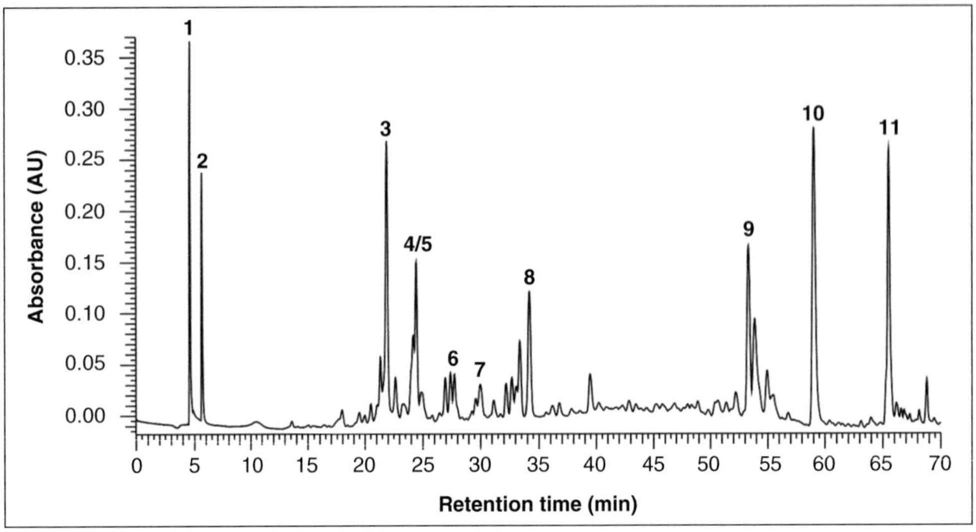

Fig. 4a HPLC – fingerprint analysis of the methanol extracts of Ramulus Mori (sample 4)

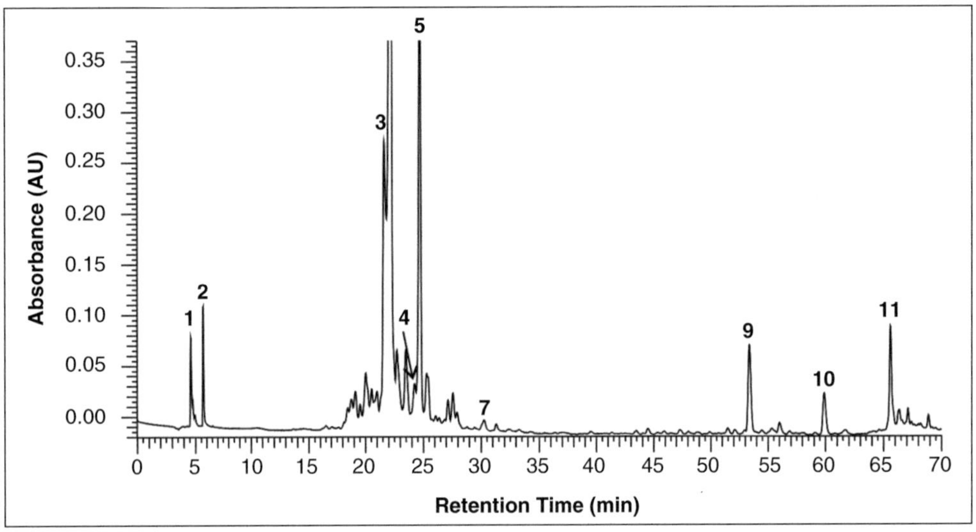

Fig. 4b HPLC – fingerprint analysis of the methanol extract of **Cortex** Mori (sample 7)

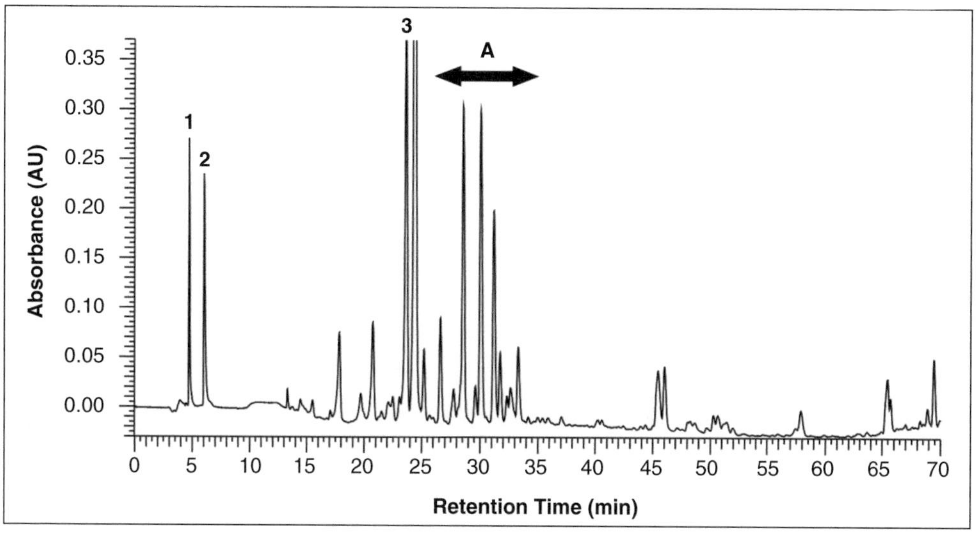

Fig. 4c HPLC – fingerprint analysis of the methanol extract of **Folium** Mori (sample 8)

4. Description of the HPLC-Figures:

   The Ramulus Mori extract sample 4 shows a fairly similar peak profile as Cortex Mori extract sample 7 but with different concentrations of the single constituents. In the extract sample 8 of Folium Mori appears in the Rt-range 26.0–34.0 (**A**) a great assembly of flavonoids and with lacking the constituents **9–11**.

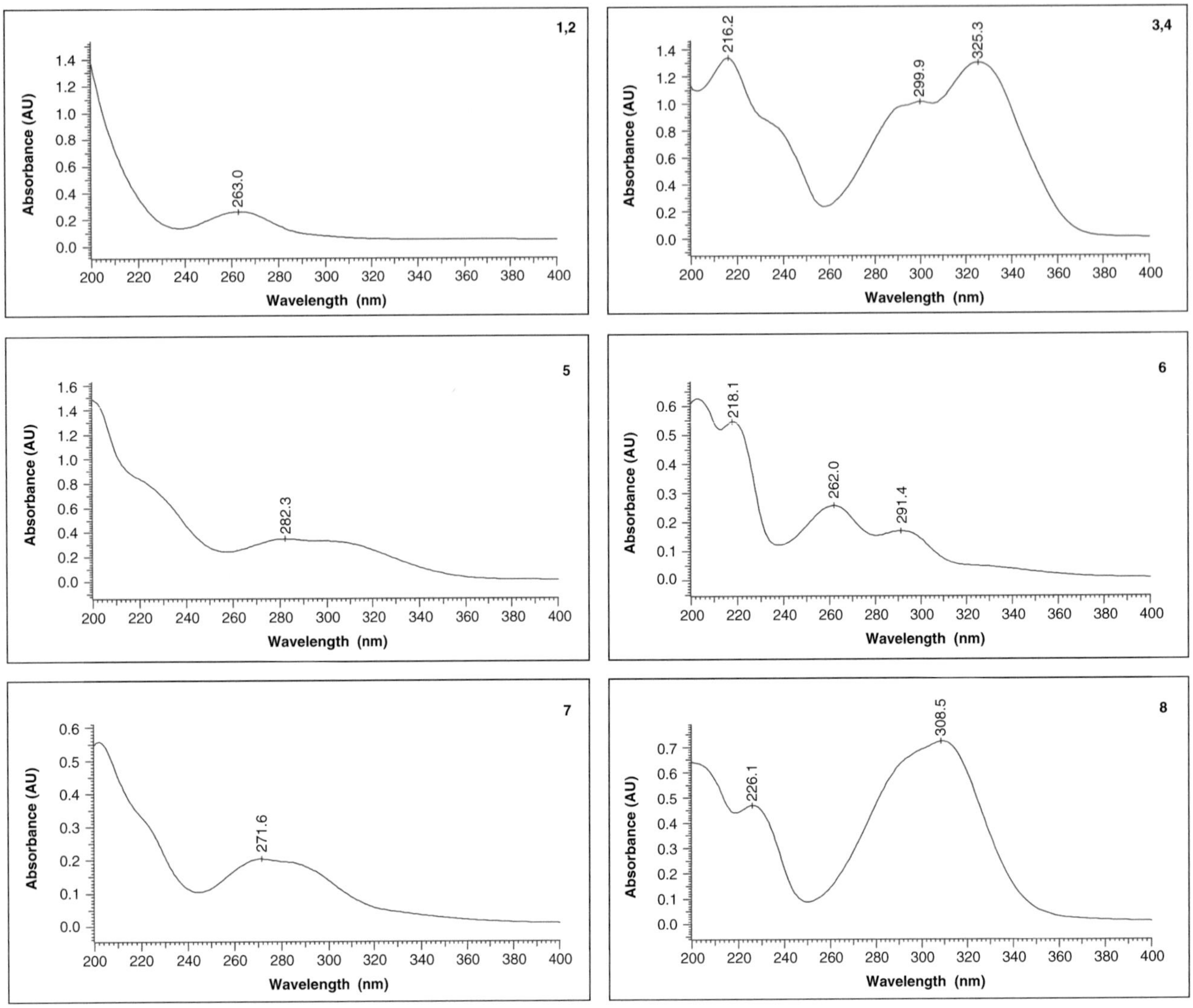

Fig. 5 On line UV-spectra of the main peaks of Ramulus, Cortex and Folium Mori

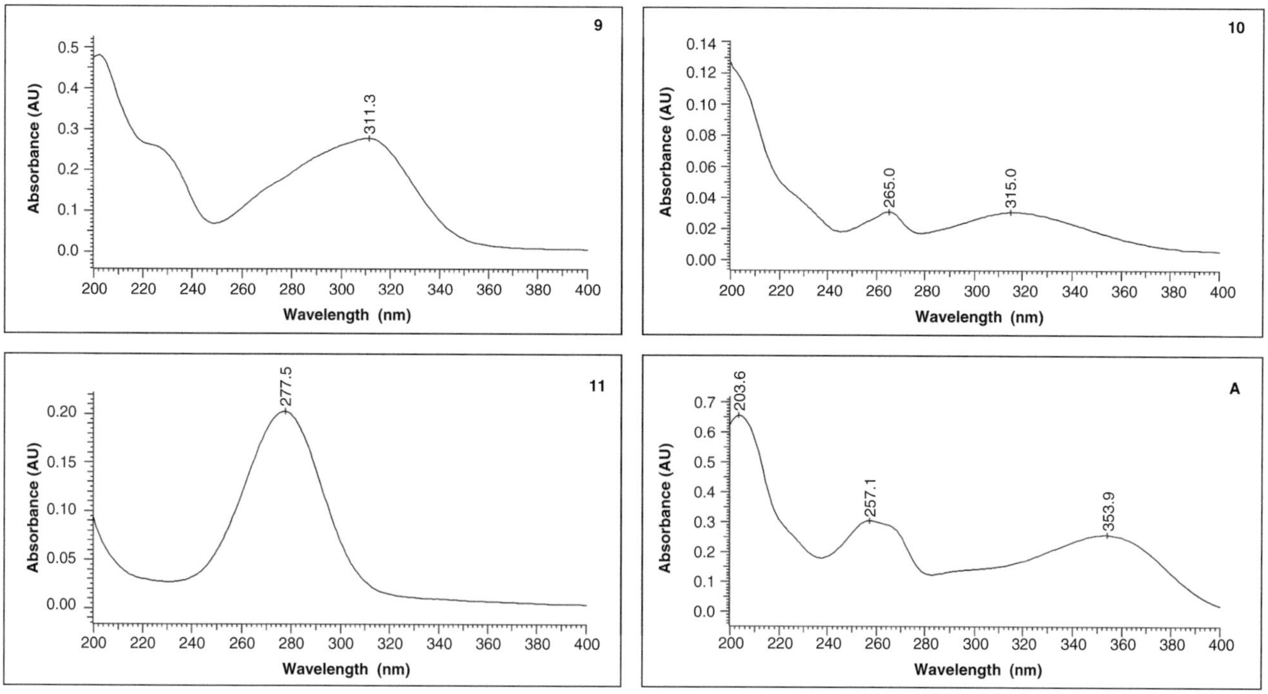

Fig. 5 (continued)

Conclusion

Ramulus Mori extracts can be easily authenticated by characteristic TLC-pictures and HPLC-graphs in contrast to Cortex and Folium Mori.

References

1. Pharmacopoeia of the people's republic of China, English edition, vol. I. People's Medical Publishing House, Beijing (2010)
2. Keys, J.D.: Chinese herbs – their botany, chemistry, and pharmacodynamics. Charles E. Tuttle Company, Rutland/Tokyo (1976)
3. Porkert, M.: Klinische Chinesische Pharmakologie, Verlag für Medizin Dr. Ewald Fischer, Heidelberg (1978)
4. Hempen, C.H., Fischer, T.: A materia medica for Chinese medicine (plants, minerals and animal products), 2nd edn. Churchill Livingstone Elsevier, New York (2009)
5. Zhang, Z., Shi, L.: Anti-inflammatory and analgesic properties of *cis*-mulberrosid A from Ramulus mori. Fitoterapia **81**(3), 214–218 (2010)
6. Stöger, E.A.: Arzneibuch der chinesischen Medizin. Deutscher Apotheker Verlag, Stuttgart (2005)
7. Zheng, Z.P., Cheng, K.W., Zhu, Q., Wang, X.C., Lin, Z.X., Wang, M.: Tyrosinase inhibitory constituents from the roots of *Morus nigra*: a structure-activity relationship study. J. Agric. Food Chem. **58**(9), 5368–5373 (2010)
8. Choi, S.W., Jang, Y.J., Lee, Y.J., Leem, H.H., Kim, E.O.: Analysis of functional constituents in mulberry (*Morus alba* L.) twigs by different cultivars, producing areas, and heat processings. Prev. Nutr. Food. Sci. **18**(4), 256–262 (2013)
9. Galindo, I., Hernáez, B., Berná, J., Fenoll, J., Cenis, J.L., Escribano, J.M., Alonso, C.: Comparative inhibitory activity of the stilbenes resveratrol and oxyresveratrol on African swine fever virus replication. Antiviral Res. **91**(1), 57–63 (2011)
10. Zhang, Z., Jin, J., Shi, L.: Protective function of *cis*-mulberroside A and oxyreveratrol from Ramulus mori against ethanol-induced hepatic damage. Environ. Toxicol. Pharmacol. **26**(3), 325–330 (2008)

11. Shi, Y.W., Wang, C.P., Wang, X., Zhang, Y.I., Liu, L., Wang, R.W., Ye, J.F., Hu, L.S., Kong, L.D.: Uricosuric and nephroprotective properties of Ramulus Mori ethanol extract in hyperuricemic mice. J. Ethnopharmacol. **143**(3), 896–904 (2012)
12. Roh, S.S., Seo, B.I., Kim, Y.U., Ku, S.K., Seo, Y.B.: Preventive and therapeutic effects Ramulus Mori extract on collagen-induced Arthritis in mice via suppression of inflammatory T cells. J. Health Sci. **57**(2), 142–152 (2011)
13. Yu, Z., Fong, W.P., Cheng, C.H.K.: Morin (3,5,7,2′,4′-Pentahydroxyflavone) exhibits potent inhibitory actions on urate transport by the human urate anion transporter (hURAT1) expressed in human embryonic kidney cells. Drug Metab. Dispos. **35**(6), 981–986 (2007)
14. Yu, Z., Fong, W.P., Cheng, C.H.K.: The dual actions of morin (3,5,7,2′,4′-Pentahydroxyflavone) as a Hypouricemic agent: uricosuric effect and xanthine oxidase inhibitory activitiy. J. Pharmacol. Exp. Ther. **3016**(1), 169–175 (2006)
15. Chang, L.W., Juang, L.J., Wang, B.S., Wang, M.Y., Tai, H.M., Hung, W.J., Chen, Y.J., Huang, M.H.: Antioxidant and antityrosinase activity of mulberry (*Morus alba* L.) twigs and root bark. Food Chem. Toxicol. **49**(4), 785–790 (2011)
16. Wagner, H., Bladt, S.: Plant drug analysis: a thin layer chromatography atlas, 2nd edn. Springer, Berlin (1996)
17. Wagner, H., Bauer, R., Melchart, D., Xiao, P.G., Staudinger, A.: Chromatographic fingerprint analysis of herbal medicines: thin layer and high performance liquid chromatography of chinese drugs, Volume 2; (a) Radix et Rhizoma Rhei; (b) Folium Mori, Springer, Wien (2011)

Semen Celosiae – *Qingxiang*

Pharmacopoeia: [1]	Pharmacopoeia of the People's Republic of China, English Edition Vol. I, 2010
Official drug: [1]	Feather Cockscomb Seed is the dried ripe seed of *Celosia argentea* L. (Fam. Amaranthaceae).
Origin: [2, 3]	Southern China, south Asia, Sri Lanka, Africa, America
Description of the drug: [1]	Oblate, a few rounded- reniform, 1–1.5 mm in diameter. Externally black or reddish-black, lustrous, somewhat raised in center, with a hilum on the slightly dented lateral side. Testa thin and brittle. Odour, slight; tasteless.
Pretreatment of the raw drug: [1]	The Plant is cut up or the infructescence is picked up in autumn when the fruit is ripe, dried in the sun, and the seed is gathered and removed from foreign matters.
Medicinal use: [2, 4]	It is used for eye and hepatic diseases, treatment of jaundic, inflammation, metrorhagia, gonorrhoea, healing of wounds and injuries.

Effects and indications of Semen Celosiae according to Traditional Chinese Medicine [1, 3–5]

Taste:	Bitter
Temperature:	Mild cold
Channels entered:	*Orbis hepaticus, o. felleus, o. renalis*
Effects (functions):	To clear the liver and purge fire, improve vision and remove nebule, Treatment of hepatitis, caligo corneae and hypertension, sarcoptidosis. Antipyretic, antispasmodic, anticancer, diuretic, anti-inflammatory and antibacterial
Symptoms and indications:	Red eyes caused by liver heat, nebule, blurry vision, dizziness caused by liver fire

H. Wagner et al. (eds.), *Chromatographic Fingerprint Analysis of Herbal Medicines*, Vol. 4,
DOI 10.1007/978-3-319-32328-2_17, © Springer International Publishing Switzerland 2016

Main constituents: - **Triterpenoids with Saponins** [2–4, 6, 7, 17, 18]

Celosins (A-G), oleanolic acid, celosin (H-J), 3-O-β-D-xylopyranosyl-(1→3)-β-D-glucuronopyranosyl-polygalagenin 28-O-β-D-glucopyranosyl ester, 3-O-β-D-glucuronopyranosyl-medicagenic acid 28-O-β-D-xylcopyranosyl-(1→4)-α-L-rhamnopyranosyl-(1→2)-β-D-fucopyranosyl ester, and 3-O-β-D-glucuronopyranosyl-medicagenic acid 28-O-α-L-arabinopyranosyl-(1→3)-[β-D-xylopyranosyl-(1→4)]-α-L-rhamnopyranosyl-(1→2)-β-D-fucopyranosyl ester

- **Bicyclic peptides** [4, 6, 10, 11, 13, 14, 18]

Moroidin, celogentin A, celogentin B, celogentin C

- **Cyclic nonapeptide:**

Celogenamide A

- **Phenolic glycoside** [8]

4-O-β-D-Apifuranosyl-(1→2)-β-D-glucopyranosyl-2-hydroxy- 6-methoxyacetophenone

-**Di-and Poly-Saccharides** [7]

Saccharose, celosian

Minor constituents: Kaempferol, quercetin, β-sitosterol, palmatic acid, stigmasterol, daucosterol, 2-hydroxyoctadecanoic acid, stearic acid, palmatic acid and 4-hydroxyphenethyl alcohol

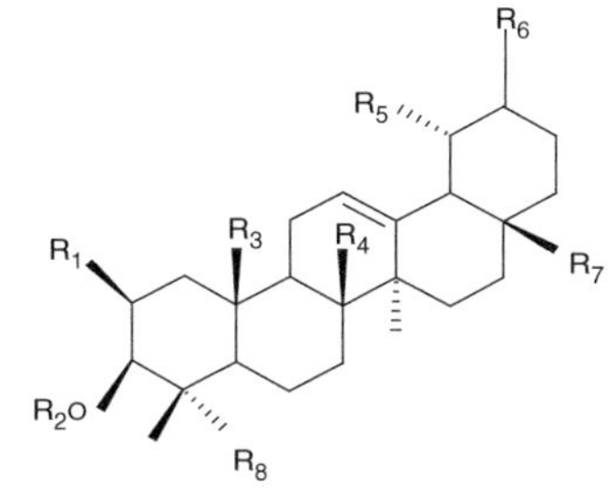

	R_1	R_2	R_3	R_4	R_5	R_6	R_7	R_8
Celosin A	OH	glcA-ara	CH_3	CH_3	H	$(CH_3)_2$	COOH	CHO
Celosin B	OH	glcA-gal	CH_3	CH_3	H	$(CH_3)_2$	COOH	COOH
Celosin C	H	H	COOH	CH_3	H	$(CH_3)_2$	COO-fuc-rha-glc-xyl	COOH
Celosin D	H	H	COOH	CH_3	H	$(CH_3)_2$	COO-fuc-rha-glc-xyl-ara	COOH
Celosin E	OH	glcA	CH_3	COOH	H	$(CH_3)_2$	CH_3	COOH
Celosin F	OH	xyl	CH_3	COOH	CH_3	$=CH_2$	COOH	COOH
Celosin G	H	glc	CH_3	CH_3	H	$(CH_3)_2$	COO-fuc-rha-glcA	COOH
Cristatin (*Celosiacristatea*)	OH	glcA	CH_3	CH_3	H	$(CH_3)_2$	COOH	COOH

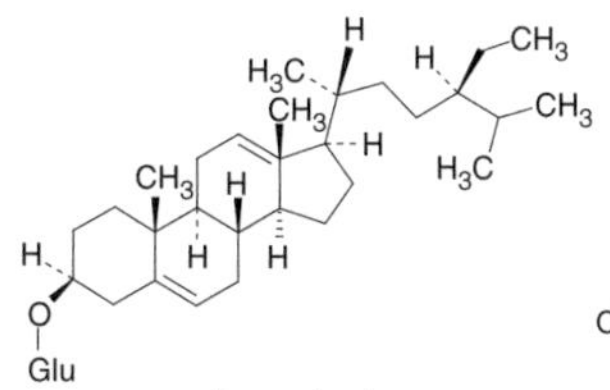

Daucosterol

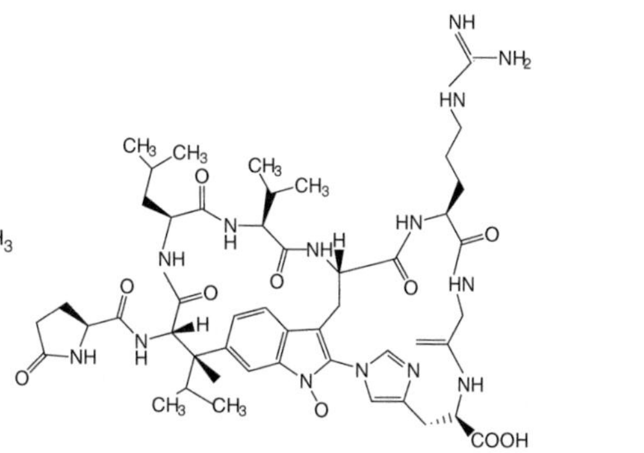

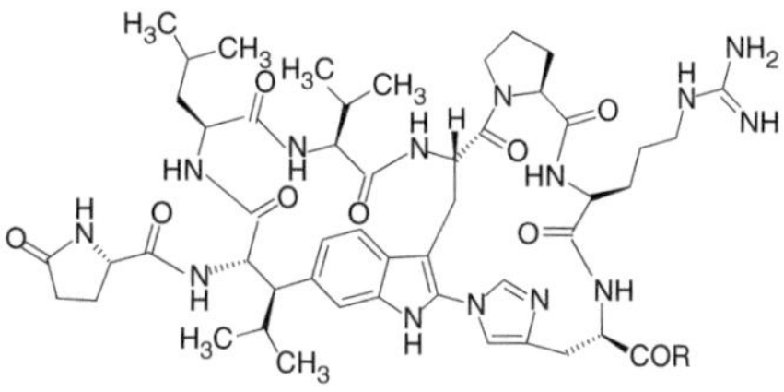

Celogentin C

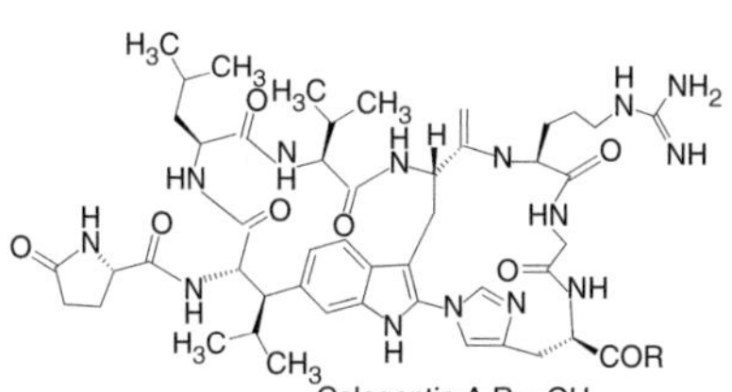

Celogentin A R = OH
Celogentin B R = His

Fig. 1 Formulae of the main constituents of Semen Celosiae [4, 7, 12, 14]

Reported pharmacological activities:

The crude EtOH extract in vitro [2–4, 8, 9, 16]

- anticancer activity
- anti-inflammatory activity
- antidiabetic activity
- antipyretic activity
- antispasmodic activity
- diuretic activity
- antibacterial activity

The aqueous extract in vitro

- exhibition of antimetastatic effects
- the acidic polysaccharide celosian was reported to exert hepatoprotective effects in mice [7, 8]

TLC Fingerprint Analysis

	Drug samples	Origin
1	Semen Celosiae / *Celosia argentea*	Sample of commercial drug, obtained from HerbaSinica (origin: ???)
2	Semen Celosiae / *Celosia argentea*	Sample of commercial drug, obtained from China Medica
3	Semen Celosiae / *Celosia argentea*	Sample of commercial drug, obtained from TCM-Clinic Bad Kötzting (Charge: K07.01.2003)

1. **TLC-fingerprint analysis of Triterpenoid saponins and Disaccharide:**

Reference compounds of Fig. 2a		R*f*
T 1	Oleanolic acid	0.98
T 2	Daucosterol	0.91
T 3	Saccharose	0.32

1. Extraction: 2.0 g powdered drug with 10 ml 80 % ethanol are sonicated for 1 h. The extract is filtered and evaporated to dryness. The residue is diluted with 0.5 ml methanol and filtered over Chromafil® Type 0.20 µm.
2. Reference compounds: Each 0.5 mg is dissolved in 0.5 ml MeOH
3. Separation parameters:

Plate:	HPTLC Silica gel 60 F_{254}, Merck
Applied amounts:	Celosiae Semen extracts: 40 µl each
	Reference compounds: 20 µl each
Solvent system:	Chloroform + methanol + water (6.4 + 5 + 1)
Detection:	<u>Liebermann- Burchard reagent</u>
	5 ml acetic anhydride and 5 ml concentrated sulphuric acid are added carefully to 50 ml absolute ethanol while cooling in ice. The reagent must be freshly prepared.
	The plate is sprayed with the solution and heated at 110 °C for 5–10 min. The evaluation is carried out under UV 366 nm and VIS.

4. Description of Fig. 2a, b:

Fig 2a: The ethanol extracts of Semen Celosiae show at R*f*=0.32 the dominant black colour zone of saccharose (**T3**), the **7–8** pink/grey zones above saccharose (**T3**) comprise e.g. oleanolic acid at R*f*=0.91 (**T2**) followed by the triterpene- di-and triglycosides of celosins A- G till the R*f*- range above saccharose.

In Fig 2b appear the same Celosia- constituents with light grey or pink, white and brown fluorescent zones, maybe partly superimposed by phenolic glycosides (see main constituent p. 171).

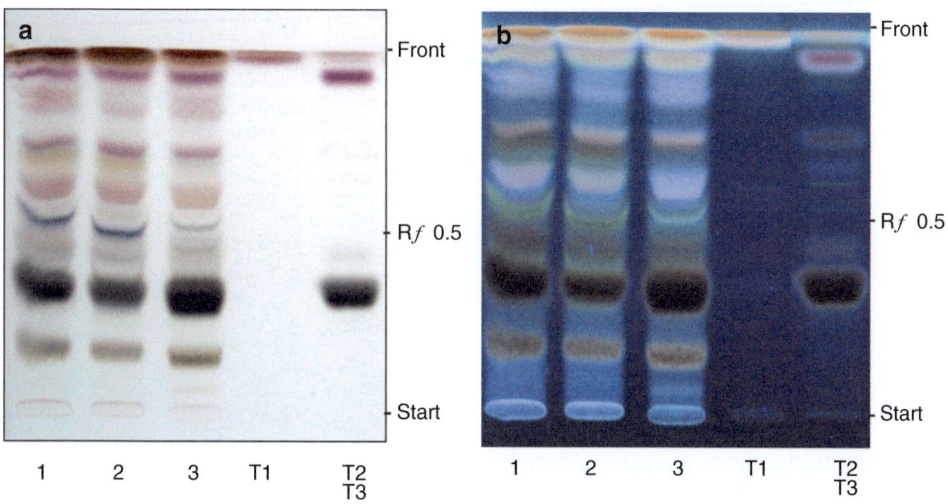

Fig. 2a, b Thin layer chromatogram of the 80 % ethanol extracts of Semen Celosiae sprayed with Liebermann-Burchard reagent **a**) under VIS and **b**) at 366 nm

2. **TLC-fingerprint analysis Cyclo peptides:**
   1. Extraction: The same extract used for TLC (1)
   2. Reference compounds: not applied
   3. Separation parameters:
      Plate: HPTLC Silica gel 60 F_{254}, Merck
      Applied amounts: Semen Celosia extracts: 30 µl each
      Solvent system: n-Butanol + ethanol + glacial acetic acid + water (6 + 2 + 2 + 2)
      Detection: Ninhydrin:
      Dissolve 2 g Ninhydrin in 100 ml ethanol.
      The plate is sprayed with the solution and heated at 100 °C for 5–10 min. The evaluation is carried out under Vis.
      The plate is sprayed with this solution, heated for 5 min at
      105 °C and evaluated in VIS.
   4. Description of Fig. 2c:
      With ninhydrin appear the Peptides between R*f*=0.20 and R*f*=0.65 with violet colour. An assignment of the different zones to the various dipeptides was not possible, because of lacking reference compounds.

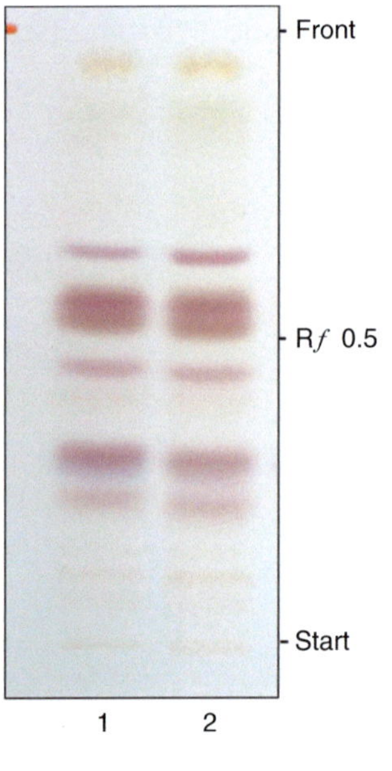

Fig. 2c Thin layer chromatogram of the 80% ethanol extracts of Semen Celosiae (Cyclic peptides) sprayed with Ninhydrin and detected under VIS

HPLC-Fingerprint Analysis

1. Sample preparation: The same extracts are used as for the TLC (see above).
2. Injection volume: Semen Celosiae extracts: 50 µl each
3. HPLC parameter:

Apparatus:	MERCK HITACHI D-6000 A Interface MERCK HITACHI L-4500 A Diode Array Detector MERCK HITACHI AS-2000 Autosampler MERCK HITACHI L-6200 A Intelligent Pump
Separation column:	LiChroCART® 250-4 LiChrospher® 100 RP-18 (5 µm), Merck
Precolumn:	LiChroCART® 4-4 LiChrospher® 100 RP-18 (5 µm), Merck
Solvent System:	A: 0.0001 H_3PO_4 (Millipore Ultra Clear UV plus® filtriert) B: Acetonitril (VWR)
Gradient:	0–40 % B in 40 min, 40–100 % B in 25 min, total run time: 65 min
Flow:	1 ml/min
Detection:	210 nm

Retention times of the main peaks

peak	Rt (min)	compound
1	2.6-6.8	Amino acids, peptides or non-cyclic peptides
2	9.9	
3	12.6	
4	36.0-43.0	Not identified
5	45.3	
6	49.0-52.0	
7	54.4	Daucosterol
8	59.9	Oleanolic acid

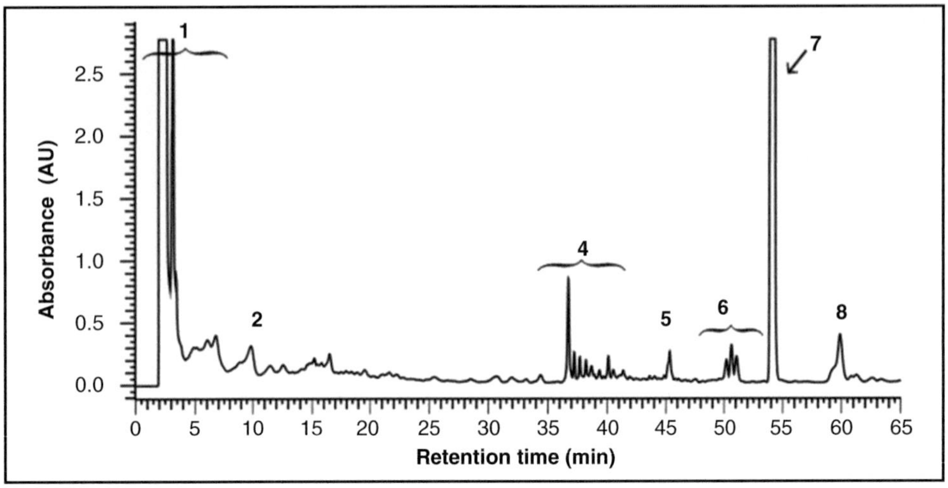

Fig. 3a HPLC-fingerprint analysis of the 80 % ethanol extract of Semen Celosiae (sample 1)

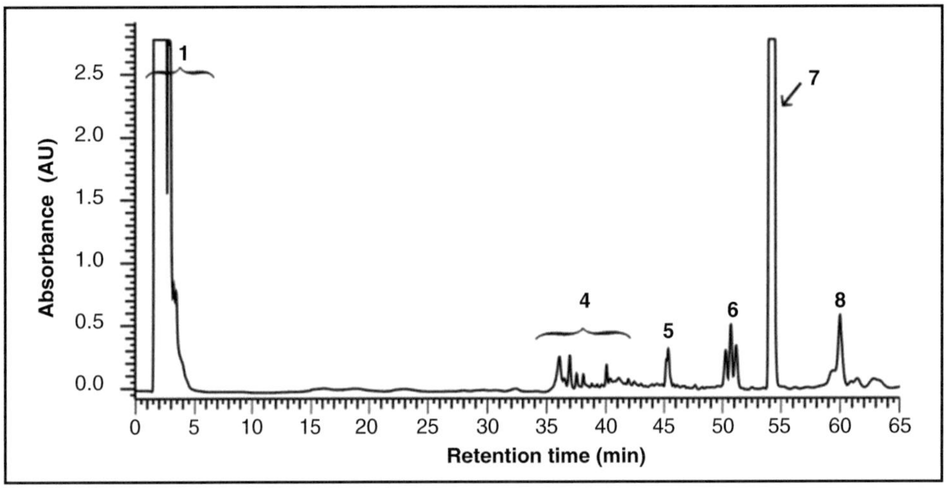

Fig. 3b HPLC-fingerprint analysis of the 80 % ethanol extract of Semen Celosiae (sample 2)

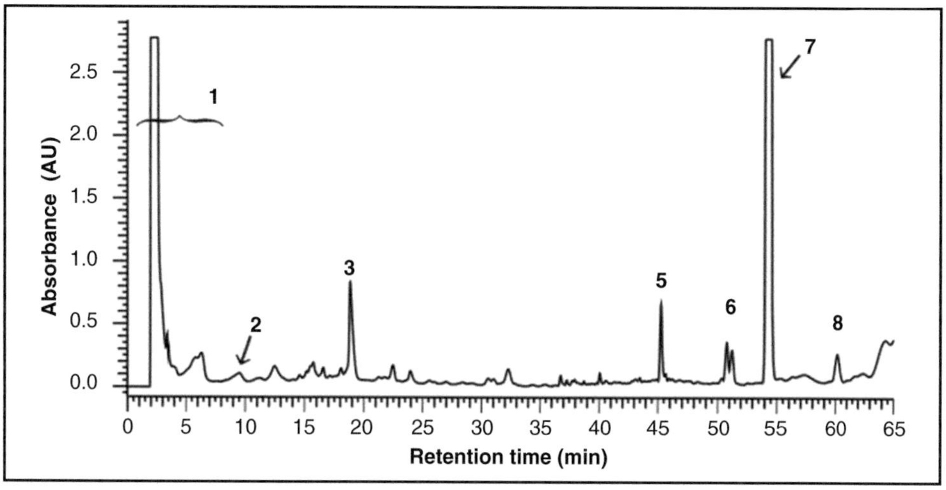

Fig. 3c HPLC-fingerprint analysis of the 80 % ethanol extract of Semen Celosiae (sample 3)

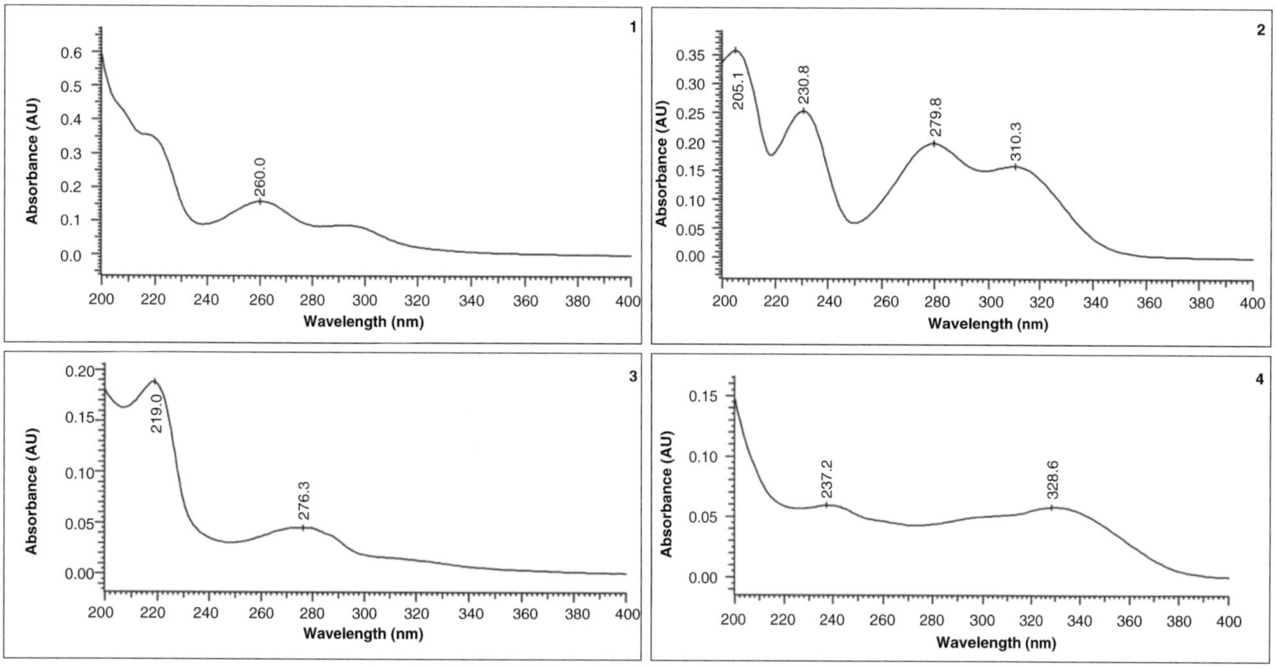

Fig. 4 On line UV-spectra of the mean peaks of Semen Celosiae

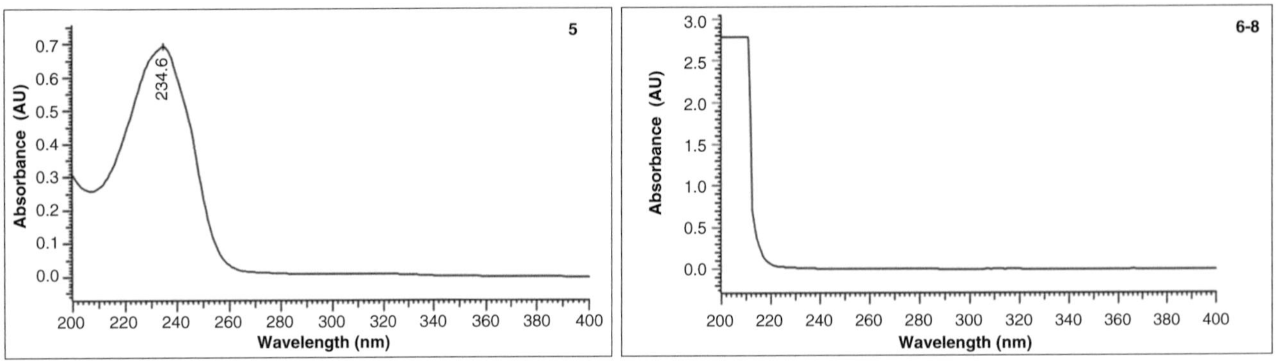

Fig. 4 (continued)

4. Description of the HPLC- fingerprints of Semen Celosiae- ethanol extracts:

   The peak profiles with peak No. 1–3 show the presence of amino acids or cyclopeptides, whereas the peak 7 and 8 can be identified as daucosterol and oleanolic acid.

Conclusion

For the authentication of Semen Celosiae ethanol extracts the TLC – fingerprints method has definite preference. For the exact HPLC- identification of the cyclopeptides and minor compounds, a precedent fractionation of extract would be necessary.

References

1. Pharmacopoeia of the People's republic of China, English edition, vol. I. People's Medical Publishing House, Beijing (2010)
2. Keys, J.D.: Chinese herbs. Charles E. Tuttle Company, Rutland/Tokyo (1976)
3. Hempen, C.-H., Fischer, T.: Leitfaden Chinesische Phytotherapie, 2nd edn. Urban & Fischer, Munich (2007)
4. Tang, W., Eisenbrand, G.: Handbook of Chinese medicinal plants. Wiley-VCH Verlag, Wheinheim (2011)
5. Paulus, E., Ding, Y.-H.: Handbuch der traditionellen chinesischen Heilpflanzen, Karl F. Haug Verlag, Heidelberg (1987)
6. Sun, Z.L., Wang, Y., Gue, M.L., LI, Y.X.: Two new hepaprotective saponins from Semen celosiae. Fitoterapia **81**(5), 375–380 (2010)
7. Kadota, S.H., Hase, K., Basnet, P., Takahashi, T., Namba, T.: Protective effect of celosian, an acidic polysacchride on chemically and immmunologically induced liver injuries. Biol. Pharm. Bull. **19**(4), 567–572 (1996)
8. Hase, K., Basnet, P., Kadota, S.H., Namba, T.: Immunostimulating activity of Celosian, an Antihepototoxic Polysaccharide isolated from *Celosia argentea*. Planta Med. **63**(3), 216–219 (1997)
9. Vetrichelvan, T., Jegadeesan, M., Devi, B.A.: Anti-diabetic activity of alcoholic extract of *Celosia argentea* LINN. Seeds in rats. Bio. Pharm. Bull. **25**(4), 526–528 (2002)
10. Moritta, H., Shimbo, K., Shegemori, H., Kobayashi, J.: Antimitotic activity of Moroidin, a bicyclic peptide from the seeds of *Celosia argentea*. Bioorg. Med. Chem. Lett. **10**(5), 469–471 (2000)
11. Morita, H., Suzuki, H., Kobayashi, J.: Celogenamide A, a new cyclic peptide from the seeds of *Celosia argentea*. J. Nat. Prod. **67**(9), 1628–1630 (2004)

12. Kahn, S.D., Booth, P.M., Waltho, J.P., Williams, D.H.: Computer assisted structure determination. structure of the peptide Moroidin from Laportea moroides. J. Org. Chem. **54**(8), 1901–1904 (1989)

13. Kobayashi, J., Suzuki, H., Shimbo, K., Takeya, K., Morita, H., Celogentins, A-C.: New antimitotic bicyclic peptides from the seeds of Celosiae argentea. J. Org. Chem. **66**(20), 6626–6633 (2001)

14. Suzuki, H., Morita, H., Shiro, M., Kobayashi, J.: Celogenatin K, a new cyclic peptides from the seeds of *Celosiae argentea* and X-ray structure of moroidin. Tetrahedron **60**. 2489–2495 (2004)

15. Wu, Q.B., Wang, Y., Guo, M.L.: Isolation and purification of two Saponins from semen Celosiae by high- speed counter-current chromatography coupled with ELSD. Nat. Prod. Res. Dev. **24**(4), 481–485 (2012)

16. Wu, Q., Wang, Y., Guo, M.: Triterpenoid Saponins from the seeds of *Celosiae argentea* and their anti-infalmmatory and antitumor activities. Chem. Pharm. Bull. **59**(5), 666–671 (2011)

17. Wu, Q., Wang, Y., Guo, M., Liang, L., Jiang Q., Zhang, J.J.: Novel triterpenoid saponins from the seeds of Celosia argentea L. Nat. Prod. Res. **27**(15), 1353–1360 (2013)

18. Ramesh, B.N., Mahalakshmi, A.M., Mallappa, S.: Toward a better understanding of an updated ethnopharmacology of *Celosia argentea* L. Int. J. Pharmaceutical. Sci. **5**(3), 54–59 (2013)

Semen Nigrum Sesami – *Heizhima*

Pharmacopoeia: [1]	Pharmacopoeia of the People's Republic of China, English Edition Vol. 1 2010
Official drug: [1]	Black Sesame is the dried ripe seed of *Sesamum indicum* L. (Fam. Pedaliceae). The plant is collected in autumn when the fruits are ripe, dried in the sun. The seed is gathered, removed from foreign matter and dried in the sun.
Origin: [2]	Chinese Provinces Henan, Shandong, Hubei, Anhui, Sichuan, Jiangxi, and Hebei
Description of the drug: [1]	Flattened- ovoid, about 3 mm long, and 2 mm wide. Externally black, smooth or reticulately wrinkled. A dotted brown raphe at the apex. Testa thin, cotyledons 2, white, oily. Odour, slight; taste sweet, slightly oily aromatic.
Pretreatment of the raw drug: [1]	The drug is eliminated from foreign matter, washed clean, dried in the sun, and broken into pieces before use.
Medicinal use: [1, 3]	Used as a general tonic, for the treatment of dizziness, tinnitus, impaired hearing, and also as laxative against constipation.

Effects and indications of Semen Nigrum Sesami according to Traditional Chinese Medicine [1–4]

Taste:	Sweet
Temperature:	Neutral
Channels entered:	*Orbis hepaticus, orbis renalis, orbis intestini crassi*
Effects (functions):	To tonify the liver- kidney, replenish essence and blood and moisten the intestines
Symptoms and indications:	Deficiency of essence and blood, dizziness and blurred vision, tinnitus and deafness, premature graying, hair loss after recovering from an illness, constipation caused by intestinal dryness

H. Wagner et al. (eds.), *Chromatographic Fingerprint Analysis of Herbal Medicines*, Vol. 4,
DOI 10.1007/978-3-319-32328-2_18,

Main constituents [3–6]

- **Lignans:** Sesamin, sesamolin, episesamin, sesamolinol, sesaminol, and *epi*-sesaminol, sesanoglin
- **Lignans glucosides:** Sesaminol mono-, di-, and triglycosides, pinoresinol mono-, di-, and triglycosides
- **Sterols:** β-Sitosterol and spinasterol
- **Sesam oil:** Composed mainly of glycerol esters of oleic acid and linoleic acid, eicosenoic acid
- **Protein:** Albumin, α- globulin, glutelin, protamine

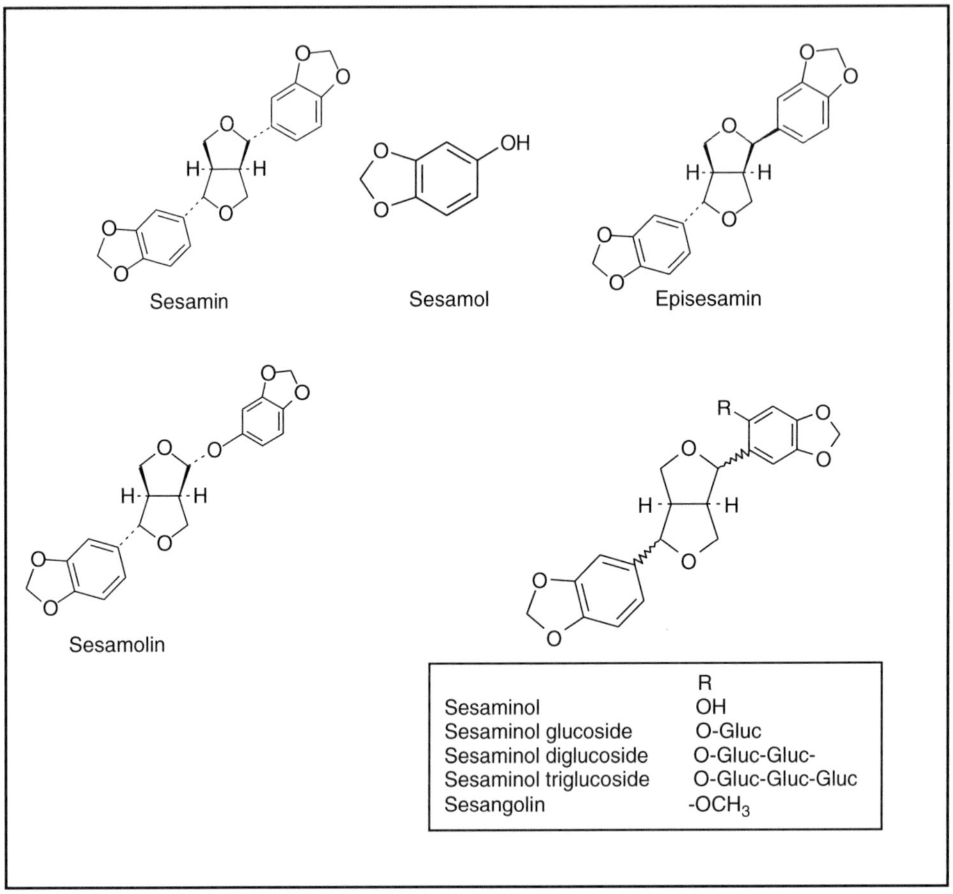

Fig. 1 Formulae of the main constituents of Semen Nigrum Sesami [1, 3, 5, 7–9]

Reported pharmacological effects:
- anti-hypertensive effect [10]
- inhibitory activity for cholesterol absorption and biosynthesis [10]
- antioxidation effect [10]
- laxative effect (sesame oil) [7]
- hypoglycemic effect (flavonoids) [7]
- inhibition of malignant melanoma (linoleate in triglyceride form) [7]
- cancer preventive (meristic acid) [7]
- neuroprotective effect [11, 12]
- chondroprotective effect (sesamin) [13]
- anti-inflammatory effect [13]
- anti-proliferation and anti- angiogenic effect on cancer cells [14]

TLC Fingerprint Analysis

	Drug samples	Origin
1	Semen nigrum Sesami / *Sesamum indicum*	Sample of commercial drug, obtained from firm China Medica, Guangyuan Sichuan, China
2	Semen nigrum Sesami / *Sesamum indicum*	Sample of commercial drug, obtained from firm Herba sinica Hillsboro GmbH, Anhui, China
3	Semen nigrum Sesami / *Sesamum indicum*	Guangxi, China
4	Semen nigrum Sesami / *Sesamum indicum*	Vietnam
5	Semen nigrum Sesami / *Sesamum indium*	Yunnan, China
6	Semen nigrum Sesami / *Sesamum indium*	Sample of commercial drug, obtained from TCM-Clinic Bad Kötzting (Charge:56513072004)

Reference compounds of Fig. 2		R*f*
T1	Sesamin	0.81
T2	β-sitosterol	0.73
T3	Linoleic acid	0.67
T4	Saccharose	0.16
T5	Oleic acid	0.69

1. Extraction: 2 g powdered drug are sonicated with 10 ml 80 % ethanol for 1 h, filtrated and evaporated to dryness. The residue is dissolved in 1 ml ethanol.
2. Reference compounds: Each 1.0 mg is dissolved in 1 ml ethanol
3. Separation parameters:

   Plate: HPTLC Silica gel 60 F_{254}, Merck (20 × 20)

   Applied amounts: Semen nigrum Sesami extracts: 20 µl each

   Reference compounds: 10 µl each

   Solvent system: 1) Chloroform + Methanol (10 + 10)

   2) Chloroform + Diethyl ether (18 + 2)

   The plate is developed in solvent system 1 about 6 cm, then dried in air and developed in solvent system 2 approx. 15 cm.

   Detection: 20 % ethanolic sulphuric acid

   The plate is sprayed with 15 ml 20 % sulphuric acid, heated for 10 min at 120 °C and detected in VIS.

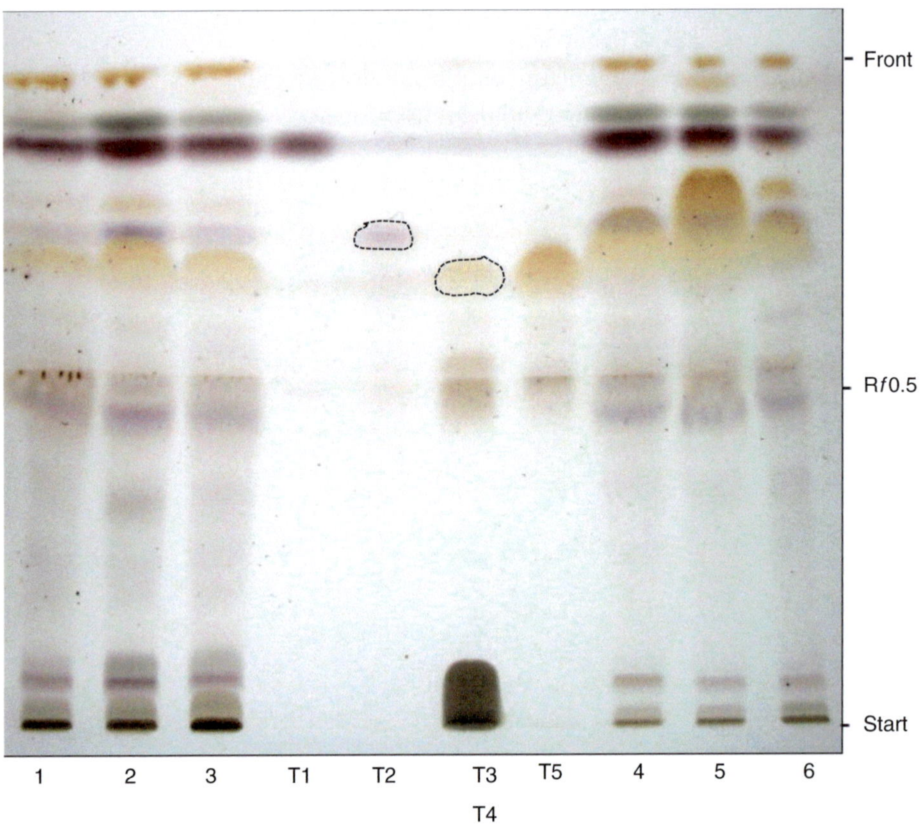

Fig. 2 Thin layer chromatogram of the 80 % ethanol extracts of Semen Sesami nigrum sprayed with 20 % ethanolic sulphuric acid (VIS)

4. Description of Fig. 2:

   The Semen Sesami ethanol extract samples 1 – 6 show in the R*f*- range from 0.75 till the solvent front 3 brown spots with the dominant (main) sesamin (**T1**), episesamin at R*f*=0.87 and at R*f*=0.91 which might be assignable to another constituent (e.g. sesamolin).

   In the R*f*- range from R*f* 0.60 up to R*f*=0.80 appear with yellow brown colour linoleic acid (**T3**), oleic acid (**T5**) and the phytosterol β- sitosterol (**T2**). Saccharose was identified at R*f*=0.16 (**T4**). In the R*f*- range R*f*=0.15 till R*f*=0.50 the three dark brown and pink brown spots might be identical with mono-, di- and triglycoside of sesaminol.

HPLC-Fingerprint Analysis

1. Sample preparation: 2 g powdered drug are sonicated with 10 ml 80% ethanol for 1 h, filtered and evaporated to dryness. The residue is dissolved in 1 ml ethanol, filtered over Chromafil®, Type 0.20 µm and injected into the HPLC apparatus.
2. Injection volume: Semen nigrum Sesami extracts: 50 µl each
   Reference compound: 20 µl
3. HPLC parameter:

Apparatus:	MERCK HITACHI D-6000 A Interface
	MERCK HITACHI L-4500 A Diode Array Detector
	MERCK HITACHI AS-2000 Autosampler
	MERCK HITACHI L-6200 A Intelligent Pump
Separation column:	LiChroCART® 250-4 LiChrospher® 100 RP-18 (5 µm), Merck
Precolumn:	LiChroCART® 4-4 LiChrospher® 100 RP-18 (5 µm), Merck
Solvent System:	A: Sodium phosphate buffer pH 2 (Millipore Ultra Clear UV plus® filtered)
	B: Acetonitrile (VWR)
Gradient:	0–15% B in 5 min.
	15–30% B in 25 min.
	30–100 %B in 25 min.
	100% B for 10 min.
	total run time: 65 min
Flow:	1 ml/min
Detection:	215 nm

Retention times of the main peaks

Peak	Rt (min)	Compound
1	11.9	Not identified
2	14.3	Not identified
3	16.9	Not identified
4	24.3	Sesaminol triglucoside
5	32.0	Sesaminol mono- or diglucoside
6	46.2	Not identified
7	47.4	Sesamin
8	48.6	Sesamolin?
9	50.8	Linoleic acid
10	62.0–64.0	Fatty acids?

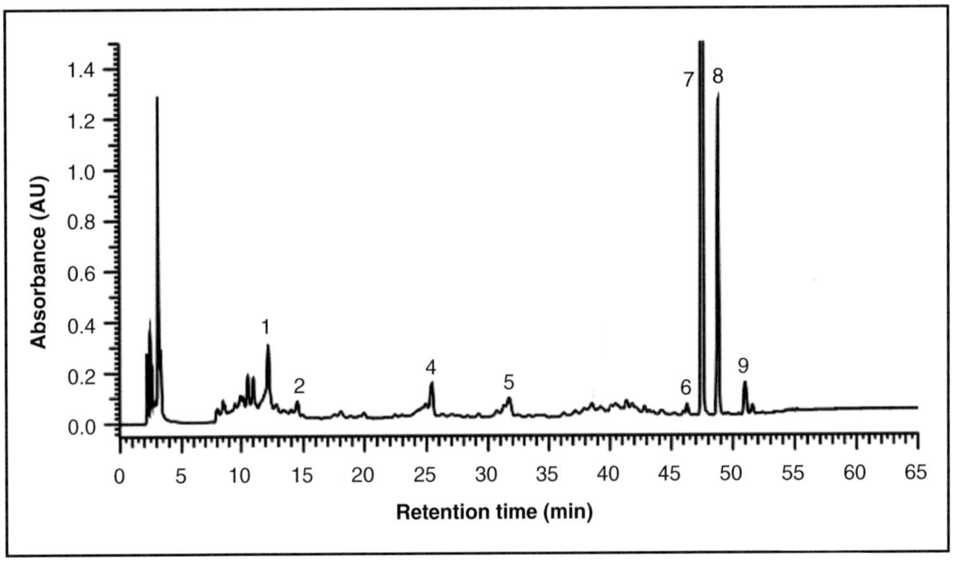

Fig. 3a HPLC-fingerprint analysis of the 80 % ethanol extract of Semen nigrum Sesami, sample 1

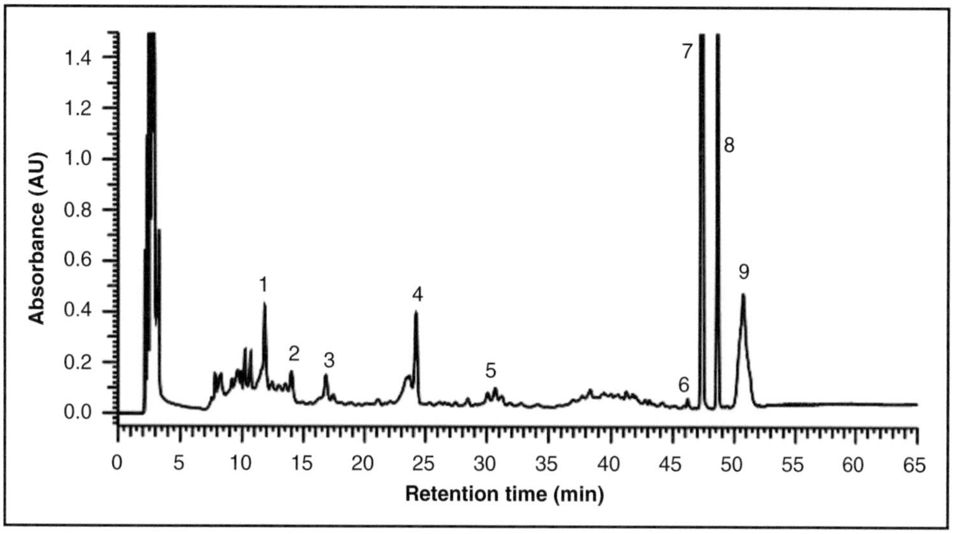

Fig. 3b HPLC-fingerprint analysis of the 80 % ethanol extract of Semen nigrum Sesami, sample 3

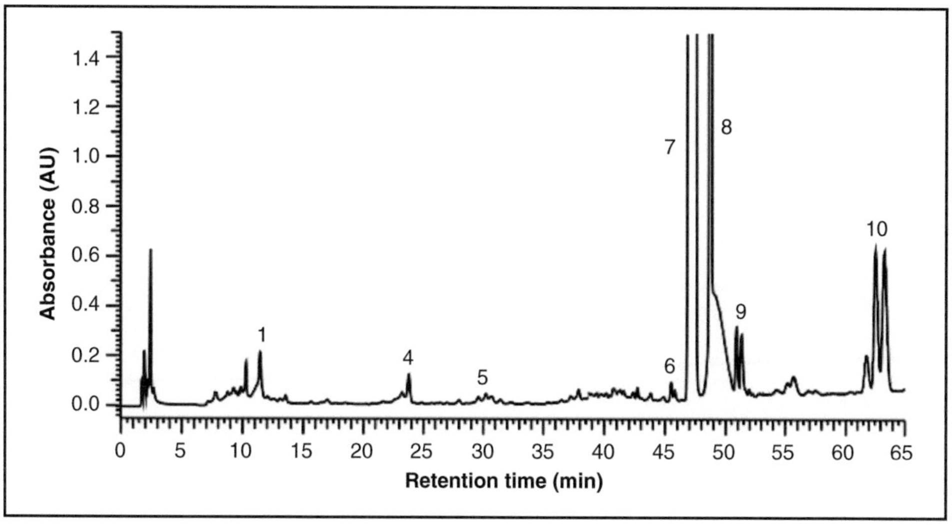

Fig. 3c HPLC-fingerprint analysis of the 80 % ethanol extract of Semen nigrum Sesami, sample 5

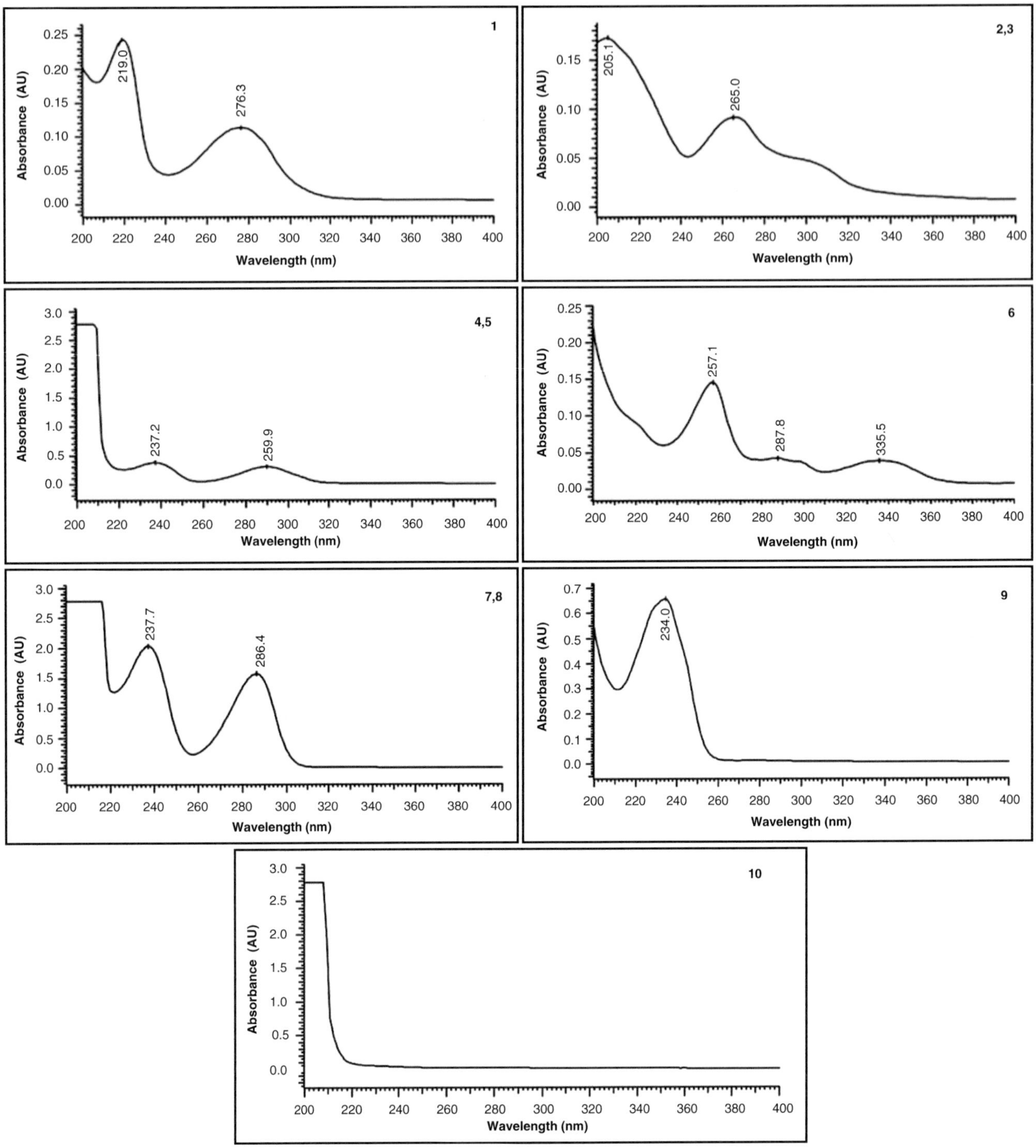

Fig. 4 On line UV-spectra of the main peaks of Semen nigrum Sesami

4. Description of Fig. 3a, b and c of HPLC- fingerprints:

   In the HPLC of the 80 % ethanol extract samples appear in opposite order to the TLC between Rt = 20.0 and 35.0 the Lignan tri-, di-, and mono-glycosides of sesaminol. The main peaks at 47.5 and Rt = 48.5 are Sesamin and Episesamin (overlapped) and Sesamolin. Between 62.0 and 64.0 appear only in the extract sample 5 double peaks which might be related to other fatty acids.

Conclusion

The authentication of Semen nigrum Sesami extracts can be definitely evidenced by the TLC-and HPLC-analysis, shown in Figs. 2 and 3a, b, c and corresponding UV- spectra of Fig. 4.

References

1. Pharmacopoeia of the people's republic of China, English edition, vol. I. People's Medical Publishing House, Beijing (2010)
2. Junying, G. et al.: Materia medica der Chinesischen Arzneimitteltherapie, Verlag für Ganzheitliche Medizin, Dr. Erich Wühr GmbH, Kötzting/Bayer. Wald
3. Tang, W., Eisenbrand, G.: Handbook of Chinese medicinal plants – chemistry, pharmacology, toxicology, vol. 1. WILEY-VCH Verlag GmbH & Co. KgaA, Weinheim (2011)
4. Moazzami, A., Andersson, R., Kamal- Aldin, A.: HPLC analysis of sesaminol glucosides in sesame seeds. J. Agric. Food Chem. **54**(3), 633–638 (2006)
5. Eldin, A., Appelqvist, L.A., Yousif, G.: Lignan analysis in seeds oils from four *Sesamum species*: comparison of different chromatographic methods. JAOCS **71**(2), 141–147 (1994)
6. Moazzami, A.A., Kamal Eldin, A.: Sesame seed is a rich source of dietary lignans. JAOCS. 83(8), 720–723 (2006)
7. Anilakumar, K.R., Khanum, F., Bawa, A.S.: Nutritional, medicinal and industrial use of sesame (*sesamium indicum* L.) seeds. Agriculturae Conspectus Seintificus **75**(4), 159–168 (2010)
8. Bhatnagar, A.S., Hemavathy, J., Krishna, A.G.G.: Development of rapid method for determination of lignans content in sesame oil. J. Food. Sci. Technol. **52**(1), 521–527 (2015)
9. Hempen, C.-H., Fischer, T.: A materia medica for Chinese medicine, english edition. Elsevier GmbH, Urban & Fischer, Munich (2009)
10. Zhu, X., Zhang, X., Sun, Y., Su, D., Sun, Y., Hu, B., Zeng, X.: Purification and fermentation in vitro of sesaminol triglucoside from sesame cake by human intestinal microbiota. J. Agri. Food. Chem. **61**(8), 1868–1877 (2013)
11. Cheng, F.C., Jinn, T.R., Hou, R.C., Tzen, J.T.: Neuroprotective effect of sesamin and sesamolin on gerbil brain in cerebral ischemia. Int. J. Biomed. Sci. **2**(3), 284–288 (2006)
12. Ohnishi, M., Monda, A., Takemoto, R., Matsuoka, Y., Kitamura, C., Ohashi, K., Shibuya, H., Inoue, A.: Sesamin supresses activation of microgali and P44/42 MAPK pathway, which confers neuroprotection in rat intracerbral hemorrhage. Neuroscience **232**, 45–52 (2013)
13. Phitak, T.H., Pothacharoen, P., Settacorn, J., Poompimol, W., Caterson, B., Kongtawelert, P.: Chondroprotective and anti- inflammatory effects of sesamin. Phytochemistry **80**, 77–88 (2012)
14. Siao, A.C., Huo, C.W., Kao, Y.H., Jeng, K.C.: Effect of sesamin on apoptosis and cell cycle arrest in human breast cancer MCF-7 cell. Asian Pac. J. Cancer Pre. **16**(9), 3779–3783 (2015)

Semen Sinapis – *Jiezi*

Pharmacopoeia: [1] Pharmacopoeia of the People's Republic of China, English Edition Vol.1 2010

Official drug: [1] Mustered Seed is the ripe seed of *Sinapis alba* L. or *Brassica juncea* (L.) Czern. et Coss. (Fam. Cruciferae). The former is known as "Bai Jiezi" (Semen Sinapis albae). the latter is known as "Huang Jiezi" (Semen Brassicae junceae).

Origin: [2] Chinese Provinces Yunnan, Fujian, Guangdong and Sichuan

Description of the drug: [1] *Bia Jiezi* Spherical, 1.5–2.5 mm in diameter. Externally greyish-white to pale yellow, finally reticulated, with an obvious point –like hilum. Testa thin and brittle, after cutting, white folded cotyledons visible, oily. Odour, slight; taste pungent.

Huang Jiezi Relatively small, 1–2 mm in diameter. Externally yellow to brownish-yellow, a few dark reddish –brown. Odour, characteristic and pungent when triturated and moistened with water.

Pretreatment of the raw drug: [1] The plant is cut up in late summer and early autumn when the fruit is ripe, and dried in the sun. The seeds are tapped out and removed from foreign matter.

Processing: [1] *Semen Sinapis (stir- baked)*

The clean drug is stir- baked as described under the method for stir-baking (Appendix II D, Chinese Pharmacopoeae) until it becomes pale yellow to dark yellow (stir- baked Baijiezi) or dark yellow to brown (stir-baked Huangjiezi) and pungent scented.

Medicinal use: [1, 3] Bronchitis, Influenza infection, urinary tract infections and external application of rheumatic pains.

Toxicity: [5] Because of its strong irritating properties, the drug should be not used topically in patients with sensitive skin. If taken in large amounts, the seeds may provoke gastrointestinal symptoms. Long-term ingestion carries the danger of nerve damage, whilst long-term topical application may lead to skin damage. Since the seeds may cause sensitization, they should be not used in children less than 6 years of age.

H. Wagner et al. (eds.), *Chromatographic Fingerprint Analysis of Herbal Medicines*, Vol. 4,
DOI 10.1007/978-3-319-32328-2_19, © Springer International Publishing Switzerland 2016

Effects and indications of Semen Sinapis according to Traditional Chinese Medicine [1–6]	
Taste:	Pungent
Temperature:	Warm
Channels entered:	*Orbis pulmonalis, o.stomachi*
Effects (functions):	To transform cold phlegm, dislodge Phlegm-wind, move and regulate qi, dissolve swelling and dislodge phlegm
Symptoms and indications:	Thin white sputum, wheezing cough, stifling sensation in the chest, respiratory complaints
	Cramps, Paresthesia, heaviness, and pain in the limbs, facial nerve paralysis, spasms, epilepsy, hemiplegia, migraine, neuralgia, arthritis
	Descends the lung qi, cough, difficult breathing. Qi movement in the channels: nodules swelling, pain
	Swelling, nodules, phlegm in the channels, joint pain, pain in the extremities
	Pain in the extremities, joint pain, swelling ulcers due to cold phlegm, nodules, pain, wheezing, cough, copious thin white sputum

Main constituents: [2, 3, 4, 7–10, 12–19]

- Glucosinolates	Sinalbin from S.alba → enzymatic reaction →4-Hydroxybenzylisothiocyanate
	Sinigrin (2-propenylglucosinolate) from Brassica juncea.
	Sinigrin → enzymatic reaction → 2-Propenyl isothiocyanate
- Phenolic Compounds	Cis- and trans-sinapine (the choline ester of sinapic acid), sinapic acid, sinapoyl glucose, Kaempferol-sinapoyl-trihexoside1-O-β-D-glucopyranosyl sinapate, Sinapoyl-hexoside, Disinapoyl hexoside, trisinapoyl-dihexoside and sinapoyl conjugate
- Further sulfur containing compounds:	By thermal degradation in aqueous solution : Di-2-propenyl sulfide, Di-2- propenyl disulphide, Di-2-propenyl trisulphide, di-2-propenyl-tetrasulfide, 2- di-propenyl thiocyanate, 3*H*-1,2-dithiolene, 2-vinyl-4*H*-1,3-dithiin, 4*H*-1,2,3-trithiin, 5-methyl-1,2,3,4-tetrathiane and *N,N*'-di-2-propenyl thiourea as major compound
- Fatty acids:	Erucic acid, oleic acid, lionoleic acid, linolenic acid, eicosenoic acid
- Proteins:	Albumin, globulin, glutelin, proteine
- Others:	β-Sitosterol

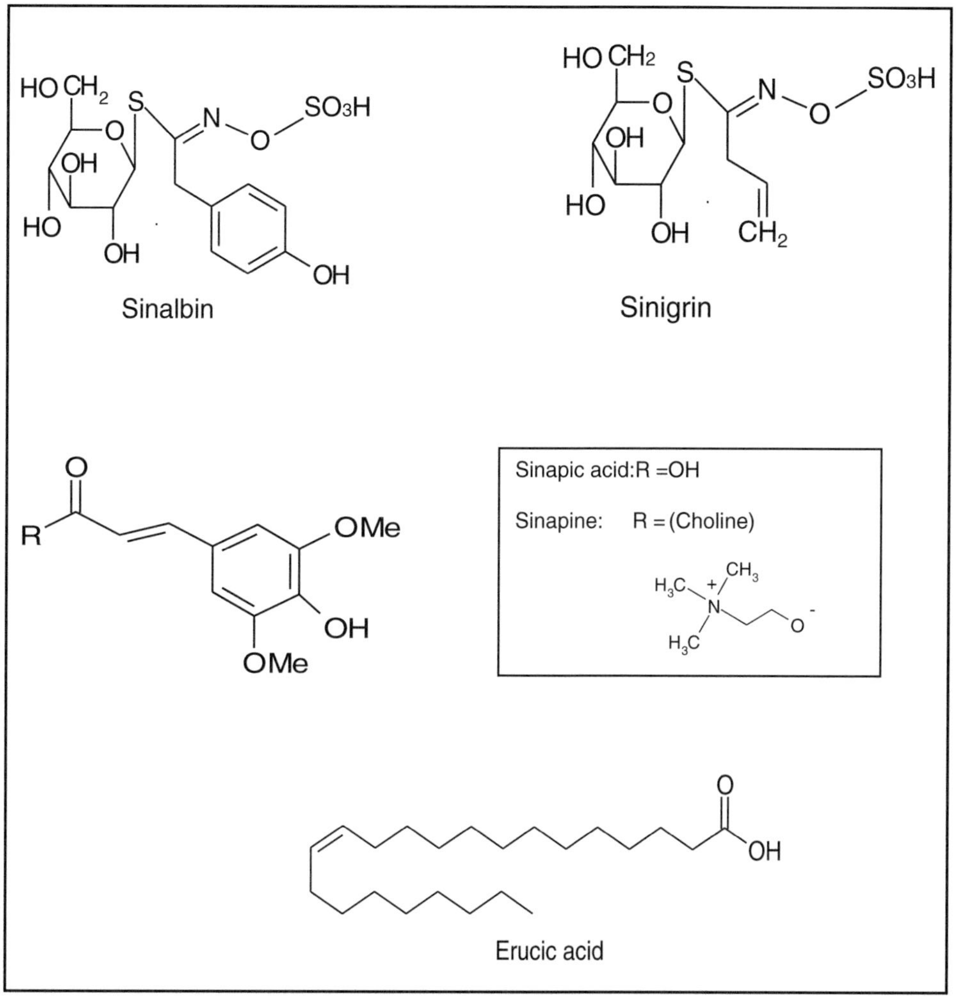

Fig. 1 Formulae of the main constituents of Semen Sinapis [2, 19])

Reported pharmacological effects [2, 7, 8, 11–16]

- radical-scavenging activity (antioxidative)
- antioxidation effect
- anti-inflammatory and anti-androgen activities (sinalbin)
- chemoprevetive efficacy (sinigrin)
- inhibition of the prostatic hyperplasia induced by testosterone propionate
- inhibition of the permeability of capillary vessel in mice skin induced by histamine
- reduction of the exacerbation frequency of chronic lung diseases in winter
- antibacterial activities

TLC Fingerprint Analysis[19]

Drug samples		Origin
1	Semen Sinapis Praep., *Sinapis alba*	Sample of commercial drug, obtained from China Medica (Germany)
2	Semen Sinapis, *Sinapis alba*	Sample of commercial drug, obtained from Cfm, (origin: Zhongjiang, Sichuan (China))
3	Semen Sinapis, *Sinapis alba*	Sample of commercial drug, obtained from TCM-Clinic Bad Kötzting (Charge: K30.12.2003)
4	Semen Sinapis, *Sinapis alba*	Sample of commercial drug, obtained from TCM-Clinic Bad Kötzting (Charge: 16.06.1995)
5	Semen Sinapis, *Sinapis alba*	Canada
6	Semen Sinapis, *Sinapis alba*	Provence Sichuan, China
7	Semen Sinapis, *Sinapis alba*	Provence Hebei, China

Reference compounds of Fig. 2a, 2b		R*f*
T 1	Sinalbin	0.58
T2	Sinigrin	0.53
(n.a)[a]	Sinapin	0.25
(n.a)	Sinapic acid	0.91

[a]Not applied

1. Extraction: 1 g powdered drug is sonicated with 50 ml methanol for 1 h, filtered and evaporated to dryness. The residue is dissolved in 2 ml methanol.
2. Reference compounds: Each 1.0 mg is dissolved in 1 ml ethanol

3. Separation parameters:

Plate:	HPTLC Silica gel 60 F_{254}, Merck
Applied amounts:	Semen Sinapis extracts: 20 µl each Reference compounds: 20 µl each
Solvent system:	n-Butanol + ethanol + glacial acetic acid + water (6 + 2 + 2 + 2)
Detection:	A) Trichloroacetic acid-potassium hexacyanoferrat-iron-III-chloride reagent (TPF) Solution (a) 25 % trichloroacetic acid in chloroform. Solution (b) 1 % aqueous potassium hexacyanoferrat mixed with an equal volume of 5 % aqueous iron-chloride. The plate is sprayed with solution (a) and heated at 100 °C for 10 min, followed with solution (b) and evaluated in VIS. B) Dragendorff reagent Solution (a) 0.85 basic bismuth nitrate in 10 ml glacial acetic acid and 40 ml water under heating. If necessary, filter. Solution (b): 8 g potassium iodide in 30 ml water Stock solution: (a)+(b) are mixed 1:1. Spray reagent: 1 ml stock solution is mixed with 2 ml glacial acetic acid, followed with 10 % aqueous sodium nitriteand evaluated in VIS.

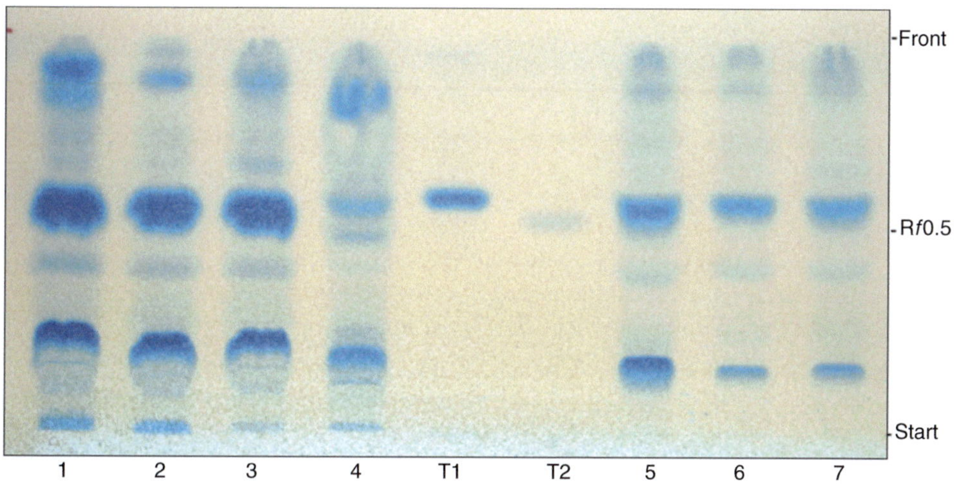

Fig. 2a Thin layer chromatogram of the methanol extracts of Sinapis Semen sprayed with Trichloroacetic acid-potassium hexacyanoferrat-iron-III-chloride reagent (TPF) (VIS)

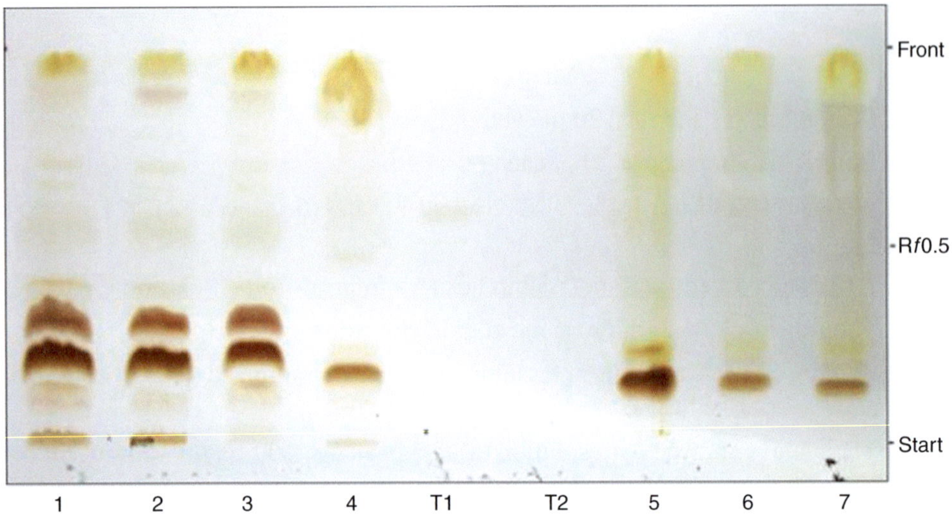

Fig. 2b Thin layer chromatogram of the methanol extracts of Semen Sinapis sprayed with Dragendorff reagent (VIS)

4. Description of Fig. 2a and 2b:

   Fig. 2a: All seven Semen Sinapis methanol extracts show in VIS 4 -5 dark blue zones with Sinalbin at R*f*=0.58 (**T1**) partly overlapped with sinigrin which in *Sinapis alba* is contained only in low concentration. In *Sinapis nigra* it is reported to be the dominant compound. The blue zone at R*f*=0.24 might be identical with sinapin the choline derivative of Sinapic acid (see also Fig. 2b). The blue zones in the front zone (R*f*=0.85 – 0.95) might be Sinapic acid and phenol carboxylic acid derivatives.

   Fig. 2b: The Dragendorff sprayed TLC shows in Sinapis extracts 1–3 strong brown zones at R*f*=0.30 and 0.20, whereas in the extracts 4, 5, 6 and 7 only one main brown zone at R*f*=0.16 could be detected. The brown zone at R*f*=0.24 might be assigned to Sinapin.

HPLC-Fingerprint Analysis

1. Sample preparation: 1 g powdered drug is sonicated with 50 ml methanol for 1 h, filtered and evaporated to dryness. The residue is dissolved in 2 ml methanol, filtered over Chromafil® Type 0.20 μm and injected into the HPLC apparatus.
2. Injection volume: Sinapis albae semen extracts: 20 μl each

   Reference compound: 20 μl

3. HPLC parameter:

Apparatus:	MERCK HITACHI D-6000 A Interface
	MERCK HITACHI L-4500 A Diode Array Detector
	MERCK HITACHI AS-2000 Autosampler
	MERCK HITACHI L-6200 A Intelligent Pump
Separation column:	LiChroCART® 250-4 LiChrospher® 100 RP-18 (5 µm), Merck
Precolumn:	LiChroCART® 4-4 LiChrospher® 100 RP-18 (5 µm), Merck
Solvent System:	A: 0.08 mol/L potassium dihydrophosphate (Millipore Ultra Clear UV plus® filtered)
	B: Acetonitrile (VWR)
Gradient:	0–40 % B in 40 min.
	40–100 % B in 20 min.
	100 % B in 5 min.
	Total run time: 65 min.
Flow:	1 ml/min
Detection:	254 and 326 nm

Retention times of the main peaks recorded at 254 and 326 nm

Peak	**Rt (min)**	**Compound**
1	6.10	Sinalbin
2	10.37	1-O-β-D-glucopyranosyl sinapate
3,6	14.1, 15.1, 20.4	Phenol carboxylic acids derivatives
4	17.9	Sinapic acid derivative
5	17.93	Sinapic acid
7	31.2	Sinapine derivative
8	34.21	*trans*-Sinapine
9	36.21	*cis*-Sinapine

4. Description of the HPLC-Fig. 3a, b, c, d:

   The HPLC-graph recorded at 254 and 326 nm is dominated by the peak (**1**) of Sinalbin and the peaks of *cis/trans* of sinapin (peak **8/9**). Differences consist in the peak intensities between Rt 13 and 32 which we have assigned to phenol carboxylic acid derivatives with conspicuous UV-maxima around 330 nm as characteristic for sinapic acid and sinapine

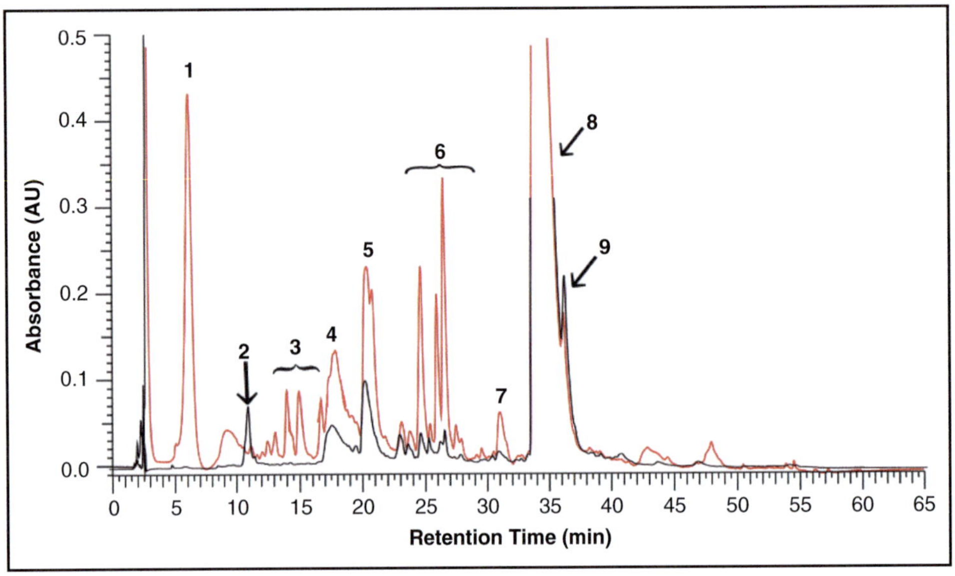

Fig. 3a HPLC-fingerprint analysis of the extract of Semen Sinapis albae (sample 1) at 254 nm (*red*) and at 326 nm (*black*)

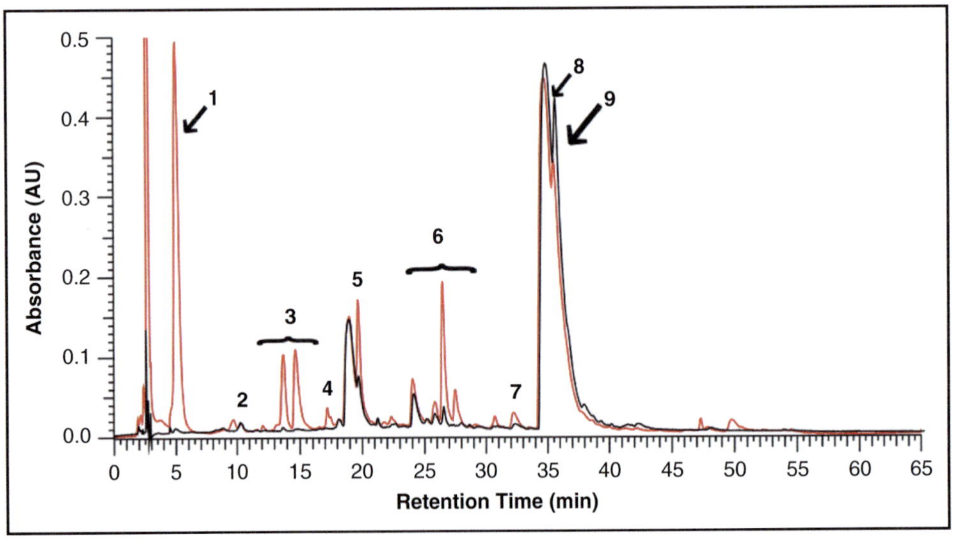

Fig. 3b HPLC-fingerprint analysis of the methanol extract of Semen Sinapis albae (sample 2) at 254 nm (*red*) and at 326 nm (*black*)

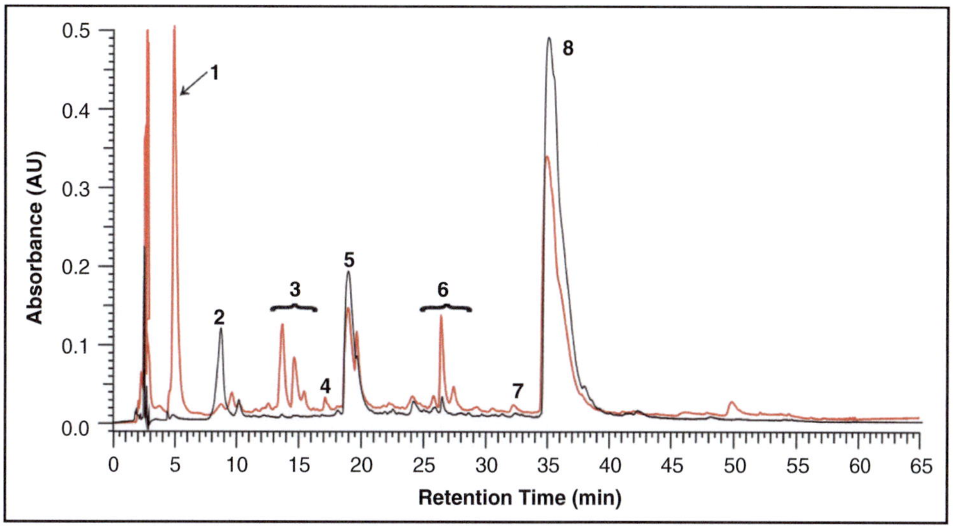

Fig. 3c HPLC-fingerprint analysis of the methanol extract of Semen Sinapis albae (sample 3) at 254 nm (*red*) and at 326 nm (*black*)

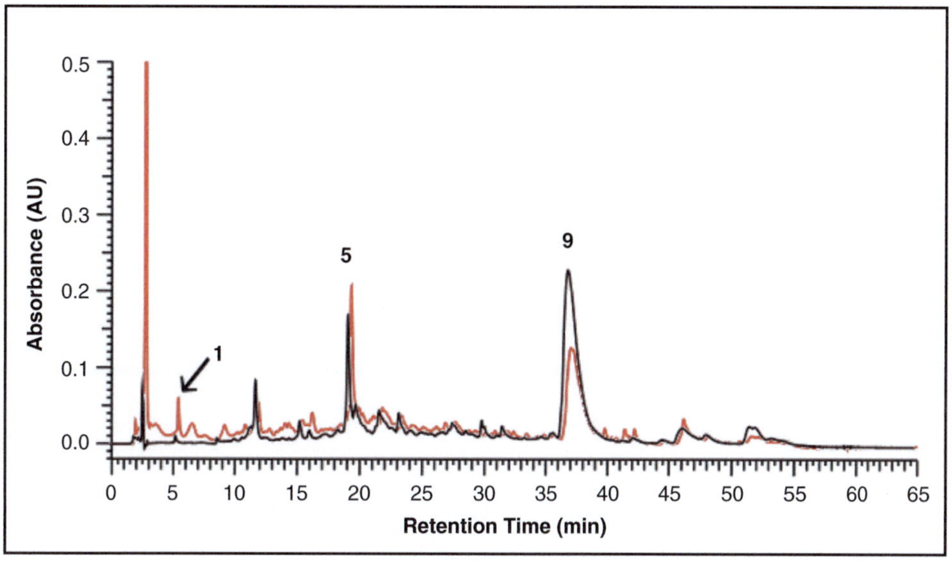

Fig. 3d HPLC-fingerprint analysis of the methanol extract of Semen Sinapis albae (sample 4) at 254 nm (*red*) and at 326 nm (*black*)

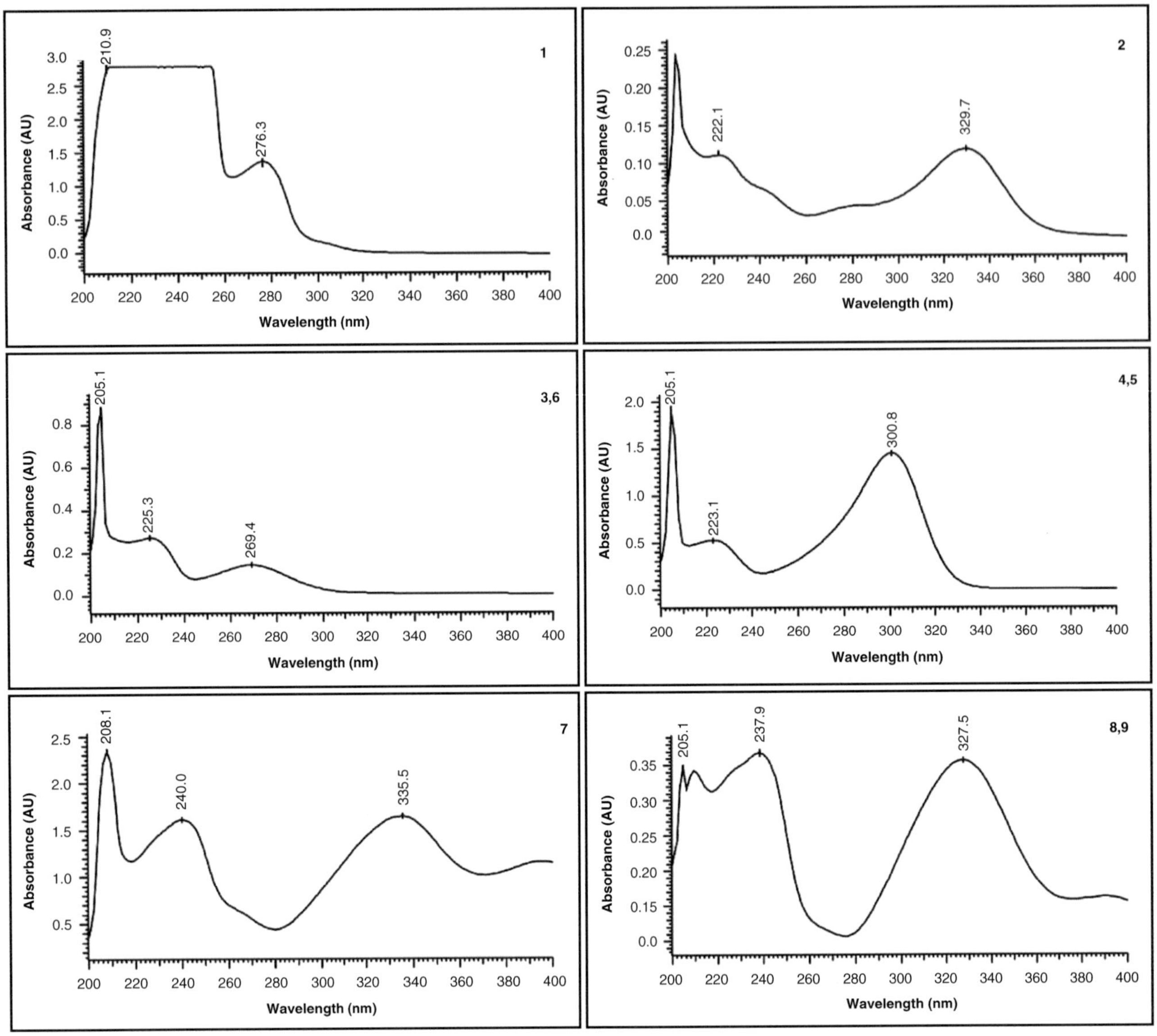

Fig. 4 UV-spectra of the main peaks of the methanol extract of Semen Sinapis

Note: The Chinese Pharmacopeia 2010 demands for Semen Sinapis a content not less than 0.4 % sinapine cyanide sulfonate, calculated with reference to the dried drug. [1]

Conclusion

The authentication of Sinapis alba Semen extracts can be definitely evidenced by the TLC-and HPLC-analysis, which is shown in Figs. 2a, b and 3a, b, c, d and corresponding UV- spectra of Fig. 4.

References

1. Pharmacopoeia of the people's republic of China, english edition, vol. I. People's Medical Publishing House, Beijing (2010)
2. Tang, W., Eisenbrand, G.: Handbook of Chinese Medicinal Plants. Wiley-VCH Verlag, Wheinheim (2011)
3. Wagner, H., Vollmar, A., Bechthold, A.: Pharmazeutische Biologie, 2, 7th edn. Wissenschftlische Verlagsgesellschaft mbH, Stuttgart (2007)
4. Hempen, C.-H., Fischer, T.: Leitfaden Chinesische Phytotherapie, 2nd edn. Urban & Fischer, Munich (2007)
5. Geng, J., Huang, W., Ren, T., Ma, X.: Materia medica der Chenesischen Arzneimitteltherapie, 2. Band, Verlag für Ganzheitliche Medizin, Dr. Erich Wühr GmbH, Kötzting/Bayer. Wald
6. Porkert, M.: Klinische Chinesische Pharmakologie, Verlag für Medizin Dr. Ewald Fischer GmbH, Heidelberg (1972)
7. Thiyam, U., Stöckmann, H., Tzm Felde, T., Schwarz, K.: Antioxidative effect of the main sinapic acid derivatives from rapeseed and mustared oil by- products. Eur. J. Lipid Sci. Technol. **3**(108), 239–248 (2006)
8. Gänzle, M.G., Engels, C., Schieber, A.: European food research and technology, sinapic acid derivatives in defatted oriental mustard (*Brassica juncea* L.) seed meal extracts using UHPLC-DAD-ESI-MS n and identification of compounds with antibacterial activity. Eur Food Res Tech. Springer, published online (2012)
9. Henning, W.: Schnelle Sinapin-Bestimmung mit Ionen-Paar HPLC aus Speisensef und Senfsaaten. Z. Lebensm. Unters. Forsch. **175**, 354–348 (1982)
10. Tan, S.H., Mailer, R.J., Blanchard, C.L., Agboola, S.A.: Extraction and residual antinutritional component in protein fraction of Sinapis alba and Beassia napus oil-free meals, 17th Ausrralian research assembly on Brassicas (ARAB), Wagga Wagga Aug (2011)
11. Peng, J., Wu, X., He, L., Fang, Y., Zi, M., Yan, S., Liu, B.: Efficacy evaluation of summer acupoint application treatment on asthma patients: a two-year follow-up clinical study, Zhong Xi Yi Yie He Xue Bao **10**(1), 39–47 (2012)
12. Wu, G.X., Lin, Y.X., Ou, M.R., Tan, D.F.: An experimental study(II) on the inhibition of prostatic hyperplasia by exert of seeds of Brassica alba. Zhongguo Zhong Yao Za Zhi **28**(7), 643–646 (2003)
13. Tesaki, S., Tanabe, S., Ono, H., Fukushi, E., Karawabata, J., Watanaba, M.: 4-Hydroxy-3-nitrophenylacetic and ainapic acids as antibacterial compounds from mustard seeds. Jpn. Soc. Biosci. Biotech. Agrochem. **62**(5), 998–1000 (1998)
14. Kim, H.W., Lee, C.H., Kim, M., Lee, H.: Antibacterial activities of phenethyl Isothiocyante and Ist derivatives against human oral pathogens. J. Korean Soc. Appl. Biol. Chem **52**(5), 555–559 (2009)
15. Lee, H., Lee, Ch., Kim, J., Pyo, S.: The inhibitory effect of sinigrin on the production of inflammatory mediatory induced by lipopolysaccharide in RAW 264.7 macrophages (1056.5). FASEB. **28**(1), (1 suppl.) 1056.5, (2014)
16. Scheffer, K.: Senföle: first- line- therapie bei Harnweginfekten. Phyto Kompass **3**, 27–29 (2014)
17. Zhang, C., Li, L., Xiao, Y., Pang, Z., Li, G.: HPLC fingerprint of processed products from Sinapis alba. Zhongguo Zhong Yao Za Zhi **35**(21), 2842–2845 (2010)
18. Yaniv, Z., Schafferman, D., Elber, Y., Ben-Moshe, Y., Zur, M.: Evaluation of *Sinapis alba*, native to Israel, as a rich source of erucic acid in seed oil. Ind. Crop Prod. **2**(2), 137–142 (1994)
19. Wagner, H., Bladt, S.: Plant drug analysis, 2nd edn. Springer, Berlin/Heidelberg (1996)

Semen Vaccariae – *Wangbuliuxing*

Pharmacopoeia: [1]	Pharmacopoeia of the People's Republic of China, English Edition Vol. I, 2010
Official drug: [1]	Cowherb Seed is the dried ripe seed of *Vaccaria segetalis* (Neck.) Garcke (Fam. Caryophyllaceae). The plant is collected in summer when the fruit is ripe but not dehiscent, and dried in the sun, the seed is tapped out, removed from foreign matter, and dried again in the sun.
Synonym: [2–4]	*Vaccaria hispanica* (Mill.) Rauschert, *Vaccaria pyramidata* (Medik)
Origin: [5]	Mainly in Chinese provinces such as Jiangsu, Hebei, Henan and Shaanxi.
Description of the drug: [1]	Spheroidal, about 2 mm in diameter. Externally black, a few in reddish-brown, somewhat lustrous, with fine and dense granular protuberances and a concave longitudinal furrow on one side. Texture hard, endosperm white, embryo curved in a ring, cotyledons 2. Odour, slight; taste, slightly astringent and bitter.
Processing: [1]	*Semen Vaccariae* (*stir-baked*) The clean Semen Vaccariae are stir-baked as described under the method for simple stir-baking (Appendix II D) until most of the seeds are burst.
Medicinal use:	Metabolic Syndrome (diabetes, hypertension) and various inflammatory diseases.

Effects and indications of Semen Vaccariae according to Traditional Chinese Medicine [1, 2, 4–13]

Taste:	Bitter
Temperature:	Neutral
Channels entered:	*Orbis hepaticus, o. stomachi*
Effects (functions):	To activate blood to unblock the meridian, promote lactation and disperse swelling, disinhibit urine and relieve stranguria
Symptoms and indications:	Amenorrhea, dysmenorrhea, agalactia, swelling and pain caused by acute mastitis, slow and painful stranguria

H. Wagner et al. (eds.), *Chromatographic Fingerprint Analysis of Herbal Medicines*, Vol. 4,
DOI 10.1007/978-3-319-32328-2_20, © Springer International Publishing Switzerland 2016

Main constituents:

- **Triterpene sapogenins** [2–4, 7]

Segetalic acid, vaccaric acid, quillaic acid, gypsogenin, gypsogenic acid, 3,4-secogypsogenic acid

- **Saponins** [2–4, 7–11, 13]

Vaccarosides A-I, vaccaroid B, dianoside G, segetoside A-I

- **Flavonoids** [2, 5, 8, 10–12]

Vaccarin, isovitexin, apigenin-6-C-arabinosyl glucoside

- **Cyclic peptides** [2, 7–11]

Cyclic pentapeptides (segetalin A-H), cyclic heptapeptide

Minor constituents:

- alkaloids, phenolic acids, steroids [2,8,10]

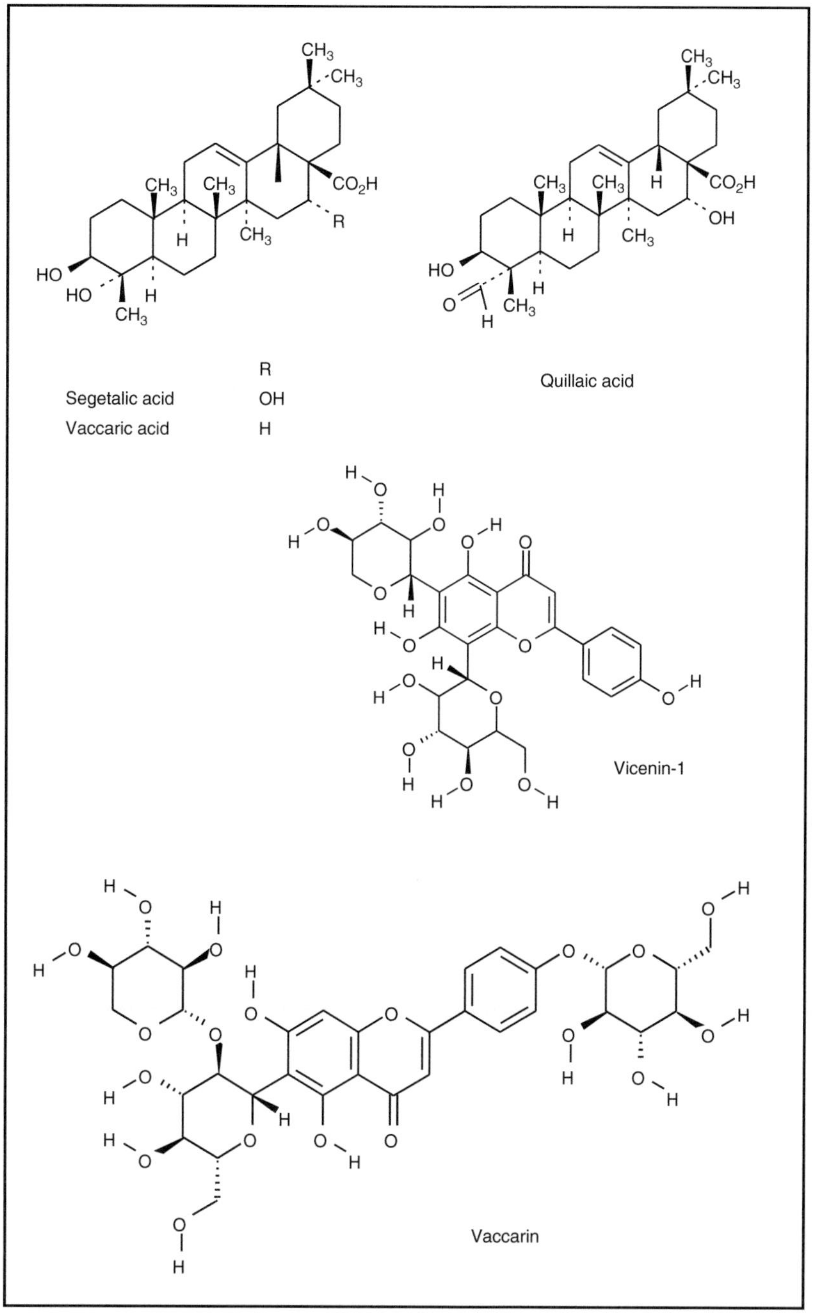

Fig. 1 Formulae of the main compounds of Semen Vaccariae (7,8)

Reported pharmacology:
- estrogen-like activity [7, 9–11]
- vasorelaxant acitivity [7, 10]
- anti-angiogenic activity [2, 12]
- growth-inhibitory activity on luteal cells, HL-60 cells and endothelial cells [2]

TLC Fingerprint Analysis

	Drug samples	**Origin**
1	Semen Vaccariae / *Vaccaria segetalis*	Sample of commercial drug obtained from TCM-Clinic Bad Kötzting (Charge: K30.08.2000)
2	Semen Vaccariae / *Vaccaria segetalis*	Sample of commercial drug obtained from TCM-Clinic Bad Kötzting (Charge: K01.09.2005)
3	Semen Vaccariae praep. / *Vaccaria segetalis*	Sample of commercial drug obtained from China Medica (origin: Neiqiu, Hebei)
4	Semen Vaccariae praep. / *Vaccaria segetalis*	Sample of commercial drug obtained from HerbaSinica (origin: Anhui)

1. TLC-fingerprint analysis of flavonoids

Reference compound of Fig. 2		**R*f***
T1	Vicenin-1[a]	0.29

[a]used as reference compound because of similar structure as vaccarin

1. Extraction: 1.5 g powdered drug is extracted under reflux with 20 ml methanol for 30 min. The extract is filtered, evaporated to dryness and the residue is dissolved in 2 ml methanol.
2. Reference compound: 1.0 mg is dissolved in 1.0 ml methanol
3. Separation parameters:

Plate: HPTLC Silica gel 60 F_{254}, Merck

Applied amounts: Semen Vaccariae extracts: each 15 µl, Reference compound: 10 µl

Solvent system: Ethyl acetate + glacial acetic acid + formic acid + water (10 + 1.1 + 1.1 + 2.6)

Detection: Natural products – Polyethylene glycol reagent (NP/PEG)

I: 1 % diphenylboric acid-*β*-ethylamino ester

(= diphenylboryloxyethylamine, NP) in methanol

II: 5 % Polyethylene glycol-4000 (PEG) in ethanol (80 %)

The plate is sprayed first with solution **I** and then with solution **II**. After 1 h the evaluation is carried out under UV 366 nm.

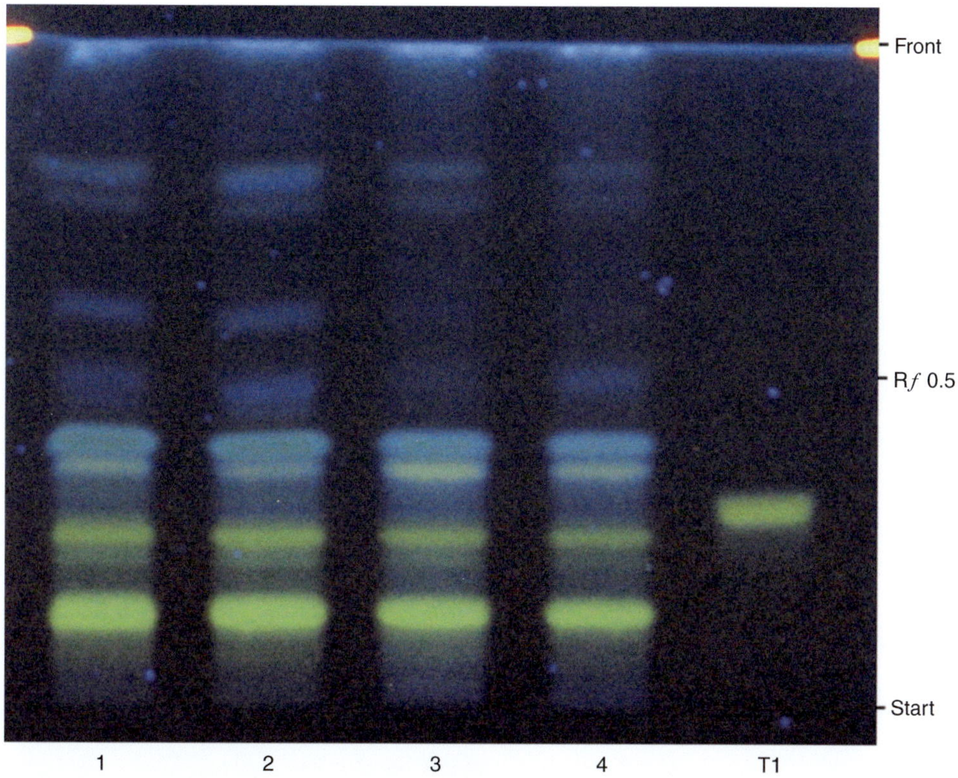

Fig. 2 Thin layer chromatogram of the methanol extracts of Semen Vaccariae, sprayed with NP/PEG reagent (UV 366 nm)

4. Description of Fig. 2:

   The Semen Vaccariae extracts samples 1–4 show in the solvent system commonly used for flavonoids a very homogenous pattern of blue and green fluorescent zones distributed over the whole TLC plate. Due to the similar structure of vicenin-1 it can be assumed that the strong green zone at R*f*=0.14 is vaccarin which contains one sugar more than vicenin-1.

2. TLC-fingerprint analysis of triterpene sapogenins and saponins

Reference compound of Fig. 3a/b		R*f*
T2	Quillaja saponin mixture	0.30

1. Extraction: 1.5 g powdered drug is extracted under reflux with 20 ml methanol for 30 min. The extract is filtered, evaporated to dryness and the residue is dissolved in 2 ml methanol.

2. Reference compound: 1.0 mg is dissolved in 1.0 ml methanol

3. Separation parameters:

Plate:	HPTLC Silica gel 60 F_{254}, Merck
Applied amounts:	Semen Vaccariae extracts: each 15 µl, Reference compound: 10 µl
Solvent system:	Chloroform + glacial acetic acid + methanol + water (6 + 3.2 + 1.2 + 0.8)
Detection:	Anisaldehyde – Sulphuric acid reagent 0.5 ml anisaldehyde is mixed with 10 ml glacial acetic acid, followed by 85 ml methanol and 5 ml concentrated sulphuric acid, in that order. The plate is sprayed with 10 ml, heated at 110 °C for 10 min and evaluated after 30 min in VIS (Fig. 3a) and under UV 366 nm (Fig. 3b). Note: The reagent has only limited stability and is no longer useable when the colour has turned to red-violet.

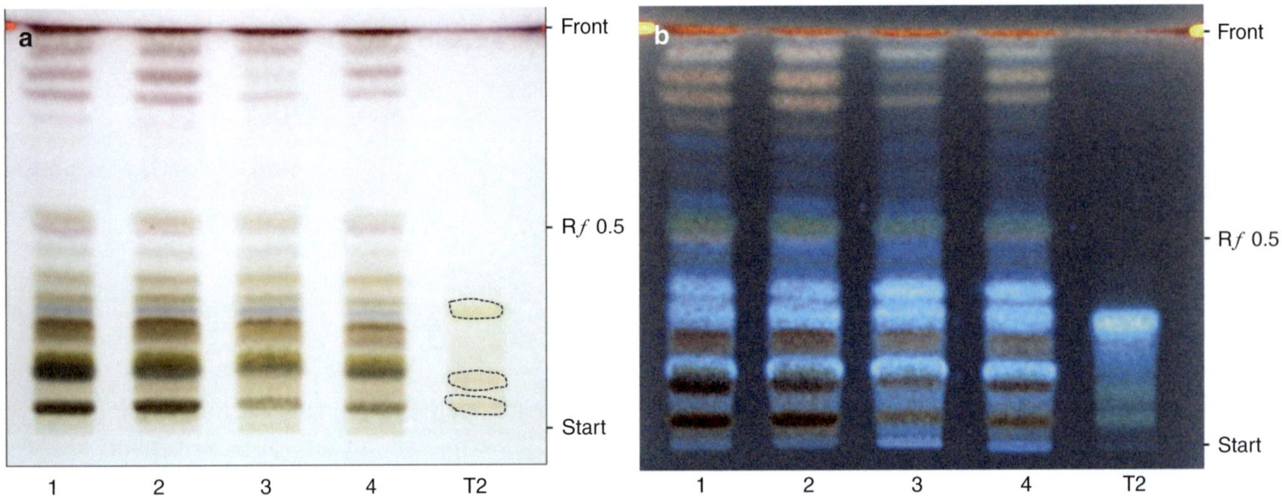

Fig. 3a/b Thin layer chromatogram of the methanol extracts of Semen Vaccariae, sprayed with Anisaldehyde – Sulphuric acid reagent (**a** = VIS, **b** = UV 366 nm)

4. Description of Fig. 3a/b:

   Fig. 3a/b: In the R*f*-range from solvent start till R*f*=0.5 the brown zones of the triterpene glycosides (saponins) of Semen Vaccariae such as vaccarosides A-I, vaccaroid B or segetoside A-I acid are visible. As reference is applied the saponin mixture of Quillaja (**T2**). In the upper TLC-range from R*f*=0.85 till the solvent front appear the corresponding sapogenins (triterpenoic acids) of Semen Vaccariae in red-violet colour.

   Fig. 3a/b: The triterpene saponins are detectable under UV 366 nm as blue fluorescent zones and the three sapogenins as auburn zones.

HPLC-Fingerprint Analysis

1. Extraction:	1.5 g powdered drug is extracted under reflux with 20 ml methanol for 30 min. The extract is filtered, evaporated to dryness and the residue is dissolved in 2 ml methanol. The extract is filtered over Chromafil®, Typ 0.20 µm.
2. Injection volume:	Semen Vaccariae extracts: each 10 µl
3. HPLC parameter:	
Apparatus:	MERCK HITACHI D-6000 A Interface
	MERCK HITACHI L-4500 A Diode Array Detector
	MERCK HITACHI AS-2000 Autosampler
	MERCK HITACHI L-6200 A Intelligent Pump
Separation column:	LiChroCART® 250-4 LiChrospher® 100 RP-18 (5 µm), Merck
Precolumn:	LiChroCART® 4-4 LiChrospher® 100 RP-18 (5 µm), Merck
Solvent system:	A: 0.05 % phosphoric acid/water (Millipore Ultra Clear UV plus® filtered)
	B: acetonitrile (VWR)
Gradient:	0 % B for 5 min,
	0–100 % B in 60 min,
	100 % B for 5 min
	total runtime: 70 min
Flow:	1.0 ml/min
Detection:	210 nm

Retention times of the main peaks

Peak	Rt (min)	Compound
1	21.8	Flavonoids
2	23.4	
3	25.0	
4a	62.8	Sapogenins
4b	64.2	
4c	65.6	

4. Description of the HPLC-Figures:

   The HPLC-peaks **1–3** of the extract samples 1 and 3 can be identified as main flavonoid-C-glycosides of Semen Vaccariae. The peaks **4a–c** are identical with the triterpenic acids (sapogenins) of Semen Vaccariae.

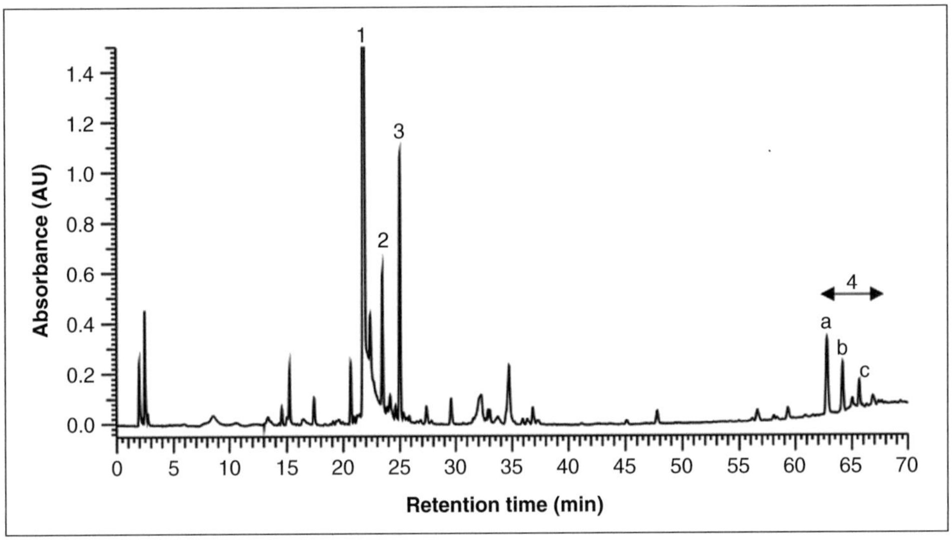

Fig. 4a HPLC-fingerprint analysis of the methanol extract of Semen Vaccariae, sample 1

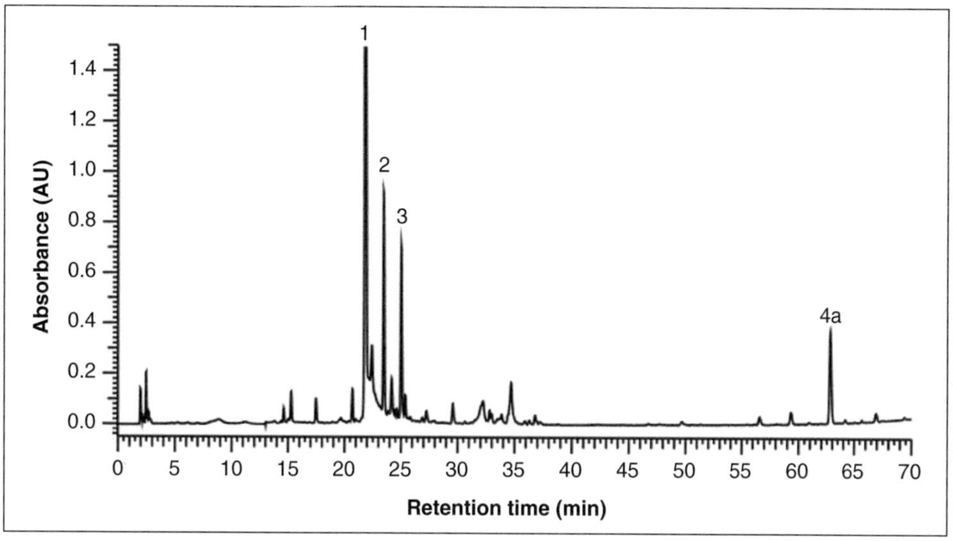

Fig. 4b HPLC-fingerprint analysis of the methanol extract of Semen Vaccariae, sample 3

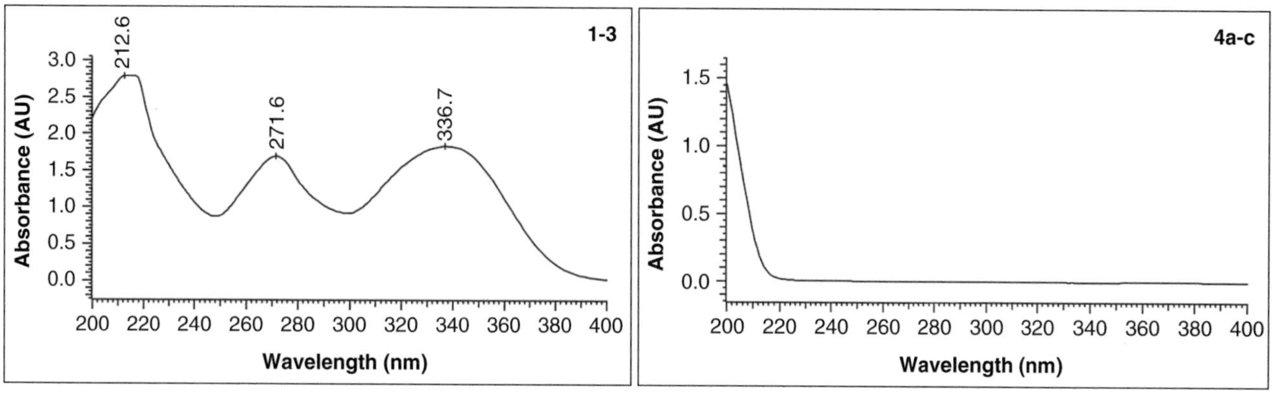

Fig. 5 On line UV-spectra of the main peaks of Semen Vaccariae extracts

Note: According to the Chinese Pharmacopeia 2010 Semen Vaccariae contains not less than 0.40 % of vaccarin, calculated with reference to the dried drug. [1]

Conclusion

The TLC-pictures and HPLC-graphs provide a significant authentication of Semen Vaccariae.

References

1. Pharmacopoeia of the people's republic of China, English edition, vol. I. People's Medical Publishing House, Beijing (2010)
2. Zhang, H, Jing, Y, Wu, G: Inhibitory effects of crude polysaccharides from semen vaccariae on benign prostatic hyperplasia in mice. J. Ethnopharmacol. **145**(2), 667–669 (2013)
3. Jia, Z, Koike, K, Kudo, M, Li, H, Nikaido, T: Triterpenoid saponins and sapogenins from *Vaccaria segetalis*. Phytochemistry **48**(3), 529–536 (1998)
4. Koike, K, Jia, Z, Nikaido, T: Triterpenoid saponins from *Vaccaria segetalis*. Phytochemistry **47**(7), 1343–1349 (1998)
5. Zhang, ZZ: An illustrated Chinese materia medica in Hong Kong. School of Chinese Medicine/Hong Kong Baptist University, Hong Kong (2004)
6. Hempen, C-H, Fischer, T: A materia medica for Chinese medicine, English edition. Elsevier GmbH, Urban & Fischer, Munich (2009)
7. Tang, W, Eisenbrand, G: Handbook of Chinese medicinal plants – chemistry, pharmacology, toxicology, vol. 2. WILEY-VCH Verlag GmbH & Co. KgaA, Weinheim (2011)
8. Qi, P, Zhang, F, Xue, R, Li, Z, Chen, M, Sun, Z, Zhu, K, Huang, C: Identification of multiple constituents from seed of *Vaccaria segetalis* with an adsorben-separation strategy based on liquid chromatography coupled to quadrupole time-of-flight mass spectrometry. Rapid Commun. Mass Spectrom. **28**(11), 1243–1257 (2014)
9. Shih, CC, Lin, CH, Lin, WL: Ameliorative effects of Vaccaria segetalis extract on osteopenia in ovariectomized rats. J. Nat. Med. **63**(4), 386–392 (2009)
10. Wang, X, Dong, H, Liu, Y, Yang, B, Wang, X, Huang, L: Application of high-speed counter-current chromatography for preparative separation of cyclic peptides from Vaccaria segetalis. J. Chromatogr. B **879**(11–12), 811–814 (2011)
11. Itokawa, H, Yun, Y, Morita, H, Takeya, K, Yamada, K: Estrogen-like activity of cyclic peptides from Vaccaria segetalis extracts. Planta Med. **61**(6), 561–562 (1995)
12. Yang, G, Zhang, N, Wang, T, Zhang, S, Xu, R, Zhu, Z, Liu, K: Liquid chromatography-tandem mass spectrometric assay for the determination of vaccarin in rat plasma: application to a pharmacokinetic study. Biomed. Chromatogr. **28**(12), 1789–1794 (2014)
13. Sang, SM, Lao, AN, Chen, ZL, Uzawa, J, Fujimoto, Y: Three new triterpenoid saponins from the seeds of Vaccaria segetalis. J. Asian Nat. Prod. Res. **2**(3), 187–193 (2000)

Semen Plantaginis – *Cheqianzi*
Herba Plantaginis – *Cheqiancao*

Pharmacopoeia: [1]

Pharmacopoeia of the People's Republic of China, English Edition Vol. I, 2010

Official drugs: [1]

Semen: Plantain Seed is the dried ripe seed of *Plantago asiatica* L. or *Plantago depressa* Willd. (Fam. Plantaginaceae).

The fruit-spike is collected in summer and autumn when the seeds are ripe and dried in the sun. The seeds are rubbed out and removed from foreign matter.

Herba: Plantain Herb is the dried herb of *Plantago asiatica* L. or *Plantago depressa* Willd. (Fam. Plantaginaceae)

The drug is collected in summer, removed from soil, and dried in the sun.

Origin: [2, 3]

China (Heilongjiang, Jilin, Liaoning, Hebei, Shandong, Shanxi, Shaanxi, Gansu, Qinghai, Inner Mongolei, Henan, Guizhou, Anhui, Jaiangxi, Sichuan)

Description of the drugs: [1]

Semen Plantaginis: Ellipsoid, irregularly oblong or triangular oblong, slightly flattened, about 2 mm long, 1 mm wide. Externally yellowish-brown to dark brown, with fine wrinkles, and a greyish-white concave pointed hilum on one side. Texture hard. Odour, slight; taste, weak.

Herb of Plantago asiatica: Roots fascicle, fibrous. Leaves basal. Long petioled; lamina crumpled, when whole, ovate-elliptical or broadly ovate, 6–13 cm long, 2.5–8 cm wide; externally greyish- green or blackish-green, with 5–7 distinct arched veins; obtuse or short acute at the apex, broadly coneate at the base, margin entire or irregularly sinuate-dentate. Spikes several, with along scape. Capsules circumscissile, calyx persistent. Odour, slightly aromatic; taste, slightly bitter.

Herb of Plantago depressa**:** Main roots straight and long. Lamina relatively narrow, long-elliptical or ellipticalanceolate, 5–14 cm long, 2–3 cm wide.

H. Wagner et al. (eds.), *Chromatographic Fingerprint Analysis of Herbal Medicines*, Vol. 4,
DOI 10.1007/978-3-319-32328-2_21, © Springer International Publishing Switzerland 2016

Pretreatment of the raw drugs: [1, 4]

Semen Plantaginis: Elemination of foreign matters.
Semen Plantaginis (processed with salt):

The clean Semen Plantaginis is stir-baked as described under the method for processing with salt-water (Appendix II D) until it cracks, sprayed with salt-water and stir-baked to dryness.

Herba Plantaginis: The drug is eleminated from foreign matter, washed clean, cut into section and dried.

Medicinal use: [1, 4]

Semen Plantaginis is used for the treatment of edema, dysuria, painful urination, cough, and ophthalmic infection. Herba Plantaginis is used for the treatment of edema with oligoriae, also for infection and painful urination, cough, hemoptysis, and carbuncle.

Effects and indications of Semen and Herba Plantaginis according to Traditional Chinese Medicine [1, 2, 4–6]

Taste:	Semen: sweet Herba: sweet
Temperature:	Semen: cold Herba: cold
Channels entered:	Semen: *Orbis pulmonalis*, *o. intestini tenuis*, *o. hepaticus*, *o. renalis* Herba: *Orbis linalis*, *o. hepaticus*, *o. pulmonalis*, *o. intestini tenuis*
Effects (functions):	Semen: promotes urination, cools heat, clears damp- heat, dislodges phlegm, stops coughing, diffuses the lung, soothes the throat and expels pus. To brighten the eyes Herba: to clear heat, disinhibit urine and relive stranguria, to dispel phlegm, cool the blood, and remove toxin
Symptoms and indications:	Semen: painful bladder dysfunction, rough scanty urination, edema, painful urination, couch and profuse sputum, oppression and discomfort in the chest, sore throat and hoarseness, lung abscess with pyemesis. Heat of the liver, visual impairment, red eyes, red eyes due to wind- heat Herba: heat stangury with slow pain, edema and small quantity of urination, diarrhea caused by summer heat dampness, phlegm- heat cough, hematemesis and epistaxis, swelling abscess, sore and toxin

Published constituents: [1, 3, 4, 7–10]

Phenylethanol glycosides:	Plantaginoside, acteoside (verbascoside), isoacteoside, calcerolioside B, leucoscoposide, martynoside, isomartynoside, plantainoside (A-F), plantasioside. Plantamajoside (Herba)
Flavonoide glycosides:	Plantaginis Herba : Homoplantaginin (hispidulin-7-O-D-glucoside), plantaginin (scutellarein 7-O-D-glucoside), luteolin, luteolin-7-glucoside, plantagoside (from Semen)
Iridoid glycosides:	Aucubin, catalpol, geniposidic acid
Triterpenoides:	Ursolic acid, palmitates of β-sitosterol, β-sitosterol, stigmasterin

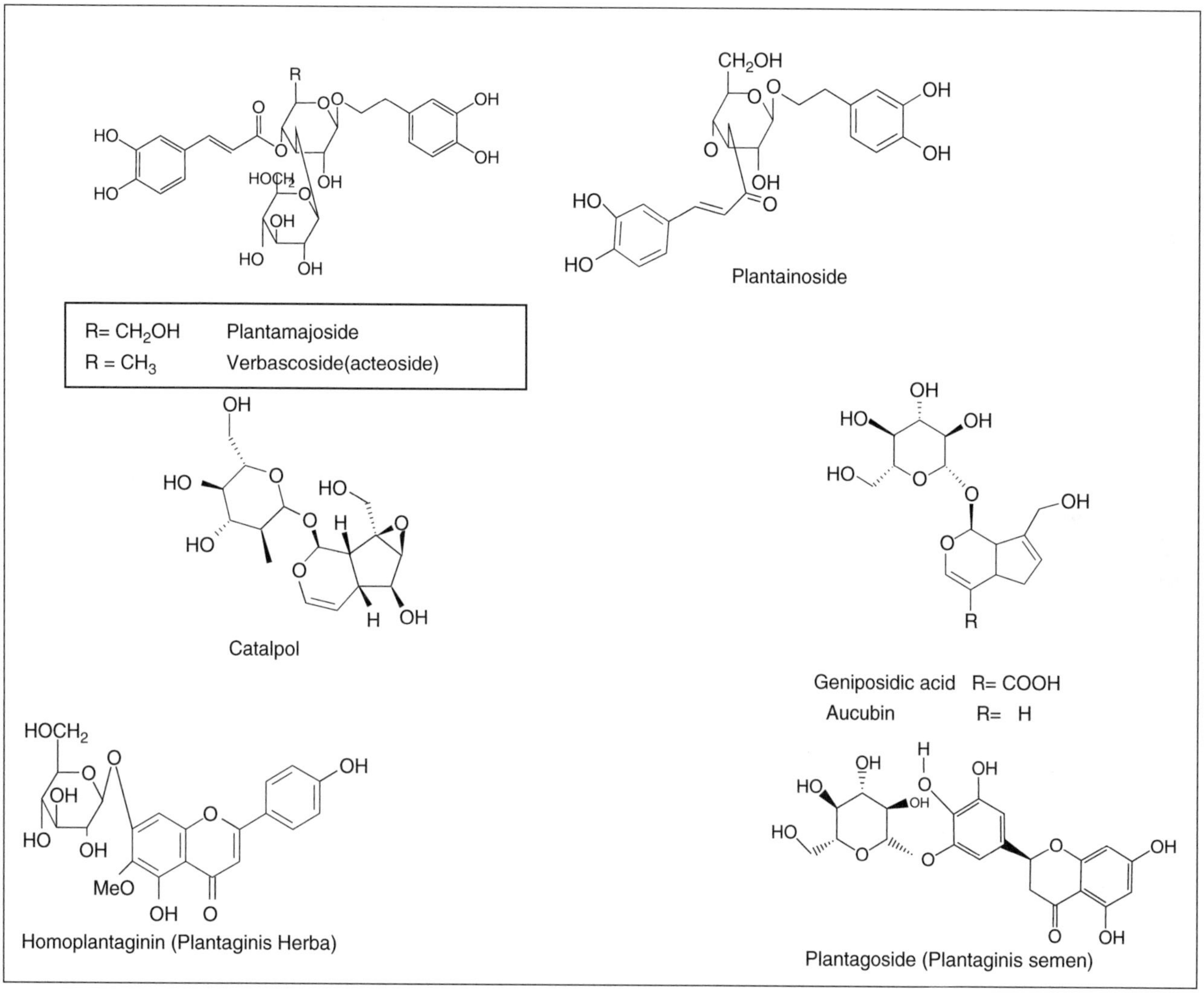

Fig. 1 Formulae of the main constituents of Semen and Herba Plantaginis [4, 8, 11]

Reported pharmacological activities:

- antioxidant with very low toxicity (*in vitro* and *in vivo*) [8, 11, 12]
- diuretic (Seeds and Herba) [4]
- expectorant (Seeds and Herba) [4]
- hemostatic agent (Herba) [4]
- anti- inflammatory
- immunomodulatory activity (Seeds) [11, 13]
- plantagoside showed a potent inhibitory activity against jack bean α-mannosidase [14]
- hepatoprotective [11]
- antibacterial [11]

TLC Fingerprint Analysis

	Drug samples	Origin
1	Semen Plantaginis / *Plantago asiatica* L. or *Plantago depressa* Willd	Sample of commercial drug, obtained from firm China Medica Province Shifang, Sichuan (China)
2	Semen Plantaginis / *Plantago asiatica* L. or *Plantago depressa* Willd	Sample of commercial drug, obtained from China Medica
3	Semen Plantaginis / *Plantago asiatica* L. or *Plantago depressa* Willd	Sample of commercial drug, obtained from TCM-Clinic Bad Kötzting (Charge: A 22.08.02)
4	Semen Plantaginis / *Plantago asiatica* L.	Sample of commercial drug, obtained from Firm Herba Sinica (origin: Hebei)
5	Semen Plantaginis / *Plantago asiatica* L. or *Plantago depressa* Willd	Sample of commercial drug, obtained from TCM-Clinic Bad Kötzting (Charge: K 30.08.2000)
6	Herba Plantaginis / *Plantagio asiatica* L. or *Plantago depressa* Willd	Province Hebei (China)
7	Herba Plantaginis / *Plantago asiatica* L. or *Plantago depressa* Willd	Province Henan (China)

1. **TLC-fingerprint analysis of Phenylethanol glycosides and iridoides:**

Reference compounds of Fig. 2a		**R*f***
T1	Aucubin	0.32
T2	Verbascoside (Semen)	0.75
T3	Catalpol	0.26

1. Extraction: 1 g powdered drug is extracted with 10 ml methanol under reflux for 20 min. and filtered. The filtrate is evaporated to dryness; The residues are dissolved in 1 ml MeOH and used as the test solution.

2. Reference compounds: Each 1 mg is dissolved in 1 ml ethanol

3. Separation parameters:

   Plate: HPTLC Silica gel 60 F_{254}, Merck

   Applied amounts: Semen, Herba Plantaginis extracts: 10 µl each
   Reference compounds: 10 µl each

   Solvent system: ethyl acetate + glacial acetic acid + formic acid + water (8 + 1 + 1 + 2)

   Detection: Anisaldehyde- sulphuric acid reagent:
   0.5 ml anisaldehyde is mixed with 10 ml glacial acetic acid,
   followed by 85 ml methanol and 5 ml concentrated sulphuric acid, in that order.
   The plate is sprayed with about 20 ml reagent and heated at 105 °C for 5–10 min.
   The evaluation is carried out under Vis and at 366 nm.

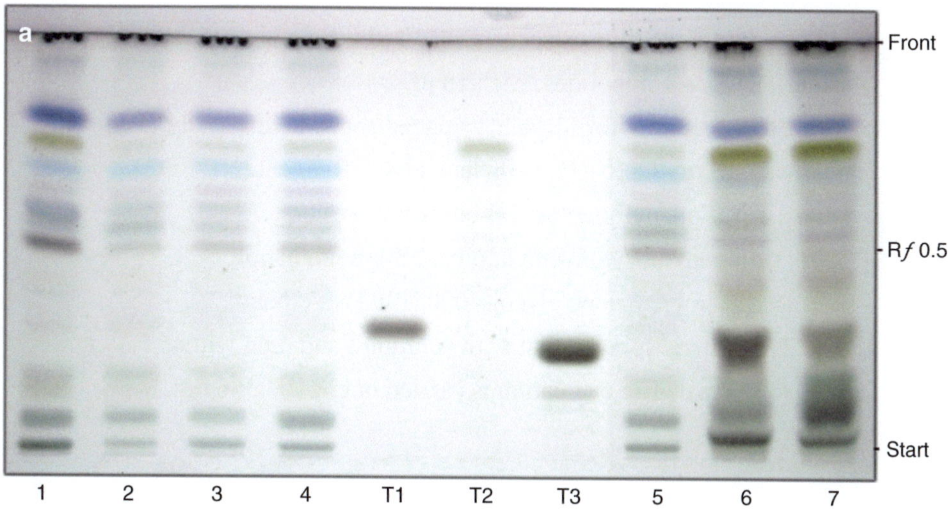

Fig. 2a Thin layer chromatogram of the methanol extracts of **Semen** and **Herba Plantaginis** sprayed with anisaldehyde- sulphuric acid reagent (VIS)

Description of Fig. 2a:

All 1–5 samples of Semen Plantaginis show in the deep R*f*- range from start till R*f*~0.25 three weak grey zones. None of them can be assigned to any defined compound. In the upper R*f* – range appear from R*f*=0.50 till R*f*=0.83 ca.5 grey zones with one green and one blue zone. The grey zone at R*f*=0.75 is identical with Verbascoside (**T2**) whereas in the deep *Rf* value one of the two grey zones on the solvent start and R*f*=0.15 might be geniposidic acid. The Herba Plantago extracts samples 6 and 7 differ from the Semen extracts by two strong zones at R*f*=0.75 (Verbascoside) and above of this at R*f*=0.78 a non-identified compound. In the deep R*f*- range of the Herba extract appear in high concentration Aucubin (**T1**), catalpol (**T3**) and again geniposidic acid analogue to the Semen extract sample.

2. TLC-fingerprint analysis of flavonoids

Reference compounds of Fig. 2b		**R*f***
T1	Luteolin	0.98
T2	Luteolin-7-O-glucoside	0.71
T3	Rutin	0.43
n.a*	Verbascoside	0.67

*not applied

1. Extraction: The same extract from TLC (1)
2. Reference compounds: Each 0.5 mg is dissolved in 0.5 ml methanol
3. Separation parameters:

Plate:	HPTLC Silica gel 60 F254, Merck
Applied amounts:	Semen Plantaginis extracts: each 25 µl
	Herba Plantaginis extracts: each 10 µl
	Reference compounds: each 10 µl
Solvent system:	Ethyl acetate + formic acid + glacial acetic acid + water (100 + 11 + 11 + 26)
Detection:	Natural products – Polyethylene glycol Reagent (NP/PEG)
	I: 1 % diphenylboric acid-β-ethylamino ester
	(= diphenylboryloxyethylamine, NP) in methanol
	II: 5 % polyethylene glycol-4000 (PEG) in ethanol
	The plate is sprayed first with solution I and then with solution II. The evaluation is carried out in UV 366 nm.

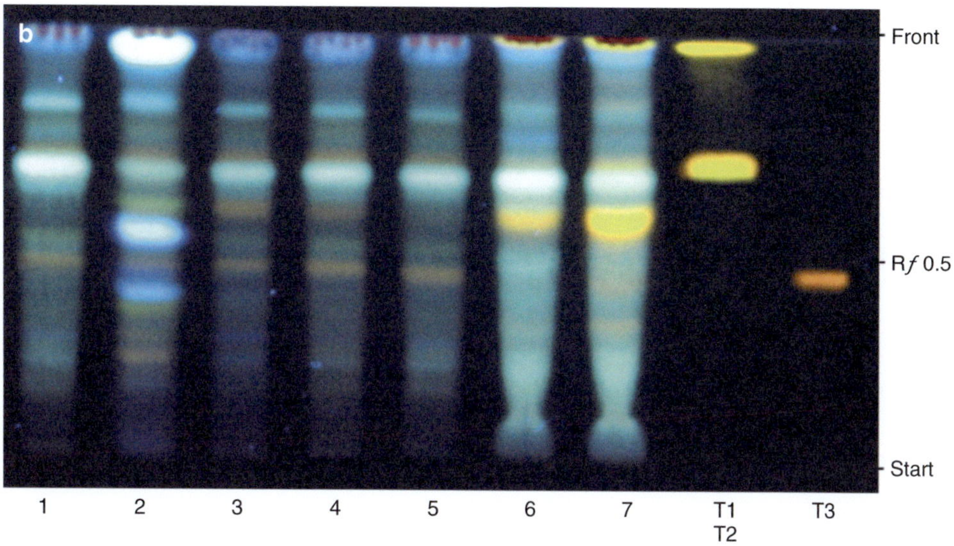

Fig. 2b Thin layer chromatogram of the methanol extracts of **Semen** and **Herba Plantaginis** sprayed with NP/PEG reagent (UV 366 nm)

Description Fig. 2b**:**

The Plantago **semen** extract samples 1–5 show with the exception of extract sample 2 a very homogeneous TLC- zone profile with green/white fluorescene zones in the R*f*-range of 0.4 up to the solvent front with the dominant zone of verbascoside at R*f*=0.67. Extract sample 2 differs from the others by a strong white fluorescent zone at R*f*=0.98 under the solvent front. The brown zones in extract samples 3, 4, and 5 could be not correctly assigned to any characteristic compound.

The Herba Plantaginis extract samples 6 and 7 are characterized again by the dominant verbascoside zone overlapped by luteolin-7-glucoside (**T2**). The strong flavonoid glycoside at R*f*=0.67 might be identical with luteolin-O-diglucoside

HPLC-Fingerprint Analysis

1. Sample preparation: 1 g powdered drug is extracted with 10 ml methanol under reflux for 20 min. The extract is cooled, filtrated and evaporated to dryness. The residue is dissolved in 1 ml methanol and filtered over Chromafil® filtration unit, type 0–20 µm/25 mm
2. Injection volume: Plantaginis Semen extracts: 30 µl each
   Plantaginis Herba extracts: 20 µl each
   Reference compounds: 10 µl each

3. HPLC parameter:

Apparatus:	MERCK HITACHI D-6000 A Interface MERCK HITACHI L-4500 A Diode Array Detector MERCK HITACHI AS-2000 Autosampler MERCK HITACHI L-6200 A Intelligent Pump
Separation column:	LiChroCART® 250 -4 LiChrospher® 100 RP-18 (5 µm), Merck
Precolumn:	LiChroCART® 4-4 LiChrospher® 100 RP-18 (5 µm), Merck
Solvent System:	A: 0.0001 % aq. H_3PO_4: (Millipore Ultra Clear UV plus® filtriert)
Gradient:	B: Acetonitril (VWR)
Flow:	1 ml/min
Detection:	210 nm

Retention times of the main peaks recorded at 210 nm

Peak	**Rt (min)**	**Compound**
1	2.5	Catalpol
2	3.2	Geniposidic acid
3	4.1	Aucubin
4	8.3	Not identified
5	17.8	Not identified
6	21.4	Phenylethanol glycoside?
7	22.5	Phenylethanol glycoside?
8	25,6	Not identified phenylethanoid from (**Herba**)
9	27.8	Verbascoside (**Semen**) or Plantamajoside (**Herba**)
10	29.6	Luteolin-7-O- glucoside? (**Herba**)
11, 12	31.0, 38.3	Not identified phenylethanoid glycoside
13	39.9	Hispidulin-7-O-glucoside
14,15,16	(40–50)	Not identified flavones

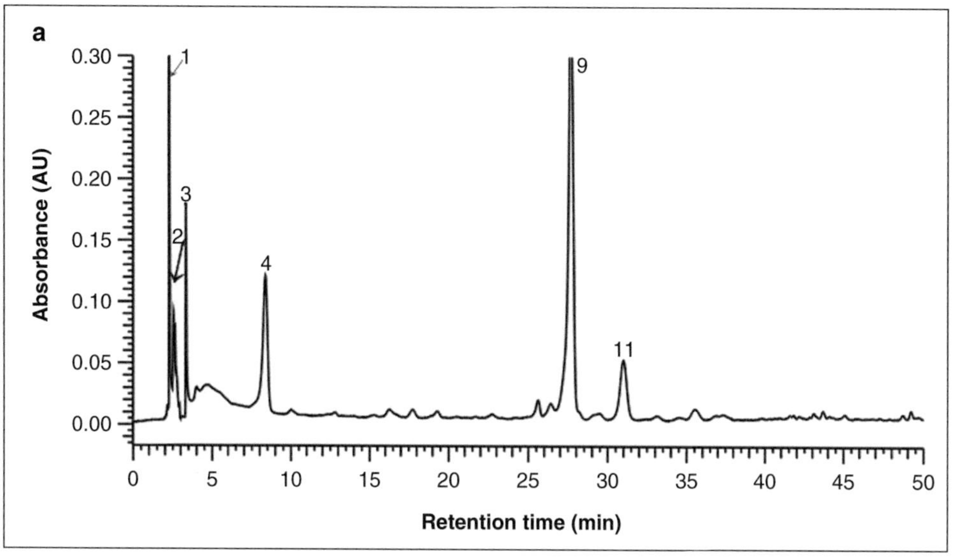

Fig. 3a HPLC-fingerprint analysis of the methanol extract of **Semen Plantaginis** sample 1

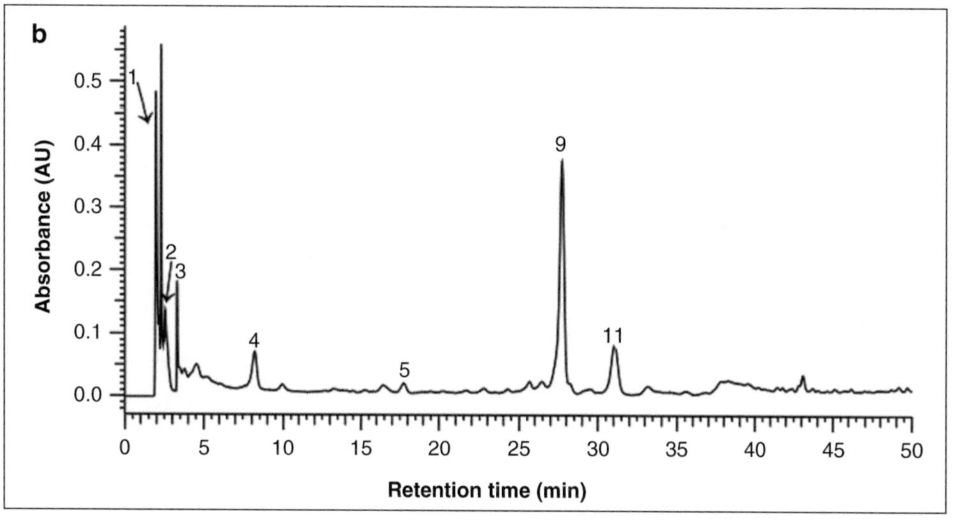

Fig. 3b HPLC-fingerprint analysis of the ethanol extract of **Semen Plantaginis** sample 3

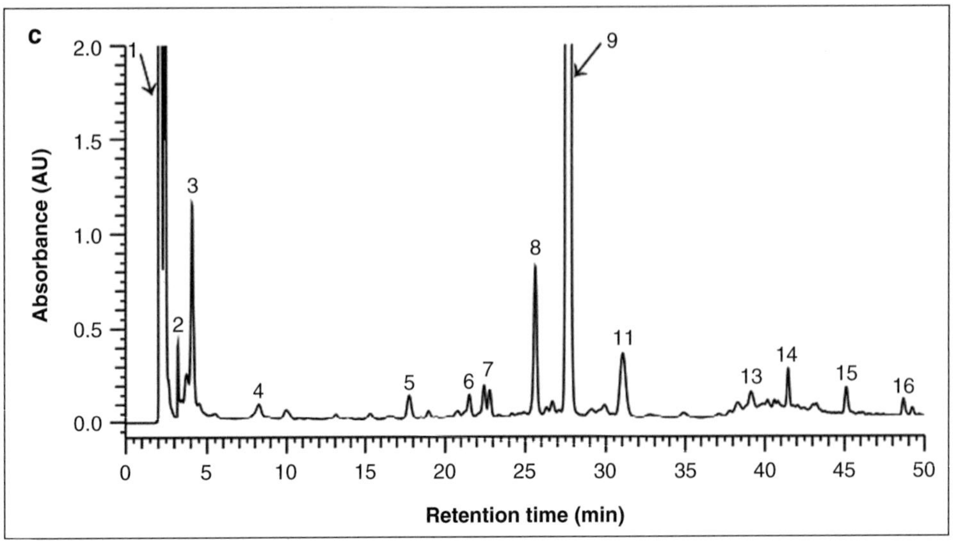

Fig. 3c HPLC-fingerprint analysis of the ethanol extract of **Herba Plantaginis** sample 6

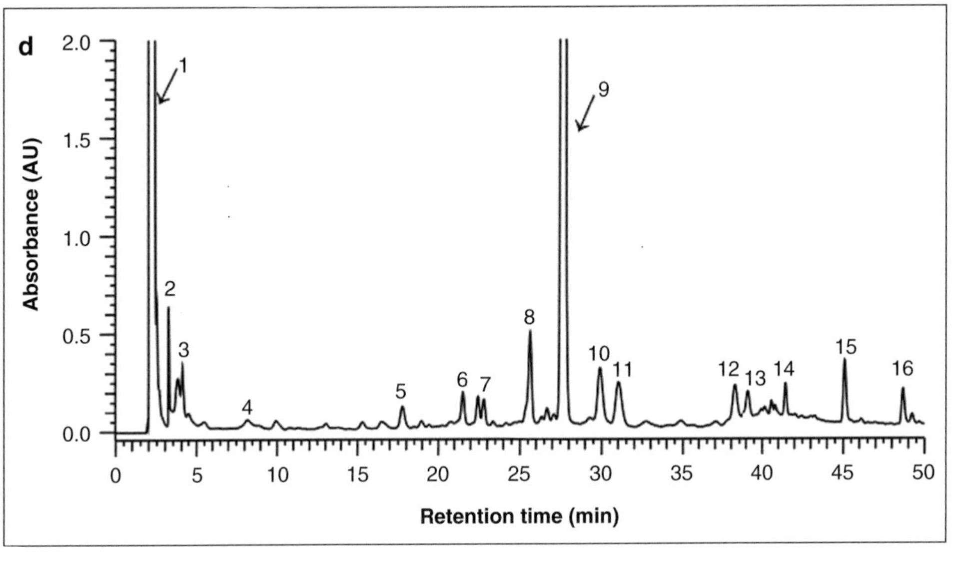

Fig. 3d HPLC-fingerprint analysis of the ethanol extract of **Herba Plantaginis** sample 7

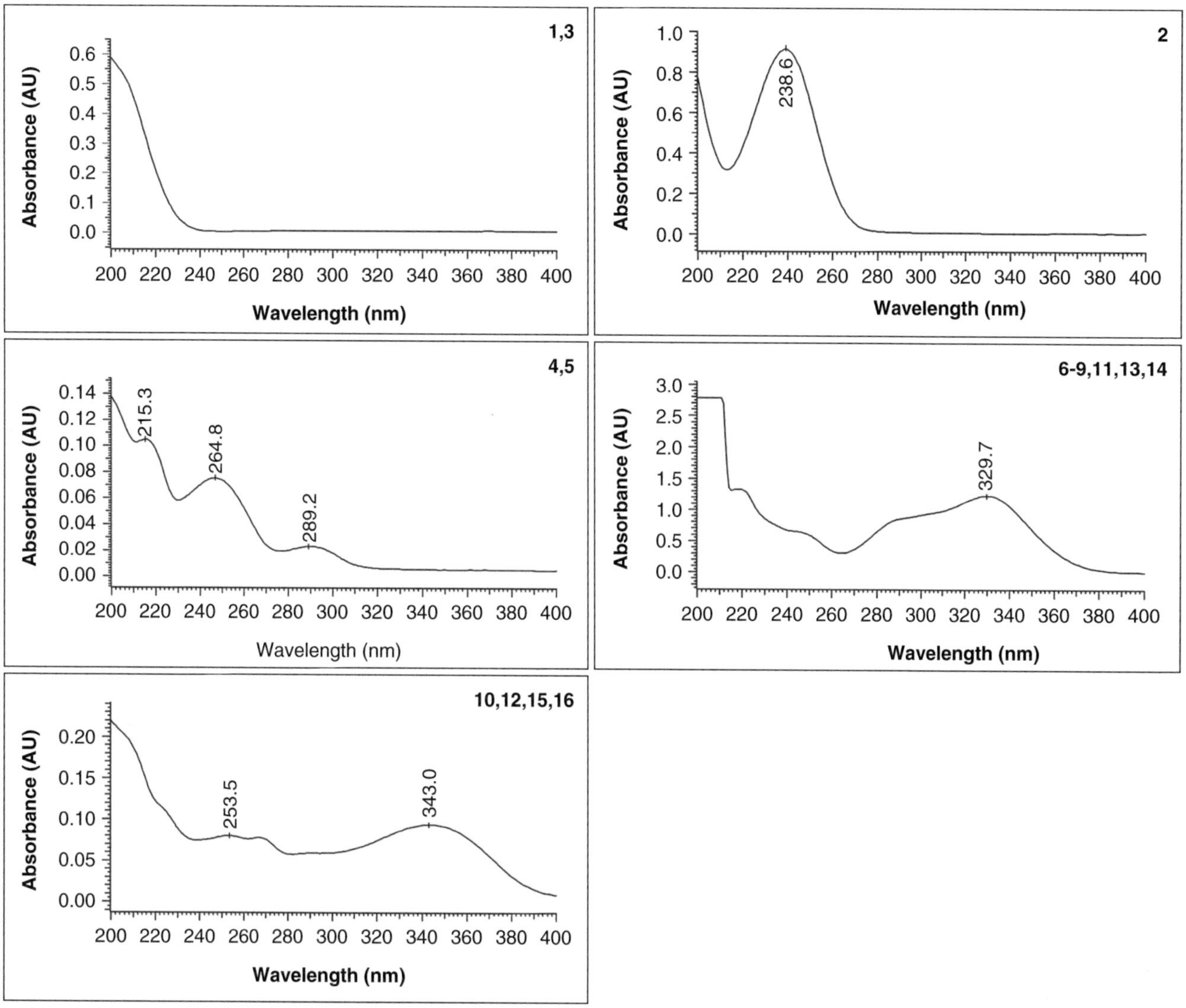

Fig. 4 On line UV-spectra of the main compounds (peaks) of **Semen**, **Herba** Plantaginis

4. Description of the HPLC fingerprints

   Fig. 3a and 3b (Plantaginis Semen):

   The HPLC- peak profile of MeOH-extract of Semen Plantaginis (extract sample 1 and 3) is characterized by the peaks **1–4** in the Rt-range 2.0–10.0 which can be assigned to Catalpol (Rt=2.50), Geniposidic acid (Rt=3.23) and (Aucubin Rt=4.11). The second Rt-range between 25.0 and 32.0 (numbered with **9** and **11)** were identified as verbascoside and another not identified phenylethanol glycoside respectively.

   Fig. 3c and 3d (Plantaginis Herba):

   The extract samples 6 and 7 show a quite different peak profile. Beside the characteristic peaks of extract sample 1 and 3 there are further small 10 peaks in the Rt-range from 17.0 to 50.0 which can be assigned analogous their UV- spectra to the class of Phenylethanol (peaks **6, 7, 8, 9, 11, 13, 14**) and flavonoids (peaks: **10, 12, 15, 16**)

Conclusion

The authentication of both drugs is possible by TLC and HPLC without any difficulties.

References

1. Pharmacopoeia of the people's republic of China, English edition, vol. I. People's Medical Puplishing House, Beijing (2010)
2. Paulus, E., Ding, Y.-H.: Handbuch der traditionellen chinesischen Heilpflanzen. Karl F Haug Verlag, Heidelberg (1987)
3. Sun, Q., Geng, F., Cheng, X., Yang, L., Wang, Z.: Qualitative and quantitative analysis of plantamajoside in Plantaginis Herba. Zhongguo Zhong Yao Za Zhi **35**(16), 2095–2098 (2010)
4. Tang, W., Eisenbrand, G.: Handbuch of Chinese medicinal plants. Wiley-VCH Verlag, Weinheim (2011)
5. Hempen, C.-H., Fischer, T.: Leitfaden Chinesische Phytotherapie, 2nd edn. Urban & Fischer, Munich (2007)
6. Porkert, M.: Klinische Chinesische Pharmakologie, Verlag für Medizin Dr. Ewald Fischer GmbH, Heidelberg (1972)
7. Gonda, S., Nguyen, N., Gyèmánt, D., Máthé, C., Vasas, G.: Detrmenation of phenylethaoid glycosides and irdroid glycosides from therapeutically used *Plantago* species by CE-MEKC. Electrophoresis **34**(17), 2577–2584 (2013)
8. Hu, L., Nie, S., Wu, Q., Li, C., Fu, Z., Gong, J., Cui, S., Xie, M.: Polysaccharide from seed of *Plantago asiatica* L. affects lipid metabolism and colon microbiota of mouse. Agric. Food Chem. **62**(1), 229–234 (2014)
9. Aritomi, M.: Homoplantaginin, a new Flavonoid Glycoside in Leaves of *plantago asiatica* Linnaeus. Chem. Pharm. Bull. **15**(4), 432–434 (1963)
10. Ravn, H., Mondolot, L., Kelly, M., Lykke, A.: Plantamajoside – a current review. Phytochem. Letters **12**, 42–53 (2015)
11. Qi, M., Xiong, A., Geng, F., Yang, L., wang, Z.: A novel strategy for target profiling analysis of bioactive phenylethanoid glycosides in *Plantago* medicinal plants using ultra-performance liquid chromatography coupled with tandem quadrupole mass spectrometry. J. Sep. Sci. **35**(12), 1470–1478 (2012)
12. Zhongzhen, Z.: An illustrated Chinese materia medica in Hong Kong, 1st edn. Chung Hwa Book Co, Hong Kong (2004)
13. Huang, D.F., Xie, M.Y., Yin, J.Y., Nie, S.P., Tang, Y.F., Xie, X.M., Zhou, C.: Immunomodulatory activity of the seeds of *Plantago asiatica* L. J. Ethnopharmacol. **3**(124), 493–498 (2009)
14. Yamada, H., Nagai, T., Takemoto, N., Endoh, H., Kiyohara, H., Kawamura, H., Otsuka, Y.: Plantagoside, a novel alpha-mannosidase inhibitor isolated from the seeds of *Plantago asiatica*, suppresses immune response. Biochem. Biophys. Res. Commun. **3**(165), 1292–1298 (1989)

Spica Prunellae – *Xiakucao*

Pharmacopoeia: [1]	Pharmacopoeia of the People's Republic of China, English Edition Vol. I, 2010
Official drug: [1]	Common Selfheal Fruit-Spike is the dried fruit-spike of *Prunella vulgaris* L. (Fam. Lamiaceae). The drug is collected in summer when the spike becomes brownish-red, removed from foreign matter, and dried in the sun.
Origin: [2–9]	Mainly in Chinese provinces (Anhui, Henan, Jiangsu, Zheijiang), Korea and Japan. Northern Hemisphere (Europe, northwestern Africa, North America)
Descriptions of the drug: [1]	Cylindrical, somewhat flattened, 1.5–8 cm long, 0.8–1.5 cm in diameter; pale brown to brownish-red. The whole spike composed of several or up to ten or more whorls of persistent calyx and bracts, each whorl with two opposite bracts, fan-shaped, apex acuminate, striations of vein distinct, the outer surface with white hairs. Each bract with three flowers, the corolla often fallen off, persistent calyx bilabiate, with four small brown ovoid nutlets, when white convex at the acute end. Texture light; Odour, slight; taste, weak.
Medicinal use: [10]	Spica Prunellae is used in western herbal medicine in the form of a hot water infusion sweetened with honey to treat inflammatory affections of mouth and throat sores.

Effects and indications of Spica Prunellae according to Traditional Chinese Medicine [1, 3, 9–16]

Taste:	Pungent, bitter
Temperature:	Cold
Channels entered:	*Orbis hepaticus*, *o. vesicalis*, *o. lienalis*, *o. felleus*, *o. pulmonalis*
Effects (functions):	To clear the liver and purge fire, improve vision, dissipate binds and disperse swelling
Symptoms and indications:	Red painful swelling eyes, eyeball pain at night, headache and dizziness, scrofula, goiter, acute mastitis, mammary hyperplasia, distending pain in the breasts As a remedy for sore throat, fever and wounds Used to treat high blood pressure, diseases of the lymphatic system and tuberculosis Widely used for human tuberculosis, thyroid gland swell, jaundice, infectious hepatitis, bacillary dysentery, pleuritic, hypertension and cancer

H. Wagner et al. (eds.), *Chromatographic Fingerprint Analysis of Herbal Medicines*, Vol. 4,
DOI 10.1007/978-3-319-32328-2_22,

Main and minor **constituents**:

Triterpenoic acids [8, 9, 13, 17–22, 27]	**ursolic acid; oleanolic acid; betulinic acid;** euscaphic acid; hydroxyursolic acid; hydroxyoleanolic acid; 2α,3α,19α,24-tetrahydroxyurs-12-en-28-oic acid-28-O-D-glucopyranoside; 2α,3β,19α,24-tetrahydroxyurs-12-en-28-oic acid-28-O-D-glucopyranoside; 2α,3α,24-trihydroxyurs-12-en-28-oic acid-28-β-D-glucopyranoside; 2α,3α,19α-trihydroxyurs-12-en-28-oic acid-28-β-D-glucopyranoside; 2α,3α,19α-trihydroxyurs-12-en-28-oic acid-28-β-D-glucopyranosyl-(1 → 2)-β-D-glucopyranoside
Phenol carboxylic acids [4, 5, 7, 9, 12, 13, 18–20, 23, 24]	**rosmarinic acid; caffeic acid; *p*-coumaric acid**; chlorogenic acid; ellagic acid; 2-hydroxy-3-(3',4'-dihydroxyphenyl) propanoic acid
Flavonoids [5, 7–9, 12–14, 20, 23]	rutin; quercetin; hyperoside; quercetin 3-O-*β*-D-glucopyranoside (miquelianin); kaempferol 3-O-*α*-L-rhamnopyranosyl (1→6)-*β*-D-glucopyranoside; kaempferol 3-O-β-D-glucopyranoside (astragalin); acacetin-7-O-β-D-glucopyranoside
Di- / Triterpenes [7, 13, 18, 20–24]	trans-phytol; vulgarisin A; β-amyrin; lupeol; lupenone; chiratenol; uvaol; tanshinone I; 2α,3β-dihydroxy-lup-20(29)-ene; 3β-13β-dihydroxyolic-11-ene-28-oic acid
Polysaccharide [6, 8, 9, 14, 15, 24]	prunellin
Anthocyanidins [12]	cyanidin delphinidin
Other [6, 8, 9, 14, 15, 24]	essential oil, bitter principles, tannins, saponins (16-oxo-17-demethyl-3β,24-dihydroxylolean-12-en-3-O-β-D-glucuronoside=prunelloside A), polyphenols, carbohydrates, coumarins, sterols, organic acids; rhein; polygalacerebroside; α-spinasterol; stigmasterol; β-sitosterol; daucosterol; β-amyrenone; stigmast-7, 22-dien-3-one; 3,4,α-trihydroxy-methyl phenylpropionate; butyl rosmarinate

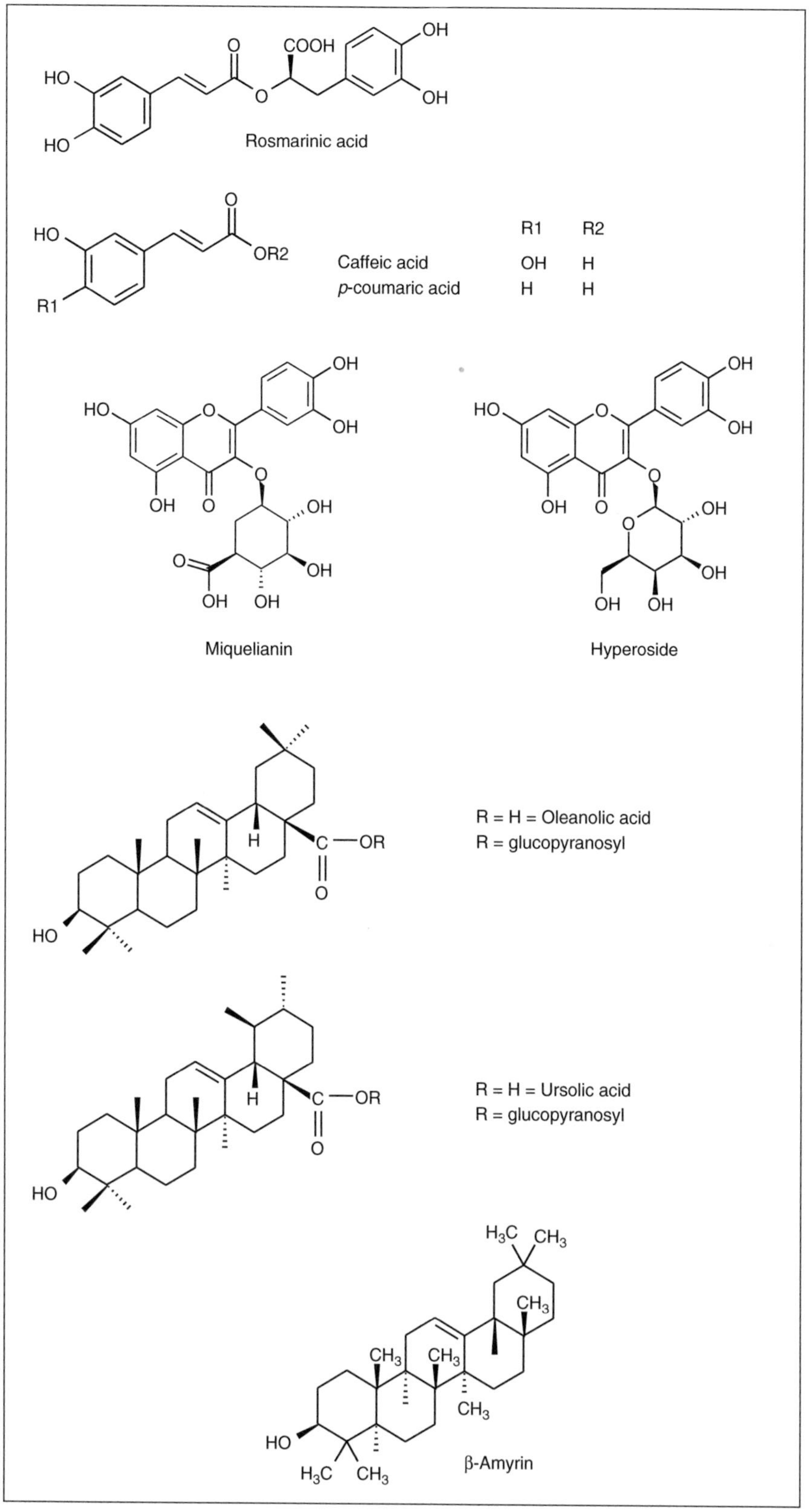

Fig. 1 Formulae of the main compounds of Spica Prunellae [4, 8]

Reported pharmacological activities:

- anti-inflammatory [5–7, 10, 12, 16, 18, 19, 21, 22, 24]
- antioxidant [7, 9, 12, 13, 15, 21, 22, 24]
- antibacterial / antibiotic [9, 12, 13, 15, 16, 18, 21–23]
- antihypertonic [10, 15, 18]
- anti-tumor activity [4, 7, 18, 19, 21]
- antiestrogenic activity [4]
- induction of apoptosis activity in tumor cells [4]
- antidiabetic / anti-hyperglycemic [5, 9, 10, 13, 18, 21, 23]
- antiviral activity against HIV-1 and herpetic keratitis [6, 7, 9, 10, 12, 13, 15, 16, 18, 19, 21–23]
- anti-ageing [6]
- antioomycete [6]
- antifungal [6, 21]
- immunomodulatory activity [9, 12, 13, 22]
- antiallergic [10, 21, 22]
- antirheumatic [15]
- antipyretic [15]
- diuretic [15, 19]
- hepatoprotective [19, 21]
- vasodilatoric actions [21]
- antimutagenic [25]

TLC-Fingerprint Analysis

Drug samples		Origin
1	Spica Prunellae / *Prunella vulgaris*	Province Anhui, China
2	Spica Prunellae / *Prunella vulgaris*	Province Zheijiang, China
3	Spica Prunellae / *Prunella vulgaris*	Sample of commercial drug (HerbaSinica, origin: Anhui)
4	Spica Prunellae / *Prunella vulgaris*	Sample of commercial drug (TCM-Clinic, Bad Kötzting)
5	Spica Prunellae / *Prunella vulgaris*	Sample of commercial drug (Caelo)
6	Spica Prunellae / *Prunella vulgaris*	Sample of commercial drug (Pharmacy, Munich, origin: Anhui)

1. TLC-fingerprint analysis of the flavonoids and phenol carboxylic acids: [25]

Reference compounds of Fig. 2		R*f*
T1	Caffeic acid	0.75
T2	Hyperoside	0.43
T3	Rosmarinic acid	0.26

1. Extraction: 2.5 g powdered drug are ultrasonicated with 35 ml ethanol (70 %) for 30 min, filtered and evaporated to dryness. The residue is dissolved in 1 ml ethanol.

2. Reference compounds: Each 0.5 mg is dissolved in 0.5 ml methanol

3. Separation parameters:

Plate:	HPTLC Silica gel 60 F_{254}, Merck
Applied amounts:	Spica Prunellae extracts: each 15 µl; Reference compounds: each 10 µl
Solvent system:	Ethyl acetate + methanol + water (20 + 2.7 + 2)
Detection:	Natural products – Polyethylene glycol reagent (NP/PEG) **I**: 1 % diphenylboric acid-*β*-ethylamino ester (= diphenylboryloxyethylamine, NP) in methanol **II**: 5 % Polyethylene glycol-4000 (PEG) in ethanol (80 %) The plate is sprayed first with solution **I** and then with solution **II**. After 30 min the evaluation is carried out under UV 366 nm.

4. Description of Fig. 2:
The ethanol extracts of Spica Prunellae are characterized in the upper R*f*-range by caffeic acid (**T1**). In the middle and lower R*f*-range appear three flavonol-glycosides [hyperoside (**T2**), rutin (R*f*=0.26), quercetin-3-glucuronid (R*f*=0.09)] and rosmarinic acid (**T3**).

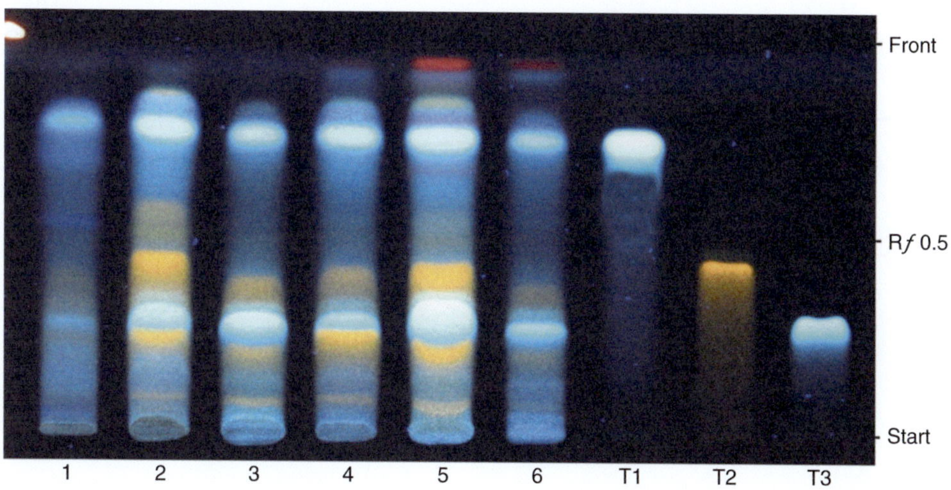

Fig. 2 Thin layer chromatogram of the ethanol extracts of Spica Prunellae, sprayed with NP/PEG (UV 366 nm)

2. TLC-fingerprint analysis of triterpenoids: [26]

Reference compounds of Fig. 3a, b		*Rf*
T4	Oleanolic acid	0.36
T5	β-Amyrin	0.66

1. Extraction: 1.0 g powdered drug are shaken with 10 ml diethyl ether for 1 h, filtered and evaporated to dryness. The residue is dissolved in 0.5 ml ethanol.
2. Reference compounds: Each 0.5 mg is dissolved in 0.5 ml methanol.
3. Separation parameters:
   Plate: HPTLC Silica gel 60 F_{254}, Merck
   Applied amounts: Spica Prunellae extracts: each 15 µl; Reference compounds: each 10 µl
   Solvent system: Petroleum ether (60–90 °C) + ethyl acetate + cyclohexane (10 + 6 + 4)
   Detection: 10 % ethanolic Sulphuric acid
   The plate is sprayed with 8 ml reagent and heated at 110 °C for 10 min. The plate is evaluated in VIS and under UV 366 nm.

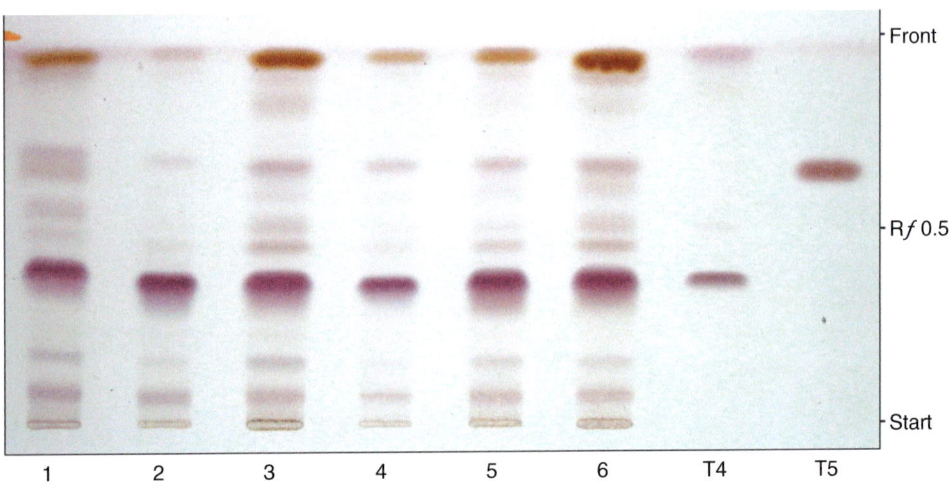

Fig. 3a Thin layer chromatogram of the ethanol extracts of Spica Prunellae, sprayed with 10 % Sulphuric acid (VIS)

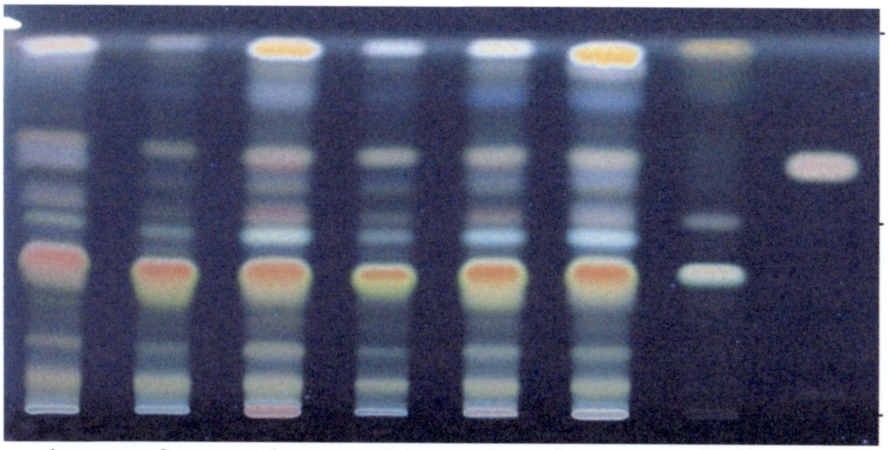

Fig. 3b Thin layer chromatogram of the ethanol extracts of Spica Prunellae, sprayed with 10% Sulphuric acid (UV 366 nm)

4. Description of Fig. 3a and b:
   The chromatogram shows in the upper R*f*-value with violet/brown colour (VIS) β-sitosterol at R*f*=0.95, β-amyrin (**T5**) at R*f*=0.66 and oleanolic acid (**T4**) at R*f*=0.36. The weak coloured 4–5 zones between β-amyrin and oleanolic acid might be other triterpenoids (e.g. betulinic acid). The zones in the lower and deep R*f*-range can be assigned to the various triterpenoic (e.g. ursolic) acids monoglycosides.

HPLC-Fingerprint Analysis

1. Sample preparation: 2.5 g powdered drug are ultrasonicated with 35 ml ethanol (70%) for 30 min, filtered and evaporated to dryness. The residue is dissolved in 1 ml ethanol and filtered over Chromafil® filtration unit, type 0–20 µm/25 mm.
2. Injection volume: Spica Prunellae extracts: each 15 µl

3. HPLC parameter:

Apparatus:	MERCK HITACHI D-6000 A Interface MERCK HITACHI L-4500 A Diode Array Detector MERCK HITACHI AS-2000 Autosampler MERCK HITACHI L-6200 A Intelligent Pump
Separation column:	LiChroCART® 250-4 LiChrospher® 60 RP-select B (5 µm), Merck
Precolumn:	LiChroCART® 4-4 LiChrospher® 60 RP-select B (5 µm), Merck
Solvent:	A: 0.1 % Phosphoric acid/Water (Millipore Ultra Clear UV plus® filtered) B: Acetonitrile (VWR)
Gradient:	0 % B for 5 min, 0–100 % B in 60 min 100 % B for 5 min,
Total runtime:	70 min
Flow:	1.0 ml/min
Detection:	210 nm

Retention times of the main peaks

Peak	**Rt (min)**	**Compound**
1	16.7	Not identified
2	17.8	Not identified
3	19.9	Not identified
4	22.2	Caffeic acid
5	24.5	Quercetin-3-glucuronide?
6	25.3	Hyperoside
7	27.9	Rosmarinic acid
8	32.0	Not identified
9	59.6	Oleanolic acid

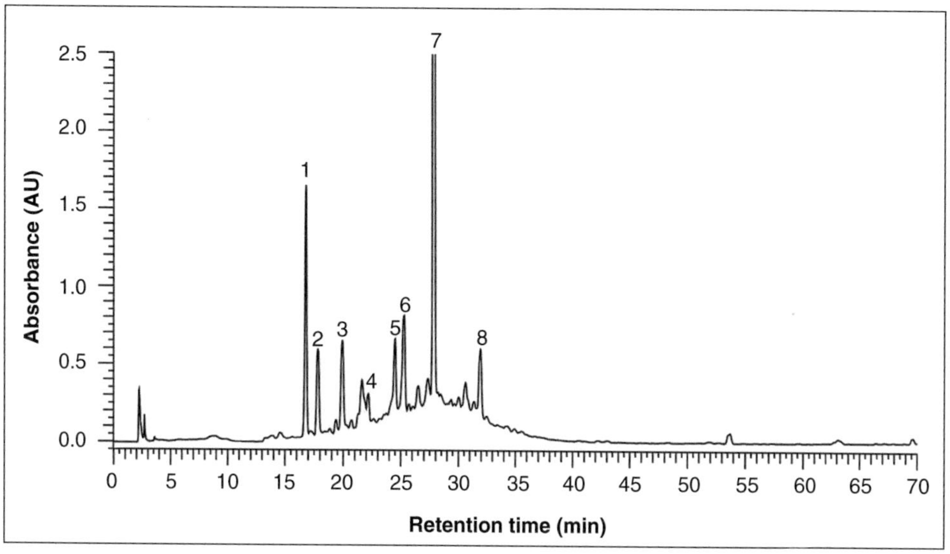

Fig. 4a HPLC-fingerprint analysis of the ethanol extract of Spica Prunellae, sample 2

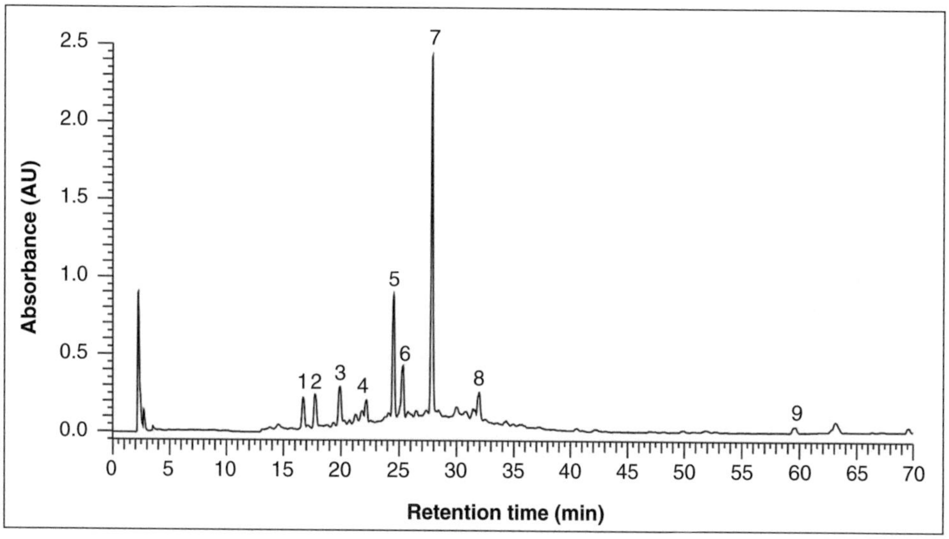

Fig. 4b HPLC-fingerprint analysis of the ethanol extract of Spica Prunellae, sample 4

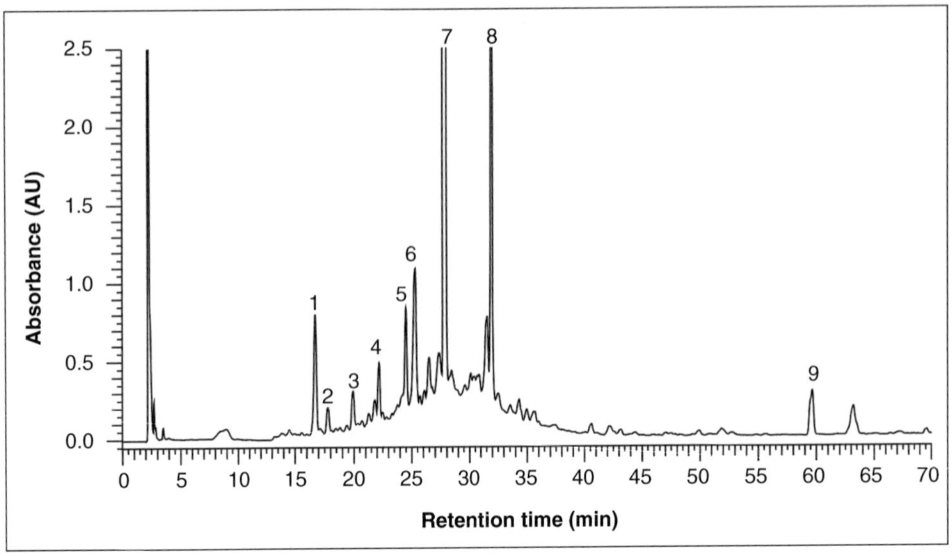

Fig. 4c HPLC-fingerprint analysis of the ethanol extract of Spica Prunellae, sample 5

4. Description of the HPLC-figures:

   All Spica Prunellae samples apart of sample 5 show a characteristic HPLC-fingerprint profile with the dominant rosmarinic acid (peak **7**). The also dominant peak **8** in sample 5 might be an isomer of rosmarinic acid.

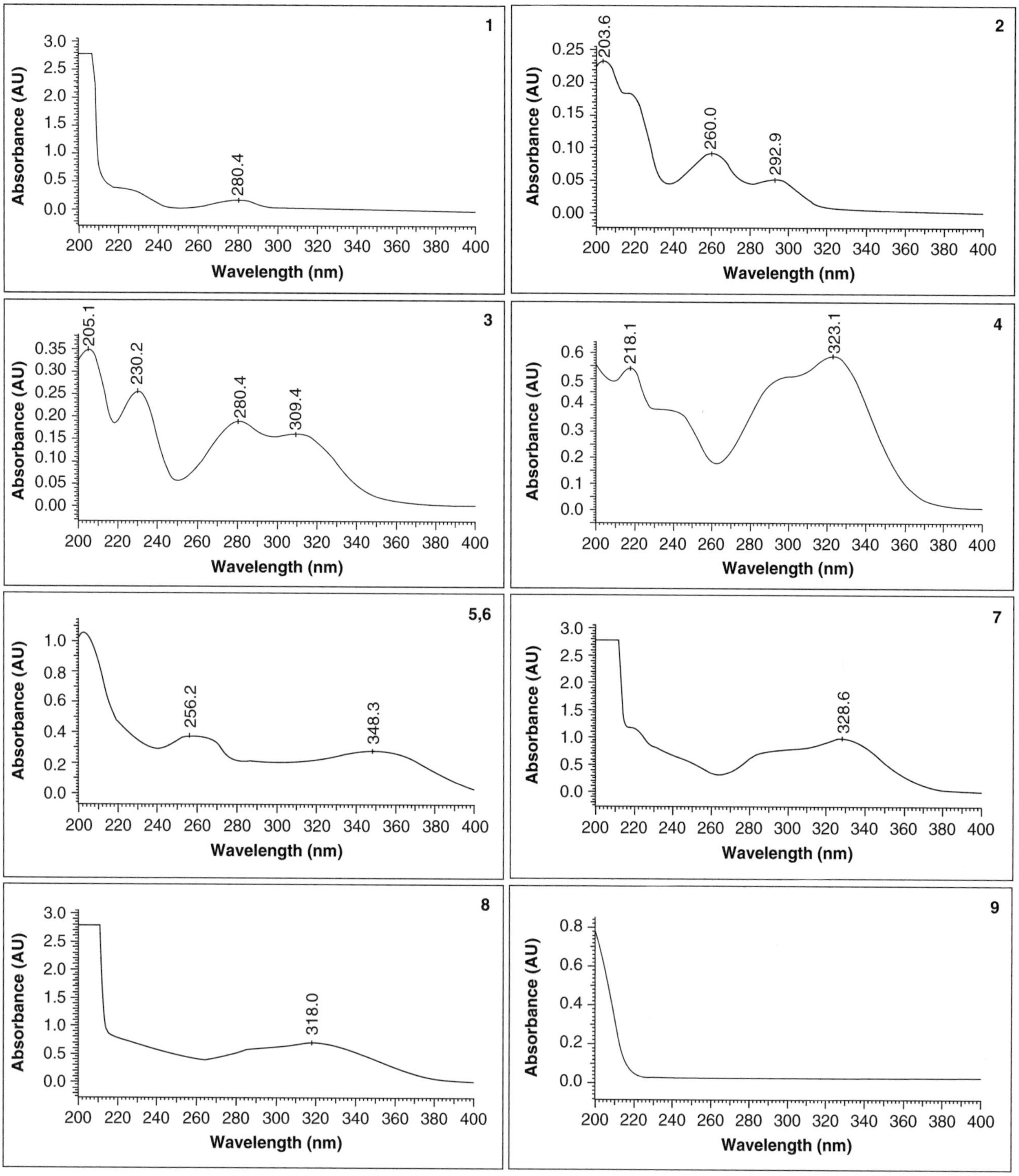

Fig. 5 UV-spectra of the main characteristic peaks of the HPLC-fingerprint of Spica Prunellae

Note: The Chinese Pharmacopeia 2010 describes for Spica Prunellae a total amount of rosmarinic acid not less than 0.20 % with reference to the dried drug. [1]

Conclusion

TLC and HPLC-fingerprints show characteristic zone- and peak profiles for authentication of Spica Prunellae.

References

1. Pharmacopoeia of the people's republic of China, English edition, vol. I. People's Medical Publishing House, Beijing (2005)
2. Zhang, Z.Z.: An illustrated Chinese materia medica in Hong Kong. School of Chinese Medicine Hong Kong Baptist University, Hong Kong (2004)
3. Keys, J.D.: Chinese herbs – their botany, chemistry, and pharmacodynamics. Charles E. Tuttle Company, Rutland/Tokyo (1976)
4. Feng, L., Jia, X., Zhu, M.M., Chen, Y., Shi, F.: Antioxidant activities of total phenols of *Prunella vulgaris* L. *in Vitro* and in tumor-bearing mice. Molecules **15**(12), 9145–9156 (2010)
5. Hwang, S.M., Lee, Y.J., Lee, Y.P., Yoon, J.J., Lee, S.M., Cha, J.D., Choi, K.M., Kang, D.G., Lee, H.S.: Anti-proliferative effect of an aqueous extract of Prunella vulgaris in vascular smooth muscle cells. Evid. Based Complement Alternat. Med. Article ID 936463 (2013)
6. Yoon, M.Y., Choi, G.J., Choi, Y.H., Jang, K.S., Park, M.S., Cha, B., Kim, J.C.: Effect of polyacetylenic acids from *Prunella vulgaris* on various plant pathogens. Lett. Appl. Microbiol. **51**(5), 511–517 (2010)
7. Wurtele, E.S., Chapepell, J., Jones, A.D., Celiz, M.D., Ransom, N., Hur, M., Rizshsky, L., Crispin, M., Dixon, P., Liu, J., Widrlechner, M.P., Nikolau, B.J.: Medicinal plants: a public resource for metabolomics and hypothesis development. Metabolites **2**(4), 1031–1059 (2012)
8. Li, H.M., Kim, J.K., Jang, J.M., Kwon, S.O., Cui, C.B., Lim, S.S.: The inhibitory effect of *Prunella vulgaris* L. on Aldose Reductase and protein glycation. J. Biomed. Biotechnol. Article ID 928159 (2012)
9. Chen, Y., Zhu, Z., Guo, Q., Zhang, L., Zhang, X.: Variation in concentrations of major bioactive compounds in *Prunella vulgaris* L. related to plant parts and phonological stages. Biol. Res. **45**(2), 171–175 (2012)
10. Sun, H.X., Qin, F., Pan, Y.J.: In vitro and in vivo immunosuppressive activity of Spica Prunellae ethanol extract on the immune responses in mice. J. Ethnopharmacol. **101**(1–3), 31–36 (2005)
11. Hempen, C.H., Fischer, T.: A materia medica for Chinese medicine (plants, minerals and animal products), 2nd edn. Churchill Livingstone Elsevier, New York (2009)
12. Sahin, S., Demir, C., Malyer, H.: Determination of phenolic compounds in *Prunella* L. by liquid chromatography-diode array detection. J. Pharm. Biomed. Anal. **55**(5), 1227–1230 (2011)
13. Cheung, H.Y., Zhang, Q.F.: Enhanced analysis of triterpenes, flavonoids and phenolic compounds in *Prunella vulgaris* L. by capillary zone electrophoresis with the addition of running buffer modifiers. J. Chromatogr. A. **1213**(2), 231–238 (2008)
14. Feng, L., Jia, X.B., Shi, F., Chen, Y.: Identification of two Polysaccharides from *Prunella vulgaris* L. and evaluation on their anti-lung Adenocarcinoma activity. Molecules **15**(8), 5093–5103 (2010)
15. Rasool, R., Ganai, B.A., Akbar, S., Kamili, A.N., Masood, A.: Phytochemical screening of *Prunella vulgaris* L. – an important medicinal plant of Kashmir. Pak. J. Pharm. Sci. **23**(4), 399–402 (2010)
16. Qu, L., Widrlechner, M.P.: Variation in the breeding system of *Prunella vulgaris* L. Hort. Sci. **46**(5), 688–692 (2011)
17. Ryu, S.Y., Oak, M.H., Yoon, S.K., Cho, D.I., Yoo, G.S., Kim, T.S., Kim, K.M.: Anti-allergic and anti-inflammatory Triterpenes from the herb of *Prunella vulgaris*. Planta Med. **66**(4), 358–360 (2000)
18. Gu, X., Li, Y., Mu, J., Zhang, Y.: Chemical constituents of *Prunella vulgaris*. J. Environ. Sci. **25**(Suppl 1), S161–S163 (2013)
19. Lee, M.K., Ahn, Y.M., Lee, K.R., Jung, J.H., Jung, O.S., Hong, J.: Development of a validated liquid chromatographic method for the quality control of Prunella Spica: determination of triterpenic acids. Anal. Chim. Acta **633**(2), 271–277 (2009)
20. Zhou, Q.X., Liu, F., Zhang, J.S., Lu, J.G., Gu, Z.L., Gu, G.X.: Effects of triterpenic acid from *Prunella vulgaris* L. on glycemia and pancreas in rat model of streptozotozin diabetes. Chin. Med. J. **126**(9), 1647–1653 (2013)
21. Chen, Y., Yu, M., Zhu, Z., Zhang, L., Guo, Q.: Optimisation of potassium chloride nutrition for proper growth, physiological development and bioactive component production in *Prunella vulgaris* L. PLoS One **8**(7), e66259 (2014)
22. Du, D., Cheng, Z., Chen, D.: A new unusual $\Delta^{11(12)}$-Oleane Triterpene and anti-complementary Triterpenes from *Prunella vulgaris* spike. Nat. Prod. Commun. **7**(4), 501–505 (2012)

23. Cheng, Q., Zhang, X., Wang, O., Liu, J., Cai, S., Wang, R., Zhou, F., Ji, B.: Anti-diabetic effects of the ethanol extract of a functional formula diet in mice fed with a fructose/fat-rich combination diet. J. Sci. Food Agric. **95**(2), 401–408 (2015)
24. Wu, H., Gao, M., Ha, T., Kelley, J., Young, A., Breuel, K.: *Prunella vulgaris* aqueous extract attenuates IL-1β-induced apoptosis and NF-κB activation in INS-1 cells. Exp. Ther. Med. **3**(6), 919–924 (2012)
25. Wagner, H., Bladt, S.: Plant drug analysis: a thin layer chromatography atlas, 2nd edn. Springer, Berlin (1996)
26. Wagner, H., Bauer, R., Melchart, D., Xiao, P.G., Staudinger, A.: Chromatographic fingerprint analysis of herbal medicines, vol. I+II. Springer, Wien-New York (2011)
27. Yu, Q., Qi, J., Wang, L., Liu, S.J., Yu, B.Y.: Pentacyclic triterpenoids from spikes of *Prunella vulgaris L.* inhibit glycogen phosphorylase and improve insulin sensitivity in 3T3-L1 adipocytes. Phytother. Res. **29**(1), 73–79 (2015)

Index

A
Acteoside (verbascoside), 215
Adenosine, 80, 84–87, 89
Alkaloids, 68, 80, 108, 109, 138, 204
Amino acids, 80, 148–150, 153, 154, 178
Apigenin, 92, 93, 116
Apigenin 6-C-α-L-arabinopyranosyl-8-C-β-D-glucopyranoside (isoschaftoside), 116
Apigenin 6-C-β-D-glucopyranosyl-8-C-α-L-arabinopyranoside (schaftoside), 116
Apigenin 6-C-β-D-glucosyl-8-C-α-L-rhamnoside (Violanthin) apigenin 6,8-di-C-α-L-arabinopyranoside, 116
Apigenin 6,8-di-C-β-D-glucopyranoside (vicenin-2), 116
Arctigenin, 20–22, 24, 25
Arctiin, 20–22, 24, 25
Artemisinin, 91, 92, 95, 96, 98–101
α-Spinasterol, 6, 226
Astragalin, 92, 96, 226
Aucubin, 215, 217, 218, 220, 223

B
Bakuchiol, 53
β-amyrin, 93, 226, 230, 231
Bavachinin, 48, 49, 51, 53
Bavachromene, 53, 54
Bicyclic peptides, 170
β-sitosterol, 2, 20, 72, 126, 170, 182, 183, 185, 192, 215, 226, 231

C
Caffeic acid, 28, 80, 81, 83, 92, 114, 226, 228, 229, 232
Casticin, 72–76, 92
Catalpol, 215, 217, 218, 220, 223
Celosins (A-G), 170, 173
Chalcones, 54, 158
Chlorogenic acid, 20, 24, 25, 80, 81, 83, 87, 89, 92, 96, 98–100, 102, 226
Coumarins, 12, 92, 93, 96, 116, 120, 122, 158, 226
Cucurbitacin B, 148, 151, 153
Cucurbitacin D, 148, 151, 153
Cyclic nonapeptide, 170
Cyclotides, 116

D
Daucosterol, 20, 126, 138, 170, 178, 226

E
Ephedrine, 108, 110, 111, 114
Esculetin, 12, 13, 15, 17, 116, 117, 119–121
Esculin, 12, 13, 15, 17, 117, 119–121
Essential oil, 48, 80, 83–84, 92, 93, 96, 100, 108, 126, 138, 226

F
Fatty acids, 20, 38, 60, 72, 83–84, 93, 186, 189, 192
Flavone(s), 2, 54, 72, 80, 116, 120, 122, 138, 220
Flavonoid glycosides, 116, 119, 210, 215, 219
Flavonoids, 12, 28, 48, 87, 89, 92, 93, 96, 111, 114, 119, 138, 158–161, 163, 165, 183, 204, 206–207, 210, 218–223, 226, 228
Fraxetin, 12, 13, 15
Fraxin, 12, 13, 15
(Furano)coumarins, 47, 48

G
Gallic acid, 28, 33
Gentiopicroside, 138, 140–142, 144
Glucosinolates, 192

H
Hispidulin-7-O-glucoside, 220
Homoplantaginin (hispidulin-7-O-D-glucoside), 215
Hyperosid, 81, 158–160, 226
Hyperoside, 28, 83, 92, 96, 98, 108, 228, 229, 232

I
Iridoid glycoside, 12, 28, 33, 215
Iridoid glycosides
Isobavachalcone, 48, 50, 51, 53, 54
Isochlorogenic acids, 80, 83, 93
Isofraxidin, 92, 96, 99, 100
Isoorientin, 116, 117, 119, 120, 138
Isopsoralen, 48, 50, 53, 56
Isopsoralenoside, 48, 53
Isoquercitrin, 28, 41, 92, 158

L
L-Argenin, 149
L-arginin, 148, 150
Lariciresinol, 12, 13, 15

H. Wagner et al. (eds.), *Chromatographic Fingerprint Analysis of Herbal Medicines*, Vol. 4,
DOI 10.1007/978-3-319-32328-2, © Springer International Publishing Switzerland 2016

L-citrullin, 148–150, 153
Lignans, 2, 3, 5, 6, 9, 12, 20, 22, 60, 62–69, 182, 189
Lignans and derivatives, 2
Linoleic acid, 20, 48, 72, 80, 83, 84, 182, 183, 185, 186
Lipids, 60, 108
Loganic acid, 28, 30, 33, 138, 140–142, 144
Loganin, 28, 30, 33, 35, 146
Luteolin, 92, 218
Luteolin 6-C-β-D-glucopyranoside (isoorientin), 116
Luteolin-7-glucoside, 120, 215, 219
Luteolin-7-O-glucosid, 117
Luteolin-7-O-glucoside, 120, 215, 218–220

M
Momordin Ic, 38, 40, 41, 43, 45
Morin, 158–161
Morusin, 158, 161, 163
Mulberroside A, 158

O
Oleanolic acid, 3, 5, 6, 28, 30, 33, 38, 40, 41, 43, 45, 138, 140, 141, 170, 172, 173, 178, 226, 230–232
Oleic acid, 60, 72, 80, 182, 183, 185, 192
Oleuropein, 12, 13, 15
Orientin, 117, 120
Oxyresveratrol, 158, 161

P
Phenol carboxylic acids, 80–83, 196–198, 226, 228
Phenolic compounds, 126, 192
Phenolic glycoside, 170, 173
Phenylethanol glycosides, 215, 217–218, 220, 223
Phenylpropane derivatives, 12
Phytosterols, 60, 69, 185
Pinoresinol, 12, 64, 65, 182
(+)-Pinoresinol, 62–64
(-)-pinoresinol, 60
Plantamajoside, 215, 220
Polysaccharides, 29, 148, 172, 226
Protein(s), 60, 62, 148, 182, 192
Psoralen, 48, 50, 51, 53, 54, 56
Psoralenoside, 48

Q
Quillaja saponin mixture, 207

R
Resveratrol, 158, 161
Rosmarinic acid, 226, 228, 229, 232, 239
Rutin, 92, 117, 119–121, 158–160, 218, 226, 229

S
Saccharides, 60, 158
Saccharose, 3, 5, 170, 172, 173, 185
Salidroside, 12, 13
Saponaretin, 116, 117, 119
Saponins, 2, 3, 5, 12, 29, 38, 40, 41, 45, 80, 170, 172, 204, 207–211, 226
Schaftoside, 116, 117, 120, 121
Scopoletin, 12, 92, 93, 96, 98–100, 117, 119–121, 158
Secoiridoid glycosides, 12, 138
Sesamin, 182, 183, 185, 186, 189
Sesquiterpenes, 80, 92, 93, 95–96
Sesquiterpens, 116
Sinalbin, 192–194, 196–198
Sinapic acid, 192, 194, 196–198
Sinapin, 194, 196, 198
Sinapine, 198, 200
cis-Sinapine, 197
trans-Sinapine, 192, 197
Sinigrin, 192–194, 196
Stigmasterin, 62, 63, 65, 69, 215
Stilbenes, 158, 161
Sweroside, 28, 30, 33, 138, 140–142, 144
Syringaresinol diglucoside, 3, 5
(–)-Syringaresinol-4-O-β-apiofuranosyl-(1→2)-β-D-glucopyranosid, 2, 6
Syringic acid, 2, 3, 6
Syringin, 12, 13

T
Triterpenoides, 215
Triterpenoids, 28, 60, 62–63, 69, 126, 138, 151, 170, 230–236
Triterponoids, 148
Tryptophan, 148, 153, 154

V
Verbascoside, 215, 217–220, 223
Vicenin-1, 117, 119, 120, 206, 207
Violanthin, 116, 117, 119–121
Vitexin, 92, 108, 117, 120, 121, 138, 204

MIX
Papier aus verantwortungsvollen Quellen
Paper from responsible sources
FSC® C105338

If you have any concerns about our products,
you can contact us on
ProductSafety@springernature.com

In case Publisher is established outside the EU,
the EU authorized representative is:
Springer Nature Customer Service Center GmbH
Europaplatz 3, 69115 Heidelberg, Germany

Printed by Libri Plureos GmbH
in Hamburg, Germany